Marine Engineering: Emerging Developments, Challenges and Applications

Marine Engineering: Emerging Developments, Challenges and Applications

Editor: Timothy Chapman

Callisto Reference
www.callistoreference.com

Callisto Reference,
118-35 Queens Blvd., Suite 400,
Forest Hills, NY 11375, USA

Visit us on the World Wide Web at:
www.callistoreference.com

ISBN: 978-1-64116-806-9 (Hardback)

Cataloging-in-Publication Data

Marine engineering : emerging developments, challenges and applications / edited by Timothy Chapman.
p. cm.
Includes bibliographical references and index.
ISBN 978-1-64116-806-9
1. Marine engineering. 2. Engineering. I. Chapman, Timothy.
VM600 .M37 2023
623.87--dc23

Table of Contents

Preface

Marine engineering is a field of study that deals with the innovation, design, construction and maintenance of navigation equipment and seagoing vessels. It is mainly concerned with the development and manufacture of internal systems for submarines, boats and ships. Marine engineering is intrinsically related to other fields such as oceanographic engineering, naval architecture, nautical sciences, and automotive and mechanical engineering. Recent advances and trends in marine engineering involve the utilization of wind and solar power as well as the usage of robots in ship manufacturing and port management. Recent developments in offshore marine engineering are focused on addressing issues such as fuel scarcity and environmental degradation. This book is a compilation of chapters that discuss the emerging developments, challenges and applications of marine engineering. It will serve as a valuable source of reference for graduate and postgraduate students.

The researches compiled throughout the book are authentic and of high quality, combining several disciplines and from very diverse regions from around the world. Drawing on the contributions of many researchers from diverse countries, the book's objective is to provide the readers with the latest achievements in the area of research. This book will surely be a source of knowledge to all interested and researching the field.

In the end, I would like to express my deep sense of gratitude to all the authors for meeting the set deadlines in completing and submitting their research chapters. I would also like to thank the publisher for the support offered to us throughout the course of the book. Finally, I extend my sincere thanks to my family for being a constant source of inspiration and encouragement.

Editor

1

Self-Excited Torsional Vibration in the Flexible Coupling of a Marine Propulsion Shafting System Employing Cardan Shafts

Myeong-Ho Song [1], Taek-Kun Nam [1] and Jae-ung Lee [2,*]

[1] Division of Marine Engineering, Mokpo National Maritime University, Mokpo 58628, Korea; smhsrs@mmu.ac.kr (M.-H.S.); tknam@mmu.ac.kr (T.-K.N.)

[2] Division of Marine Information Technology, Korea Maritime and Ocean University, Busan 49112, Korea

* Correspondence: julee@kmou.ac.kr

Abstract: The reliability of propulsion shafting systems is a major concern for ocean-going vessels because mid-ocean repairs can be time-consuming and spare parts must be available. To address this concern, vibration modeling and experimental measurements were conducted on a propulsion shafting system with a Z-drive propeller, with the objective of identifying the source of failure for the flexible rubber coupling connecting the diesel engine with the intermediate shaft. The torsional fluctuations in the flexible coupling dramatically increased and then abruptly ceased. The modeling results revealed that the frictional losses during power transmission through the universal joints could act as an excitation force for self-excited vibration. The coupling connected to the intermediate shaft did not have sufficient radial flexibility to dampen these vibrations. To avoid the effects of the self-excited torsional vibration, it is recommended that this coupling is replaced with one that is capable of absorbing the radial shaft displacement.

Keywords: Z-drive propeller; Cardan shaft; self-excited torsional vibration; flexible coupling; flexible link; frictional torque

---

1. Introduction

Certain marine propulsion applications require efficient maneuverability under extreme conditions. One solution for this issue is the use of a Z-drive propeller, which may rotate freely to alter the direction of the thrust applied to the vessel. However, such a propulsion system is known to be subject to self-excited vibrational modes, which may damage the system. In propulsion shaft systems, self-excited vibrations are produced by an unstable rotary power transmission mechanism. Such vibrations begin spontaneously, and the amplitude increases until a nonlinear effect limits any further increase. There are many sources of self-excitation, and in most cases, the self-excited vibrations pose a danger to the entire propulsion system. The force acting on the vibrating mechanism is typically external to the system and independent of the motion within the system. Friction-induced vibration [1–5] and flow-induced vibration [6–8] are examples of self-excited vibration.

Moreover, self-excited vibrations are characterized by the presence of a mechanism whereby the system vibrates at its own natural or critical frequency, mostly independent of the frequency of any external stimulus [9]. Lee et al. investigated the unstable torsional vibration resulting from the friction torque that occurred in a propulsion shafting system [10]. They observed that in the operating mode wherein the shaft generator is separated, the clutch should be entirely disengaged. However, if the clutch between the rubber coupling and the shaft generator is not fully disengaged, then friction torque occurs because the shaft generator is idling. During the torsional vibration experiment, significant vibrations occurred but were not associated with the excitation of the diesel engine. As measured, the self-excited torsional vibrations (SETVs) were oscillating at a frequency of 3.75 Hz, similar to the shaft generator's idle running speed (220–250 r/min) [11].

In another experiment, torsional vibrations developed in an electric system equipped with an inverter motor to drive a propeller [12,13]. The torsional vibration analysis identified that the resonant point was located on the flexible rubber coupling, which was employed to transmit the torque from the electric motor to the intermediate shaft and reduce the influence of the motor assembly on the propeller components. This resonance added to the increased dominant vibration at an approximately identical frequency. The alternating tension and compression action during the damping process of the coupling were investigated as the source of this phenomenon. The rubber element could sustain simultaneous compression, shear, and torsion stresses, but could not fully dissipate or absorb the generated torsion at certain speed ranges.

Furthermore, a fractured Cardan shaft was observed in the propulsion system using a Z-drive propeller [14]. This failure was not likely caused by the torque variation from the main engine, but is instead attributed to the SETV from the friction torque of the Cardan shaft. However, this was not confirmed owing to the lack of vibration measurements. In this study, the repeated failure of a flexible rubber coupling in a similar vessel with a Z-drive propeller was investigated using torsional vibration theory and global vibration and torsion measurements. The coupling is considered to have provided insufficient radial flexibility in the damping process. The experimental methodology is described in the next section. The main results are presented and discussed in Section 3. Finally, the major findings and conclusions of this study are presented in Section 4.

2. Materials and Methods

2.1. Case Study

In this study, the case of a propulsion shafting system on a 150-ton class oil-skimmer vessel was investigated. A Z-drive propeller, powered by a high-speed diesel engine through two sets of Cardan shafts, was used. A flexible rubber coupling was employed to isolate the vibratory torque excited by the diesel engine and the excess hydraulic variation of the propeller owing to external impact [10,15,16]. Hence, the diesel engine and the reduction gears could be safely operated by adapting the flexible coupling to the system [17]. However, the flexible coupling repeatedly failed during regular operation. Notably, the same type of new coupling was used, and a few external conditions were changed to improve the performance. The failed coupling is shown in Figure 1. The operating engineer reported abnormal noise, and vibration occurred immediately before the incident. Many experiments were performed, but abnormal vibrations were not observed. Further analysis was not possible until the next vessel in the same series was built, and the experiment could be repeated.

Figure 1. Damaged rubber part of flexible coupling installed between main engine and intermediate shaft.

2.2. Vibration Measurements

The schematics of the propulsion system and experimental setup are shown in Figure 2. To verify the torsional vibration, laser rotary torsional meters were installed onto both ends of the flexible coupling. Subsequently, the angular velocity was measured, and the relative torsion angle was calculated, as required. To evaluate the structural vibration, velocity sensors were installed onto the bearing housing and onto the fore and aft ends of the engine body. Then, measurements were taken while the engine accelerated and decelerated.

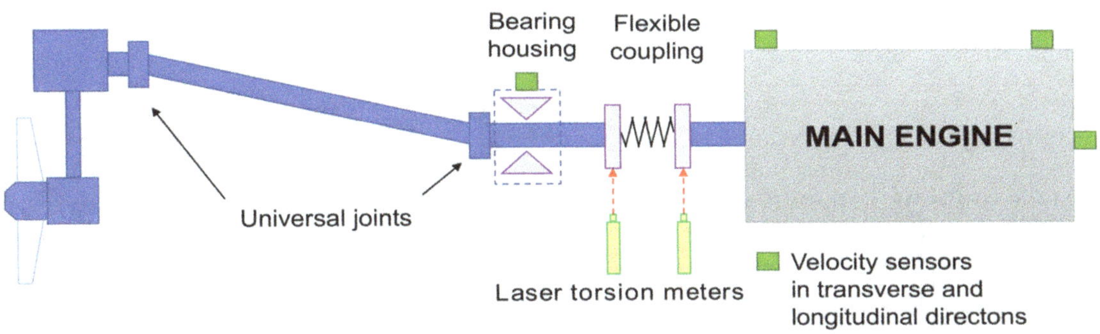

Figure 2. Equipment installation positions for global vibration measurement.

2.3. Self-Excited Torsional Vibration Theory

The model for calculating the SETV is shown in Figure 3. The model consists of an input link rotating at constant angular velocity, connected to an input shaft with a variable moment of inertia, and depending on the position of the input link. The nonlinear motion equation, considering the elasticity and damping of the input shaft, is expressed as follows [18]:

$$J(\varphi)\ddot{\varphi} + \frac{1}{2}\cdot\frac{dJ(\varphi)}{d\varphi}\dot{\varphi}^2 + c_T(\varphi - \varphi_0) + b_T(\dot{\varphi} - \dot{\varphi}_0) = 0 \tag{1}$$

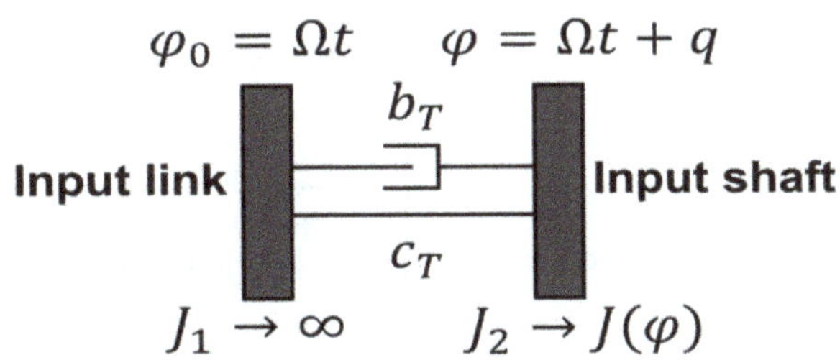

Figure 3. Model of self-excited torsional vibration mechanism.

The rotational motion is calculated as follows:

$$\varphi = \varphi_0 + q = \Omega t + q; \qquad \dot{\varphi} = \Omega + \dot{q}, \qquad \ddot{\varphi} = \ddot{q} \tag{2}$$

If the relative angle of rotation, q, is assumed to be very small, linearization is allowed. The Taylor Series expansion of the input shaft inertia moment can be expressed as follows:

$$J(\varphi) = J(\Omega t) + J'(\Omega t)q + \cdots \tag{3}$$

$$\frac{dJ(\varphi)}{d\varphi} = J'(\Omega t) + J''(\Omega t)q + \cdots \tag{4}$$

By inserting Equations (3) and (4) into Equation (1), the linear differential equation with $J = J\,(\Omega t)$ can be rewritten as follows:

$$J\ddot{q} + (b_T + J'\Omega)\dot{q} + \left(c_T + J''\Omega^2/2\right)q = -\frac{1}{2}J'\Omega^2 \tag{5}$$

In this case, all terms starting from order q^2 are ignored. The decay rate, $\delta(t)$, and natural frequency, $\omega_0(t)$, can be expressed, respectively, as follows:

$$\delta(t) = \frac{b_T + J'\Omega}{2J} \tag{6}$$

and

$$\omega_0^2(t) = \frac{c_T + \frac{J''\Omega^2}{2}}{J} \tag{7}$$

The elastic coupling with nonlinear characteristics has a significant effect on the drive system and is primarily used to lower the natural frequency range. The theoretical simulation can be represented as the numerical integration of the nonlinear motion equation.

The diagram of the drive system is shown in Figure 4a. The moment of inertia, J_1, as the prime mover (engine side) is connected to the propeller with a nonlinear flexible coupling, which has a reduction transmission ratio $i = (\varphi_{11}/\varphi_2)$. The reduced moment of inertia of the Z-drive (driven side) is expressed as follows:

$$J(\varphi_2) = J_0\left(1 + \mu\sin^2\varphi_2\right), \tag{8}$$

$$\frac{dJ}{d\varphi_2} = 2J_0\mu\sin\varphi_2\cos\varphi_2 \tag{9}$$

Therefore,

$$q = \varphi_{11} - \varphi_1 = i\varphi_2 - \varphi_1 \tag{10}$$

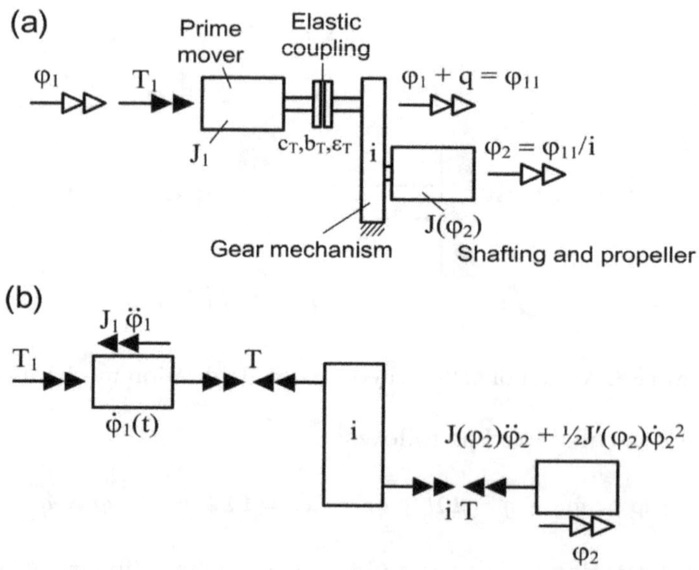

Figure 4. Drive with nonlinear coupling: (**a**) system schematic; (**b**) free-body diagram of system and occurring torques.

Then, the restoring torque, T, of the flexible coupling is calculated as follows:

$$T = c_T\left(q + \epsilon_T q^3\right) + b_T\dot{q} \tag{11}$$

The primary point of interest is the dynamic load of the flexible coupling of the starting and braking processes. From Figure 4b, the motion equation can be expressed using the torque balance of both shafts, as follows:

$$J_1\ddot{\varphi}_1 - T = T_1 \tag{12}$$

$$(\varphi_2)\ddot{\varphi}_2 + \frac{1}{2}J'(\varphi_2)\dot{\varphi}_2^2 = -iT \tag{13}$$

To solve the nonlinear differential equation for φ_1 and φ_2, Equations (8), (9), and (11) can be inserted into Equation (13) to obtain the following relationship:

$$J_0(1 + \mu\sin^2\varphi_2)\ddot{\varphi}_2 + \frac{1}{2}J_0\mu\sin 2\varphi_2\dot{\varphi}_2^2 = -i\left[c_T\left(q + \epsilon_T q^3\right) + b_T\dot{q}\right] \tag{14}$$

The nonlinear differential equation for q can be expressed as follows:

$$\ddot{q} = -\ddot{\varphi}_1 - \frac{i^2\left[c_T\left(q + \epsilon_T q^3\right) + b_T\,\dot{q}\right] - \frac{1}{2}J_0\mu\sin\left(2\frac{\varphi_1+q}{i}\right)\frac{\left(\dot{\varphi}_1+\dot{q}\right)^2}{i}}{J_0 + J_0\mu\sin^2\left(\frac{\varphi_1+q}{i}\right)} \tag{15}$$

The function for angle $\varphi_1(t)$ can be used to numerically integrate Equation (15) for the predefined initial conditions, from which $q(t)$ and $\dot{q}(t)$ are obtained. Then, the coupling torque can be calculated from Equation (11) and the angle φ_2 can be calculated from Equation (10). From φ_2, the input torque, $T_1(t)$, which is required to produce the motion from Equation (12), can be calculated.

The Cardan shaft mechanism is illustrated in Figure 5. The shaft operates at slope angle β; the angular velocity and angular acceleration on the driven side are expressed as Equations (16) and (17), respectively.

$$\Omega_2 = \frac{\cos\beta\cdot\Omega_1}{1-\sin^2\alpha_1\cdot\sin^2\beta} \tag{16}$$

$$a_2 = \frac{\cos\beta\cdot a_1}{1-\sin^2\alpha_1\cdot\sin^2\beta} + \frac{\Omega_1^2\cdot\cos\beta\cdot\sin^2\beta\cdot\sin^2\alpha_1}{\left(1-\sin^2\alpha_1\cdot\sin^2\beta\right)^2} \tag{17}$$

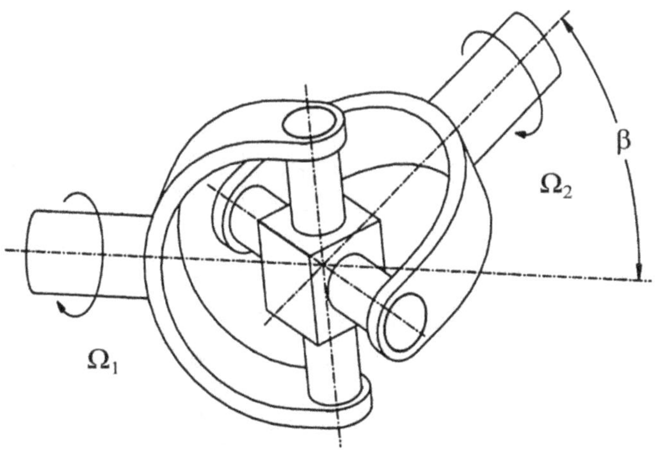

Figure 5. Schematic of Cardan shaft mechanism for rotating power transmission.

3. Results and Discussion

3.1. Torsional Vibration Calculations

In this study, the propulsion system on the 150-ton class oil-skimmer vessel was powered by a 4-stroke high-speed diesel engine as the prime mover. There were five intermediate shafts with two sets of Cardan shafts. In the stern space area, integrated reduction gears were installed onto the top of the Z-drive propeller (see Table A1 in Appendix A for system specifications).

The Vulkan flexible rubber coupling was installed between the main engine and the intermediate shaft. Its outer ring was directly connected to the flywheel, while its inner ring was coupled to the intermediate shaft. Based on the technical data provided by the coupling manufacturer [19], the propulsion shafts were modeled as 27 concentrated masses (Figure 6), and the torsional vibration calculation was performed by assuming that the coupling stiffness was 100% of the engine load. The analysis was carried out using our proprietary software. The calculation results are presented in Table 1. The torsional vibration characteristic at the flexible coupling with all cylinders firing normally is presented in Figure 7. The theoretical calculation results for the torsional vibration presented no issues compared with the measurement results.

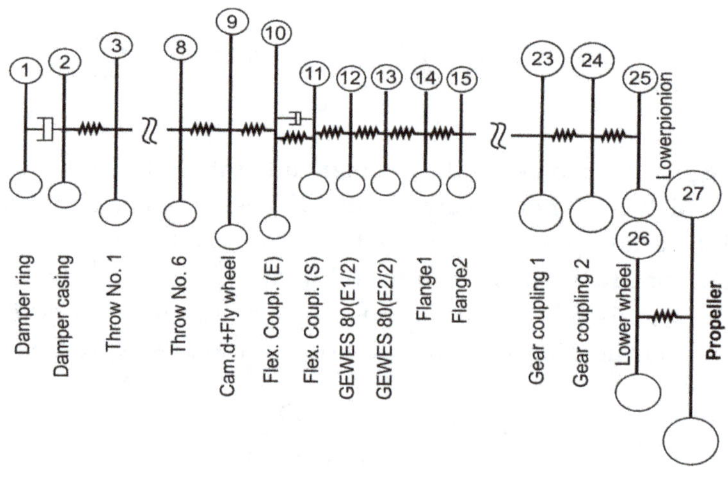

Figure 6. Mass elastic model used in theoretical calculation of torsional vibration.

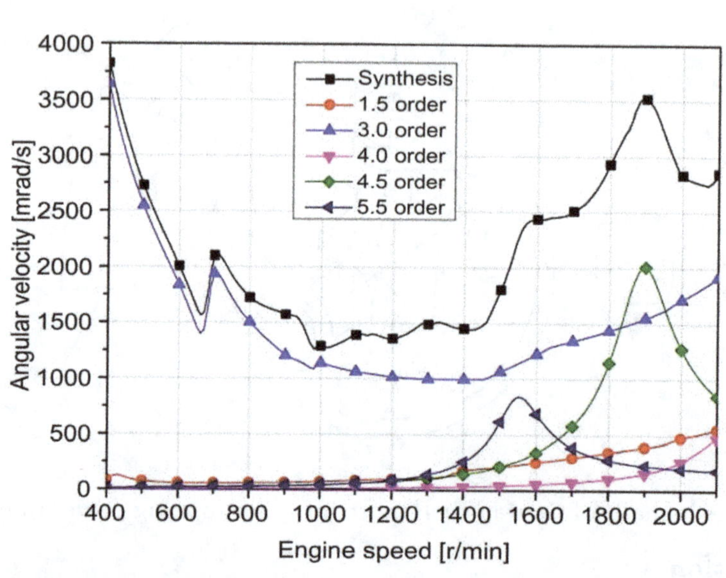

Figure 7. Calculated angular velocity on engine side of flexible coupling with all cylinders operating under normal firing condition.

Table 1. Calculated and measured natural frequencies in cycles per minute.

Mode	Calculated	Measured	Remarks
1-node	631	522	by self-excited vibration
2-node	2.051	1.980	by 3rd order at 660 r/min

The 1st order vibration was not observed in the calculation because each cylinder had an equal load, and the crank firing angles were arranged at equal intervals. Furthermore, the vibration of the 3rd order component (the vibration phase angle between the banks was 90°) was dominant because a V-type 12-cylinder engine was used. In the theoretical approach, all concentrated masses of the propulsion shafting were converted to the reduced model of a 2-mass system based on the elastic coupling and I node natural frequency [20]. In this case, the equivalent moment of inertia, J_1, of the diesel engine was 3.62 $kg{\cdot}m^2$, the equivalent inertia, J_0, of the propulsion system was 2.091 $kg{\cdot}m^2$, the equivalent torsional rigidity, c_T, was 5.8 $kN{\cdot}m/rad$, and the torsional attenuation, b_T, was 15.77 $N{\cdot}m{\cdot}s/rad$. Additional vibration elements were predicted by changing the values of Equation (15). Because these elements are complex and difficult to calculate, in this study, the abovementioned values of Equation (15) were replaced with the experimental results.

3.2. Vibrational Measurements

The angular velocity fluctuations during engine acceleration at the ends of the flexible coupling on the engine and Cardan shaft side are plotted in Figures 8 and 9, respectively. The time-domain indicates that the dominant SETV mode occurred in approximately 15 s, significantly increased when the engine speed exceeded 600 r/min, and then decreased above 800 r/min before disappearing. The results indicate that the dominant mode of the torsional vibration was 8.7 Hz or 522 cycles/min, which is slightly below the 1-node natural frequency of 631 cycles/min, as presented in Table 1. As illustrated in Figure 9, the dominant vibration mode was the 1st order at the natural frequency on the engine side of the coupling. The end on the side closest to the Cardan shaft was the 2.5th order mode, which is the major component, while the 1st order mode is much smaller. Therefore, by ignoring the angular velocity vibration on the shaft side of the coupling, the maximum angular acceleration was calculated as 1531 rad/ s^2 by applying the value measured on the engine side of the coupling to Equation (15).

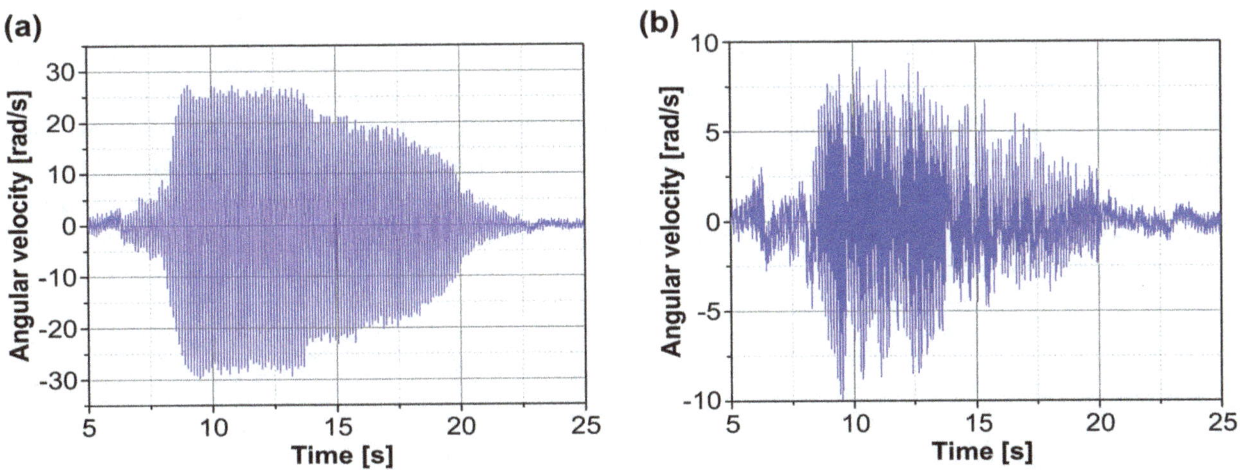

Figure 8. Time domain of angular velocity at flexible coupling during acceleration in range of 600–800 r/min (**a**) Engine side; (**b**) Shaft side.

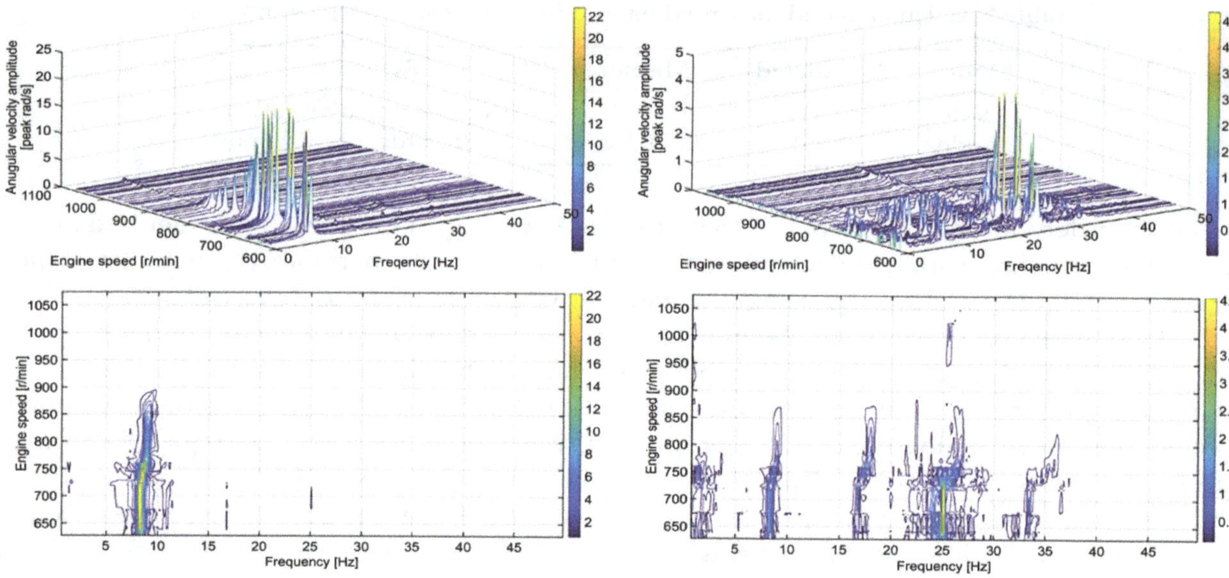

Figure 9. Frequency domain of angular velocity at flexible coupling during acceleration. Engine side (left); Shaft side (right).

The data collected during engine deceleration are presented in Figure 10. As the engine speed slowly decreased from 2200 r/min to the minimum speed, the 0.5th order resonance of 9–10 Hz occurred while the engine ran at approximately 1000–1200 r/min, which is a smaller amplitude. Additionally, the structural vibrations of the bearing housing close to the universal joint had a dominant vibration mode in the same frequency range of 9–10 Hz, as shown in Figure 11. This result confirmed that the vibration excitation frequency is a SETV that does not depend on the engine speed. The engine body vibrations were more intense when SETVs occurred. Furthermore, SETV still occurred when one cylinder misfired, but it did not bounce, unlike the last time it was recorded.

The bouncing vibration did not occur often; hence, shafting misalignment was excluded as a possible cause. Furthermore, these phenomena did not behave in the same manner as the misalignment. The universal joint consists of a cross/yolk mechanism. Based on the authors' experience, the only possible source of vibration could be the friction torque in the Cardan shaft universal joint. Friction is the source of internal damping, and this would have caused self-excited vibrations [21], but the flexible coupling did not have sufficient flexibility in the radial direction to damp it. The SETV may have occurred with a relatively small friction torque in the joint [22], which led to an increasing friction torque on the engine side of the coupling, and eventually resulted in coupling failure. The friction torque could have been detected by measuring the relative torsion angle at the universal joint, but this relative torsion angle was not appropriately measured because the equipment was inadequate. However, the bearing housing structural vibration close to the Cardan joint may reflect which part of the friction torque fluctuated.

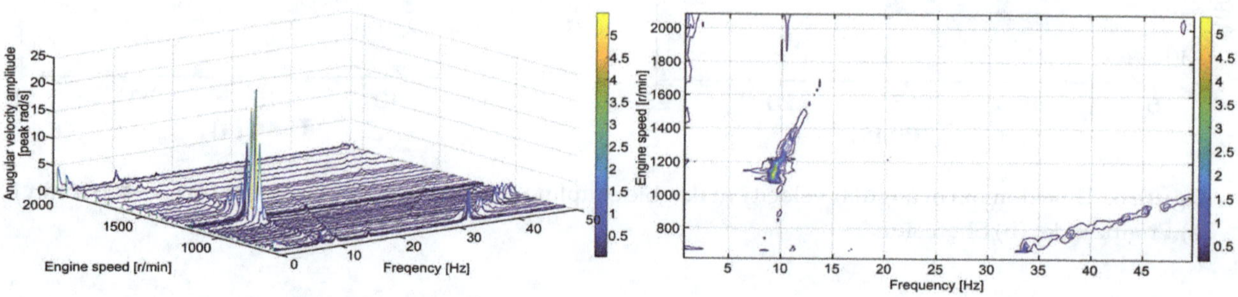

Figure 10. Frequency domain of angular velocity on engine side of flexible coupling during deceleration.

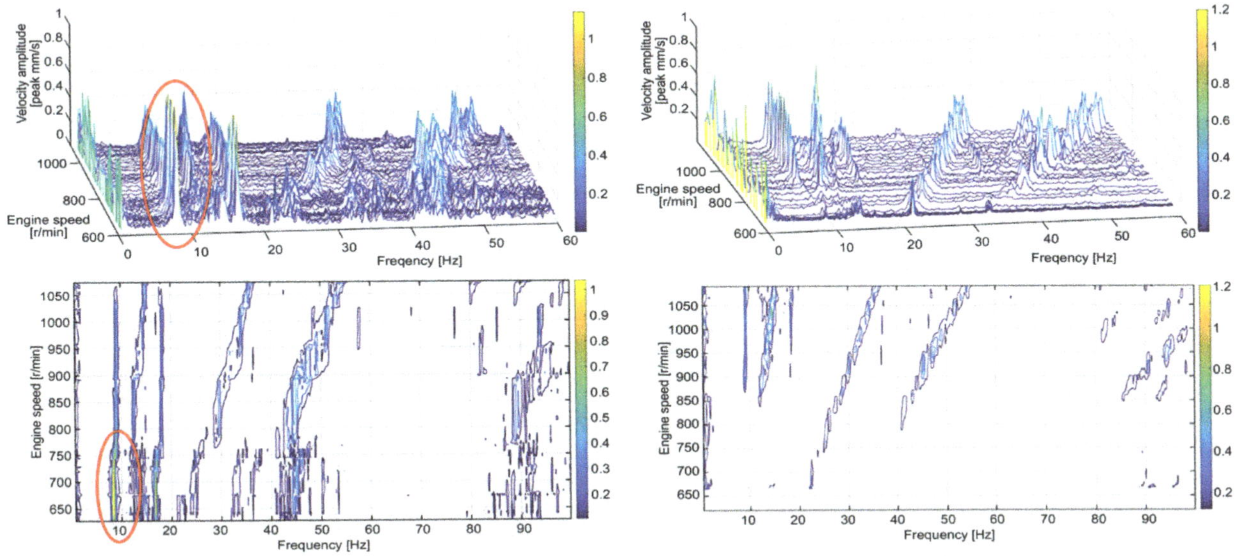

Figure 11. Velocity vibration amplitude measured at bearing housing in transverse direction. With SETV being dominant (left); Without SETV being dominant (right).

To prevent failure, the rubber coupling was replaced with the redesigned Centax elastic coupling, as shown in Figure 12. The new coupling had the same torsional stiffness as the previous coupling, but could also absorb radial shaft displacement through a flexing action. All angular, axial, and radial shaft displacements on the new elastic coupling were controlled by flexible links to avoid SETV. With this new coupling, SETV or failures have not been recorded to date.

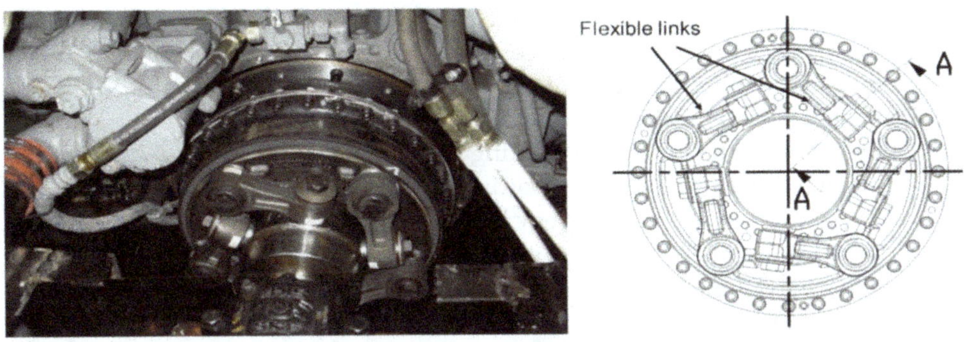

Figure 12. Installation of redesigned Centax elastic coupling with flexible links.

4. Conclusions

In this study, the occurrence of SETV in the propulsion system of a 150-ton class oil-skimmer vessel with a Z-drive propeller was theoretically investigated and verified through measurements. The results are summarized as follows:

(1) The torsional vibration calculation was performed to find out the resonance frequencies with estimated amplitudes but nothing abnormal was issued. It confirms the possibility of the SETV which began spontaneously and independently of the engine speed. The flexible coupling was repeatedly broken during operation, and the SETV amplitude rapidly increased. The dominant vibration mode occurred for approximately 15 s and then disappeared. During the occurrence of SETVs, the engine body's structural vibrations were also intense.

(2) The SETV is attributed to the friction torque from the Cardan shaft universal joint. The friction loss during power transmission acted as an excitation force, which resulted in SETV on the driven

side; then, failure occurred. Therefore, it is recommended that the friction effects on Cardan shafts are considered in the design stage of propulsion shafting system development.

(3) For a power transmission system using Cardan shafts, such as the propulsion system considered in this study, an elastic coupling should be used because it offers more radial flexibility for inhibiting the SETV effects. For the vessel considered herein, the new flexing coupling provides a high degree of control to all angular, axial, and radial shaft displacements. After replacement, dominant SETV modes were not observed.

(4) Owing to inadequate measurement, friction torque was only detected through the structural vibration of the bearing housing. In future work, appropriated experiments should be conducted for the Cardan shaft joint. The friction torque can be investigated using the relative torsion angle achieved by the torques or angular velocities measured on the two sides of the universal joint.

In summary, the results and findings of this study may be useful in addressing a significant concern for ocean-going vessels, namely the reliability of the propulsion shafting system. The proposed variations of the shafting system, whose purpose is to avoid the effects of SETV, should be further investigated to develop more reliable propulsion systems.

Author Contributions: Conceptualization, M.-H.S.; Methodology, M.-H.S., T.-K.N., and J.-u.L.; Software, M.-H.S.; Validation, T.-K.N., and J.-u.L.; Formal Analysis, T.-K.N.; Investigation, M.-H.S.; Resources, M.-H.S.; Data Curation, T.-K.N.; Writing—Original Draft Preparation, M.-H.S.; Writing—Review and Editing, T.-K.N. and J.-u.L.; Visualization, M.-H.S.; Supervision, J.-u.L.; Project Administration, M.-H.S.; Funding Acquisition, M.-H.S. All authors have read and agreed to the published version of the manuscript.

Nomenclature

a_1	angular acceleration at driving yoke of Cardan shaft (rad/s^2)
a_2	angular acceleration at driven yoke of Cardan shaft (rad/s^2)
b_T	torsional damping constant (N·m·s/rad)
c_T	torsional stiffness constant (N·m/rad)
J_0	moment of inertia for driven shafting (kg·m^2)
J_1	moment of inertia for driving shafting (kg·m^2)
$J(\varphi)$	reduced moment of inertia (kg·m^2)
i	transmission ratio of reduction gear
q	relative twist angle (rad)
T	restoring torque of elastic coupling (N·m)
T_1	input torque of driving side (N·m)
α_1: $(= \Omega_1 \mathrm{t})$	angular displacement (rad)
B	operation angle of Cardan shaft (rad)
δ	decay rate (decays/s)
ϵ_T	characteristic parameter of nonlinearity
η: $(= \Omega/\omega_0)$	frequency ratio
μ	pulsation depth of moment of inertia
t	time (s)
φ	rotational motion (rad)
Ω	constant angular velocity (rad/s)
Ω_1	angular velocity at driving yoke of Cardan shaft (rad/s)
Ω_2	angular velocity at driven yoke of Cardan shaft (rad/s)
ω_0: $(= i\sqrt{c_T/J_0})$	natural circular frequency of liner undamped oscillator (rad/s).

Appendix A

Table A1. System Specifications for 150-ton Class Oil-skimmer Ship (by courtesy of shipyard).

	Type	Viscous
Damper	Diameter × Width Effective inertia Ring inertia Weight	330 × 51 mm 0.2372 kg × m^2 0.2344 kg × m^2 22 kg
Flexible Coupling	Type Torsional stiffness (0–100% load) Permissible maximum & vibratory torque Constant kappa Nominal torque	VULASTIK L2611S 6.9–19.5 kN × m/rad 3.95 and 1.0 kN × m 0.18 3.15 kN × m
Engine	Type Cylindrical Bore × stroke Power at MCR Pmi at full load Nominal torque Reciprocating mass Firing order Diameter of crank shaft Connection ratio (r/l) Minimum speed Weight	D2842 LE405 128 × 142 mm 662 kW × 2100 r/min 20.4 bar 3.02 kN × m 4.66 kg/cyl 1-12-5-8-3-10-6-7-2-11-4-9 90 mm 0.277 600 r/min 1.79 ton
Intermediate Shaft	Diameter Length	80 mm Approximately 11 m
Propeller	Type Reduction ratio Diameter No. of blade M.O.I.	Azimuth thruster 4.7143:1 1.850 mm 4 ea. 52.0 kg × m^2 (in air)

References

1. Spelsberg-Korspeter, G. Structural optimization for the avoidance of self-excited vibrations based on analytical models. *J. Sound Vib.* **2010**, *329*, 4829–4840. [CrossRef]
2. Wu, C.; Chen, F.; Long, X. The self-excited vibration induced by friction of the shaft-hull coupled system with the water-lubricated rubber bearing and its stick-slip phenomenon. *Ocean Eng.* **2020**, *198*, 107002. [CrossRef]
3. Zhao, X.N.; Chen, G.X.; Lv, J.Z.; Zhang, S.; Wu, B.W.; Zhu, Q. Study on the mechanism for the wheel polygonal wear of high-speed trains in terms of the frictional self-excited vibration theory. *Wear* **2019**, *426*, 1820–1827. [CrossRef]
4. Cui, X.; Chen, G.; Zhao, J.; Yan, W.; Ouyang, H.; Zhu, M. Field investigation and numerical study of the rail corrugation caused by frictional self-excited vibration. *Wear* **2017**, *376*, 1919–1929. [CrossRef]
5. Charroyer, L.; Chiello, O.; Sinou, J.-J. Self-excited vibrations of a non-smooth contact dynamical system with planar friction based on the shooting method. *Int. J. Mech. Sci.* **2018**, *144*, 90–101. [CrossRef]
6. Yan, F.; Luo, C.; Zhu, R.; Wang, Z. Experimental and numerical study of a horizontal-vertical gas-solid two-phase system with self-excited oscillatory flow. *Adv. Powder Technol.* **2019**, *30*, 843–853. [CrossRef]
7. Chen, Z.; Tse, K.T.; Kwok, K.C.S.; Kim, B.; Kareem, A. Modelling unsteady self-excited wind force on slender prisms in a turbulent flow. *Eng. Struct.* **2020**, *202*, 109855. [CrossRef]
8. Purohit, A.; Darpe, A.K.; Singh, S.P. Experimental investigations on flow induced vibration of an externally excited flexible plate. *J. Sound Vib.* **2016**, *371*, 237–251. [CrossRef]
9. Davim, J.P. *Modern Machining Technology: A Practical Guide*, 1st ed.; Woodhead Publishing: Cambridge, UK, 2011.

10. Lee, D.C.; Barro, R.D.; Kim, J.S. A Case Study for the Resonance of Marine Propulsion Shaft System Excited by Diesel Engine. In *Volume 4: Dynamics, Control, and Uncertainty, Parts A and B, Proceedings of the ASME 2012 International Mechanical Engineering Congress and Exposition, Houston, TX, USA, 9–15 November 2012*; ASME: New York, NY, USA, 2012. [CrossRef]
11. Lee, D.C.; Yu, J.D. Transient and Unstable Torsional Vibrations on a 4-Stroke Marine Diesel Engine. In *Design, Application, Performance, and Emissions of Modern Internal Combustion Engine Systems and Components, Proceedings of the ASME 2003 Internal Combustion Engine Division Spring Technical Conference, Salzburg, Austria, 11–14 May 2003*; ASME: New York, NY, USA, 2003. [CrossRef]
12. Lee, D.C.; Barro, R.D. Self-Excited Vibration in a Specialized Electric Propulsion System. *MTZ* **2014**, *4*, 54–59. [CrossRef]
13. Lee, D.C.; Vuong, Q.D.; Nam, T.K. Torsional Vibration Phenomenon due to Pulse Torque of Variable Speed Induction Motor on Rotating Systems. *Trans. Korean Soc. Noise Vib. Eng.* **2015**, *25*, 414–419. [CrossRef]
14. Korean Register. *Broken Propulsion Shaft Investigation Report on the Ship Name 'Haeryong'*; Korean Register Technical Report; Korean Register: Busan, Korea, 2003.
15. Al-Hussain, K.M. Dynamic Stability of Two Rigid Rotors Connected by a Flexible Coupling with Angular Misalignment. *J. Sound Vib.* **2003**, *266*, 217–234. [CrossRef]
16. Hylarides, S.; van Oossanen, P. Some Hydrodynamic Considerations of Propeller-Induced Ship Vibrations. In Proceedings of the SNAME Symposium on Ship Vibration, Arlington, VA, USA, 16–17 October 1978.
17. Ghoneim, H.; Lawrie, D.J. Dynamic Analysis of a Hyperbolic Composite Coupling. *J. Sound Vib.* **2007**, *301*, 43–58. [CrossRef]
18. Dresig, H.; Holzweißig, F.; Grosskopfand, W.; Esche, S. *Dynamics of Machinery: Theory and Applications*; Springer: Heidelberg, Germany; New York, NY, USA, 2010.
19. Vulkan Couplings. *VULASTIK L Technical Data*; VULKAN Kupplungs- und Getriebebau Bernhard Hackforth GmbH & Co. KG: Herne, Germany, 2011.
20. Hyundai Heavy Industries, Co. Ltd. Engine & Machinery Division. Dynamic Characteristic and Performance of Tuning Torsional Vibration Damper for Hyundai-MAN B&W, Two Stroke Low-Speed Diesel Engine. In *MAN B&W Diesel A/S- 14th Meeting of Licensees*; Hyundai Heavy Industries, Co. Ltd. Engine & Machinery Division: Copenhagen, Denmark, 1993.
21. Gang, C. *Handbook of Friction-Vibration Interactions*, 1st ed.; Woodhead Publishing: Cambridge, UK, 2014; pp. 153–305. [CrossRef]
22. Saigo, M.; Okada, Y.; Ono, K. Self-excited Vibration Caused by Internal Friction in Universal Joints and Its Stabilizing Method. *J. Vib. Acoust.* **1997**, *119*, 221–229. [CrossRef]

2

Free-Surface Effects on the Performance of Flapping-Foil Thruster for Augmenting Ship Propulsion in Waves

Evangelos S. Filippas, George P. Papadakis and Kostas A. Belibassakis *

School of Naval Architecture & Marine Engineering, National Technical University of Athens, 15780 Zografos, Greece; evanfilippas@gmail.com (E.S.F.); papis@fluid.mech.ntua.gr (G.P.P.)

* Correspondence: kbel@fluid.mech.ntua.gr

Abstract: Flapping foils located beneath or to the side of the hull of the ship can be used as unsteady thrusters, augmenting ship propulsion in waves. The basic setup is composed of a horizontal wing, which undergoes an induced vertical motion due to the ship's responses in waves, while the self-pitching motion of the wing is controlled. Flapping foil thrusters can achieve high level of thrust as indicated by measurements and numerical simulations. Due to the relatively small submergence of the above biomimetic ship thrusters, the free-surface effects become significant. In the present work, the effect of the free surface on the performance of flapping foil thruster is assessed by means of two in-house developed computational models. On one hand, a cost-effective time-domain boundary element method (BEM) solver exploiting parallel programming techniques and general purpose programming on graphics processing units (GPGPU) is employed, while on the other hand a higher fidelity RANSE finite volume solver implemented for high performance computing (HPC) is used, and comparative results are presented. BEM and RANSE calculations present quite similar trends with respect to mean submergence depth, presenting 12%, 28%, and 18% of differences concerning the mean values of lift, thrust, and moment coefficients, respectively. The latter differences become very small after enhancement of the BEM model to include viscous corrections. Useful information and data are derived supporting the design of the considered biomimetic thrusters, for moderate submergence depths and conditions characterized by minor flow separation effects.

Keywords: flapping foil thruster; augment ship propulsion in waves; free surface effects

1. Introduction

Biomimetic systems such as flapping foils, located on the hull of the ship, are examined for the exploitation of energy in wave induced motions by direct conversion to useful propulsive power. The kinematics of a flapping thruster involve two oscillatory motions heave and pitch with appropriately defined phase difference; see, e.g., [1–3]. With appropriate selection of the pivot point, the most energy demanding foil oscillatory motion is the heaving motion. On the other hand, the most frequent situation of ship operation is in a wavy sea condition, causing in many cases responses of significant amplitude. In particular, the vertical motion of the ship stations especially near the bow (and perhaps also at the stern) can be found exploitable to provide the vertical oscillation of an unsteady flapping foil propulsion system free of cost with concurrently enhancing ship stability and reducing unwanted responses. At the first stage of research, passive flapping wings-systems, attached beneath the ship hull, have been considered in order to produce thrust, exploiting energy that is stored in ship motions; see, e.g., [4,5].

More recently, the replacement of passively controlled systems by actively enforced foil self-pitching motion, based on the (irregular) history of wing's vertical oscillation, has been studied by various authors; see, e.g., [6,7]. In order to produce thrust, a flapping foil performs a complex motion that can be decomposed to a linear and a rotational oscillation, while it advances with the ship's forward speed. The vertical foil oscillation is provided by ship's irregular motion in waves and is obtained by the combined effect of vessel's heaving and pitching at the station alongside the ship where the biomimetic thruster is arranged. Moreover, in the context of the project BIOPROPSHIP [8] and in the works [6,9], a method for the coupling of ship dynamics with unsteady flapping wing hydrodynamics is presented as obtained by using linear seakeeping analysis in conjunction with unsteady lifting-line theory and nonlinear 3D panel methods. Numerical results reveal that significant wave energy can be extracted from ship motions by the flapping-foil thruster and could be used for augmenting ship propulsion in waves, with concurrently reducing ship's responses, enhancing the dynamic stability of the combined system with positive effects on safety, passenger and crew comfort, and general behavior in a seaway. In addition, it is shown that the actively controlled self-pitching motion of the thruster, which is tuned on the basis of the ship vertical motion at the station where the foil is arranged, is very efficient and is characterized by very low cost since the power required is found to be only a very small fraction of the developed thrust power.

Focusing on the importance of the effects associated with the interaction of the foil with the sea waves and the free-surface boundary, in [10] the authors developed a two-dimensional time-domain boundary element method (BEM) to study the performance of a hydrofoil that performs unsteady motions in the proximity of the free surface, in harmonic waves. For the treatment of the dynamical system, the BEM is coupled to the dynamical equations and is integrated in the time domain by appropriate schemes. The above method is validated demonstrated through comparison of calculations against linear theory, RANSE solvers, and experiments. In addition, the significance of the effects of the freely moving boundary over the foil is underlined, and it is shown that the system's performance in harmonic-wave conditions is very promising. Furthermore, aiming to investigate the performance of the combined system (ship in waves and flapping-foil thruster) in more realistic conditions representing specific sea states, in [9], the performance of the system is studied in irregular waves obtained through simulation from frequency spectra. The effect of various parameters of the ship-foil system operating in realistic sea conditions have been presented and discussed, and the results indicate that single or multiple arrangements of oscillating foils attached to the ship hull could enhance the performance in a broad range of parameters.

The increased interest in wave propulsion systems by flapping foils is indicated by recent research on the topic. In fact, results from experimental research in ship models equipped with a foil below the hull demonstrated 40–45% reduction of vessels' heaving and pitching responses with simultaneous 60% reduction of the resistance in waves [11]. The study of Eco-Ship with active foils for wave-empowered propulsion including free-running model test experiments [12], showed speed improvements up to 6.2%. Additionally, applications concerning wave propulsion systems for unmanned wave glide vehicle (UWGV) are presented in [13], and the design of a flapping-foil wave powered vessel using numerical and experimental techniques for wave propelled unmanned surface vehicles (USVs) is studied in [14] in order to control forward speed and recover wave energy. Moreover, in [15], a general cargo ship with retractable foils is studied, reporting important fuel savings at full speed. In [16], the propulsion characteristics of flexible biomimetic propulsors in regular waves are studied, demonstrating that flexibility could enhance the propulsive performance and increase the wave energy extraction; see also [17,18]. A more detailed review can be found in [19,20]. In particular, a 3D nonlinear BEM is proposed in [20] for the flapping foil beneath the free surface in oblique incident waves, and a computational code is developed exploiting parallel programming techniques and general purpose programming on graphics processing units (GPGPU). In complementarity to applications concerning wave propulsion, flapping foils are also studied for the development of hydrokinetic energy devices [21–23] and hybrid devices [24], exploiting combined wave and tidal energy resources.

In the present work, the effect of the free surface on the performance of flapping foil thruster in irregular wave conditions is assessed by means of two in-house developed computational models.

On one hand, a cost-effective time-domain BEM solver exploiting parallel programming techniques and general purpose programming on graphics processing units (GPGPU) is employed, while on the other hand, a higher fidelity RANSE finite volume solver implemented for high performance computing (HPC) is used. Comparative results are presented providing qualitative and quantitative insight on the effects of viscosity, even in cases where the foil operates beneath the free surface performing general motions. Useful information and data are derived exploited for the enhancement of the BEM model and supporting the design of the considered biomimetic thrusters, for moderate submergence depths and conditions characterized by minor flow separation effects.

2. Augmentation of Ship Propulsion in Waves by Flapping Thruster-Problem Definition

We consider a ship in head waves advancing at constant forward speed U; see Figure 1. In the present approach, we use the equations of motion of the ship, derived in the body-fixed frame of reference and linearized by assuming small oscillatory amplitudes and small-slope incident waves; see, e.g., [25]. Standard linear seakeeping method in the frequency domain is used, to calculate the coupled heaving (ξ_3) and pitching (ξ_5) responses of the system (ship and foil) in the vertical plane; see [6,26]. The flapping thruster is located at the bow, at forward station (x_{wing}), and operates in random wave conditions, gaining its vertical motion from the ship heave ξ_3 and pitch ξ_5 responses. Thus, the vertical heaving motion of the foil is $\xi_3 - x_{wing}\xi_5$ and the controlled self-pitching motion is denoted by θ.

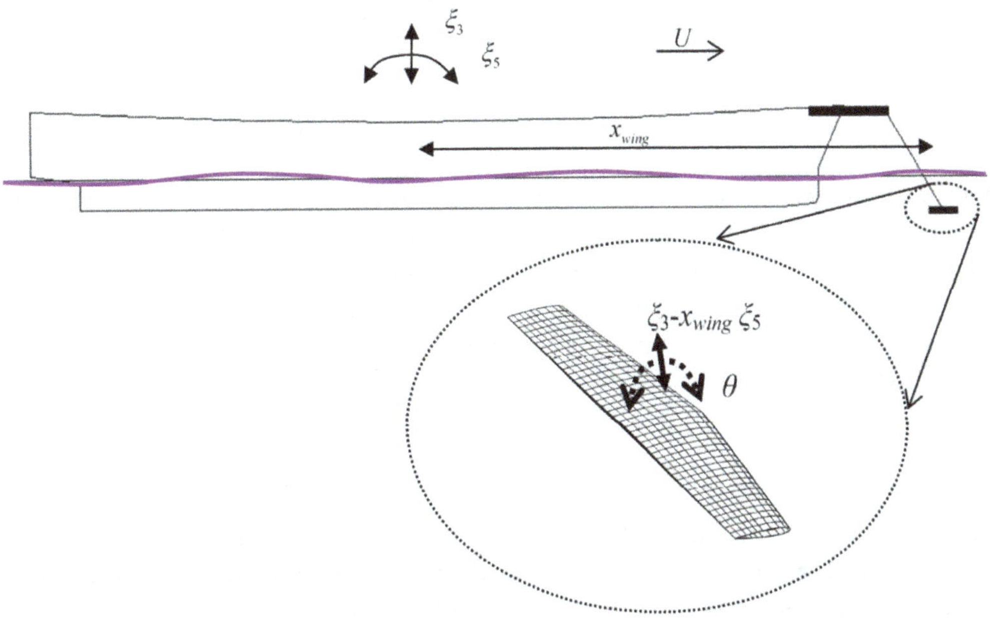

Figure 1. Ship hull, advancing in head waves, equipped with a flapping thruster located at the bow, in a forward station (x_{wing}) augmenting ship propulsion in waves. The biomimetic thruster operates in random wave conditions, gaining its vertical motion from ship heave ξ_3 and pitch ξ_5 responses. The vertical heaving motion of the foil is $\xi_3 - x_{wing}\xi_5$ and the controlled self-pitching motion is denoted by θ.

2.1. *Ship Dynamics Coupled with Unsteady Flapping Foil Thruster*

The system of the equations that describe the dynamics of the system is as follows [6]:

$$D_{33}\xi_{30} + D_{35}\xi_{50} = F_{30} + X_{30} \& \; D_{53}\,\xi_{30} + D_{55}\xi_{50} = F_{50} + X_{50} \tag{1}$$

where ξ_{30}, ξ_{50} are the complex amplitudes of ship heaving and pitching oscillations,

$$D_{33} = \left(-\omega_{en}^2(m + a_{33}) + i\omega_{en}b_{33} + c_{33}\right),\; D_{35} = \left(-\omega_{en}^2(a_{35} + I_{35}) + i\omega_{en}b_{35} + c_{35} + p\right) \tag{2}$$

$$D_{53} = \left(-\omega_{en}^2(a_{53} + I_{53}) + i\omega_{en}b_{53} + c_{53}\right), \; D_{55} = \left(-\omega_{en}^2(a_{55} + I_{55}) + i\omega_{en}b_{55} + c_{55}\right) \tag{3}$$

a_{jk} and b_{jk}, $j,k = 3,5$, are the added mass and damping coefficients, m is the total mass of the ship and the wing, and $p = -i\omega_{en}Um$ is a Coriolis term. Concerning the elements of the hydrostatic matrix, $c_{33} = \rho g A_{WL}$, $c_{35} = c_{53} = -\rho g\left(x_f A_{WL}\right)$ and $c_{55} = m\ g\ GM_L$, where A_{WL} is the waterline area, x_f is the longitudinal center of flotation, GM_L the longitudinal metacentric height, and g the acceleration of gravity. Additionally, $I_{55} = mR_{yy}^2$ and $I_{35} = I_{53} = -m\ X_G$ are inertia coefficients, where X_G is the longitudinal position of the center of gravity and R_{yy} the radius of gyration with respect to the transverse axis associated with the ship pitching motion. In this work, bow waves are considered for the excitation of motions, and the encounter frequency is $\omega_{en} = \omega + kU$, where $k = \omega^2/g$ is the wavenumber of the incident waves and ω is the absolute frequency. Additionally, with $F_{j0}, j = 3,5$, we represent the Froude-Krylov and diffraction force and moment complex amplitudes, respectively.

In the above model, the coupling with the flapping thruster dynamics is achieved through the terms $X_{j0}, j = 3,5$, that denote complex amplitudes of excitation due to the wing and are functions of heaving (ξ_3) and pitching (ξ_5) responses of the ship. The latter vertical force and moment consist of two parts: $X_{j0} = X_{j0,A}(\xi_3, \xi_5) + X_{j0,B}(\varphi_{INC})$, one part dependent on the oscillatory ship amplitudes and one dependent on the incoming wave potential $\varphi_{INC} = \Lambda \exp(kz + ikx)$, where Λ is a constant coefficient involving the incident wave height and frequency.

In the sequel, we will briefly describe the approach that is used in order to analyze the wing forces and moments and more details can be found in [9]. Beginning with the kinematics of the flapping foil, the effective angle of attack is approximately given by

$$\alpha(t) \approx -\xi_5(t) + \varepsilon(t) + \frac{1}{U}\frac{\partial \varphi_{INC}}{\partial z} - \theta(t) \tag{4}$$

where $\theta(t)$ denotes the angle of rotation of the foil. In Equation (4) the effect of wing's vertical oscillation on the angle of attack is $\varepsilon(t) \approx \frac{1}{U}\frac{dh(t)}{dt}$, where $h(t) = -\xi_3(t) + x_{wing}\xi_5(t)$ denotes the oscillatory motion of the ship at the longitudinal position x_{wing} of the foil; see Figure 1. Following the works [6,27], wing self-pitching is selected to be a linear function of the angle ε, i.e., $\theta(t) = w\ \varepsilon(t)$, and the multiplier w stands for the pitch proportional control parameter. Therefore, we finally obtain the following expression for the angle of attack.

$$\alpha(t) = -\xi_5(t) + (1 - w)\ \varepsilon(t) + \frac{1}{U}\frac{\partial \varphi_{INC}}{\partial z}, \tag{5}$$

If we assume a large aspect ratio, unsteady lifting line theory can be used to analyze foil excitation. In the present work, under the additional assumption of small Strouhal numbers, the calculation of forces is based on a quasi-steady approximation and spanwise integration of sectional lift forces, resulting in the following expressions [6].

$$X_{30} = \frac{1}{2}\rho U^2 S_w\ C_L^{3D} \approx \frac{1}{2}\rho U^2 S_w\ \frac{AR}{AR+2} C_L^{2D} = G\alpha_0 + \frac{Gc_R h_0}{4U^2}\omega^2 + X_{30,B}(\varphi_{INC}) \tag{6}$$

$$X_{50} = \frac{1}{2}\rho U^2 c_R S_w\ C_M^{3D} - \frac{1}{2}\rho U^2 S_w C_L^{3D} x_{wing} \approx -x_{wing} X_{30} \tag{7}$$

where c_R denotes the root chord of the foil, S_W is the wing planform area, AR its aspect ratio, ρ the water density, and h_0 and a_0 are the amplitude of heaving motion and the complex amplitude of the angle of attack, respectively. Furthermore, C_L^{3D}, C_M^{3D}, and C_L^{2D} are the 3D and 2D lift and moment coefficients. For more details see, e.g., [6], Equations (14)–(17). Moreover, $X_{30,B}(\varphi_{INC})$ is the contribution of the incoming wave and $G = \chi\ \pi\rho U^2 S_w\ \frac{AR}{AR+2}$. In the latter expression, χ denotes an empirical correction factor calculated by comparing the quasi-steady lifting line estimations with a posteriori predictions by more accurate time domain panel methods for the responses of the oscillatory wing, following the same combined forward and oscillatory motion.

Applying the aforementioned analysis in the dynamic system Equations (1)–(3), we introduce the following variations to the system coefficients at the left hand side, due to the operation of the flapping foil:

$$\delta a_{33} = -Gc/\left(4U^2\right),\ \delta a_{35} = \delta a_{53} = x_{wing}Gc/\left(4U^2\right),\ \delta a_{55} = -x_{wing}^2 Gc/\left(4U^2\right) \tag{8}$$

$$\delta b_{33} = G(1-w)/U,\ \delta b_{35} = \delta b_{53} = -G(1-w)x_{wing}/U,\ \delta b_{55} = G(1-w)x_{wing}^2/U \tag{9}$$

$$\delta c_{35} = G,\ \delta c_{55} = -G\, x_{wing} \tag{10}$$

Furthermore, we obtain the next set of expressions concerning the incident wave-field dependent forces:

$$X_{30,B} = G\ (i\omega/U)\ \exp(-kd+ikx),\ X_{50,B} = -G\, x_{wing}\ (i\omega/U)\ \exp(-kd+ikx) \tag{11}$$

where the vertical location of the foil beneath the free surface is at $z = -d$, where d is the foil submergence depth and c the foil root chord length.

2.2. *Flapping Thruster Performance in Irregular Waves*

Using the responses of the system, the kinematics of the flapping wing are obtained, permitting the detailed formulation and study of its performance in random head waves. For this reason, the panel method developed in [10] is applied for the numerical simulation of the performance of hydrofoils at low submergence depth including the free surface effects. We use the assumption of waves with small amplitude in comparison with the wavelength and we work in the context of linear water wave theory. Realistic sea conditions are modeled by parametric frequency spectra and the realization of the free surface elevation and of other wave signals is implemented by using the random phase model. More details can be found in [9].

The Froude number based on the length of the ship is $F_n = U/\sqrt{gL}$ and the corresponding one based on the root chord c of the foil is $F_{foil} = U/\sqrt{gc}$. We denote with $S(\omega)$ the wave spectrum in the earth fixed frame of reference and we can obtain an expression for the spectrum according to a moving observer with the foil mean velocity U, as follows:

$$S^U(\omega_e) = S(\omega)\frac{d\omega}{d\omega_e},\ \text{with } d\omega = d\omega_{en}/\left(1+\frac{2U}{gY}\right),\ \text{and } Y = \tanh(kH) + kH\,\mathrm{sech}^2(kH) \tag{12}$$

Then we use the response amplitude operator $RAO = \xi_3 - x_{wing}\xi_5$ at the position x_{wing} of the wing, in order to obtain the spectrum of wings vertical motion:

$$S^U_{foil}(\omega_e) = S^U(\omega_e)\left|RAO_{foil}(\omega_e)\right|^2 \tag{13}$$

The vertical motion of the flapping thruster is simulated by the random-phase model by discretizing the band of encounter frequencies into a large number of consecutive bins of width $\delta\omega$ centered at discrete values $\omega_{e,n}$

$$\eta(t) = \sum_n \eta_n \cos\left(\omega_n t + kx_{wing} + \varepsilon_n\right),\ \text{where } \eta_n = \sqrt{2S^U_{foil}(\omega_{e,n})\delta\omega_e} \tag{14}$$

ε_n are random phases that uniformly distributes in $[0, 2\pi)$; see, e.g., [28]. In Equation (12), the effect of finite water depth H has been included, and the case of deep water conditions is modeled by using appropriate large values.

Additionally, the rotational motion of the flapping wing, about its pivot axis is defined as negative at the clockwise direction. The pivot point is located at a distance $x_R = c/3$ from the leading edge.

The self-pitching motion of the foil is set by a simple proportional control law based on its linear oscillation history:

$$\theta(t) = w \tan^{-1}\left(\frac{1}{U}\frac{dh(t)}{dt}\right) \tag{15}$$

where $0 < w < 1$ is the control parameter; see, e.g., [6,27].

As an example, in Figures 2 and 3 we present selected results in the case of a ship model of a fast ferry travelling at $F_n = U/\sqrt{gL} = 0.25$ with a flapping thruster operating at the bow, as schematically shown in Figure 1. The ship length to beam ratio is L/B = 7.7, the beam to draft ratio B/T = 3.71, and the block coefficient $C_b = 0.45$. Moreover, the waterplane area coefficient is $C_{WL} = 0.69$, the midsection coefficient $C_m = 0.81$, and the flotation-center $x_f/L = -0.05$ (LCF aft midship). Additionally, the metacentric radius $BM_L/L = 1.8$. The vertical position of the buoyancy center is KB/L = 0.021 (from BL), and its position on the horizontal axis is LCB = −0.019 L.

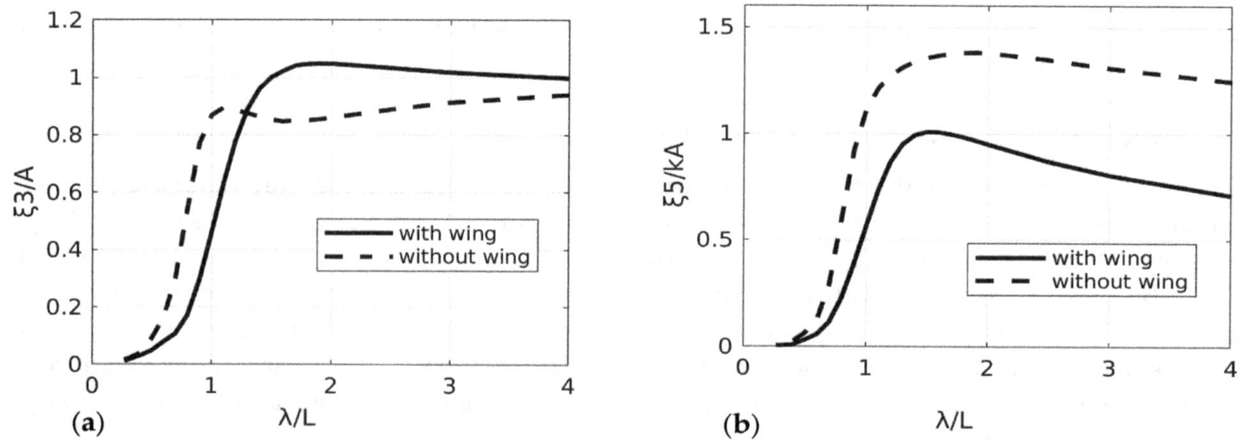

Figure 2. (**a**) Heave and (**b**) pitch response of the examined ship hull against nondimensional wavelength λ/L, for Froude number $F_n = U/\sqrt{gL} = 0.25$, $F_{foil} = U/\sqrt{gc} = 1.35$. Comparison of system responses with (bold line) and without (dashed line) the operation of the foil.

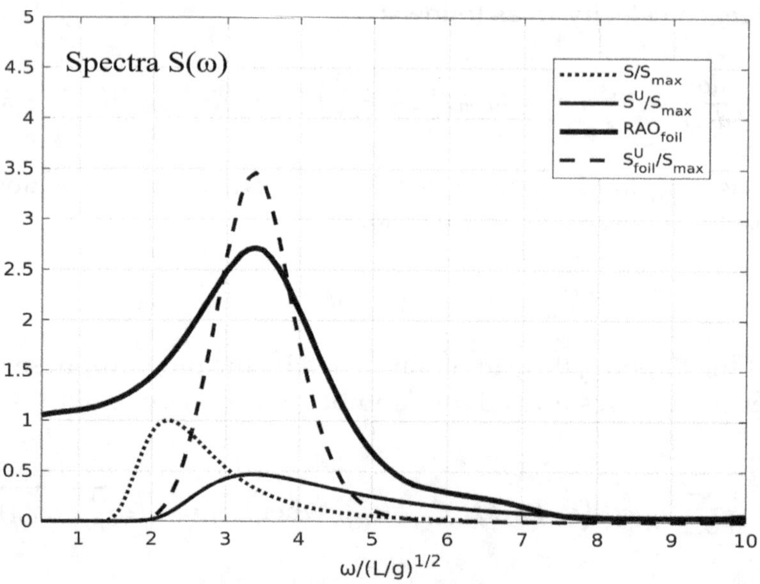

Figure 3. Sea spectrum ($H_s/L = 0.03$ and $T_pU/L = 0.7$) with respect to the inertial S (dotted line) and the moving frame S^U for $F_n = 0.25$ (solid line), for head seas ($\beta = 180°$), normalized with respect to the peak value of the incident wave spectrum S_{max}. The ship and foil system RAO modulus at the position of the foil RAO_{foil} ($x_{wing}/L = 0.6$) are shown by using a thick line and the spectrum of the foil's vertical motion S^U_{foil} by dashed line, respectively.

We assume that the horizontal position of the gravity center is the same as the location of the center of buoyancy, i.e., $X_G/L = -0.019$ (aft midship), $Y_G = 0$, and KG/T = 0.2 (from BL). Furthermore, the longitudinal metacentric height is $GM_L \approx BM_L$. Finally, the radii of gyration about the x-axis and y-axis, respectively, are taken $R_{xx}/B = 0.25$, $R_{yy}/L = 0.2$. The foil is located at a distance $x_{wing}/L = 0.6$, with respect to the midship section, and vertically at submergence d/T = 1.71 below WL. The wing planform shape is orthogonal, its chord is c/B = 0.27, and its span to chord ratio is s/c = 5, and thus, its aspect ratio is AR = 5. The wing planform area is $S_w/BT = 1.35$, and the wing sections are symmetrical NACA0012.

In particular, in Figure 2, the ship heave and pitch responses are presented against nondimensional wavelength λ/L, for Froude number. $F_n = U/\sqrt{gL} = 0.25$ and head waves ($\beta = 180°$). Additionally, the system responses are compared with (bold line) and without (dashed line) the operation of the flapping wing thruster. It is seen that the operation of the foil at a corresponding Froude number $F_{foil} = U/\sqrt{gc} = 1.35$ produces a modification of the heave responses due to changes of the resonance and has a more signifficant effect on the pitch responses, especially as frequency becomes lower than the resonant value. The latter is due to the antipitching moment generated by the foil lift and the large arm (x_{wing}/L = 0.6) due to the arrangement of the foil in front of the bow of the ship; see Figure 1. Next, in Figure 3, the frequency spectrum concerning the vertical foil motion is presented against the ones of the incident head waves in the earth-fixed system and in the ship frame of reference. The incident frequency spectrum is modeled by the Bretschneider model

$$S(\omega) = \frac{1.25}{4}\frac{\omega_m^4}{\omega^5}H_s^2 epx\left(-1.25\frac{\omega_m^4}{\omega^4}\right), \tag{16}$$

and the frequency spectrum for significant wave height $H_s/L = 0.03$ and peak period $T_p U/L = 0.7$, for head seas ($\beta = 180°$), is shown by dotted line. The corresponding wave spectrum in the ship frame of reference travelling at $F_n = 0.25$ (solid line), is shown by using a solid line, and line and the spectrum of the foil's vertical motion by dashed line, respectively. Moreover, the ship and foil system RAO modulus at the position of the foil (x_{wing}/L = 0.6) is shown by using a thick line. In the examined case, the encounter frequency spectrum is concentrated around the resonant behavior in the vertical ship motion at the bow, and we are able to observe in Figure 3 the magnification of the spectrum density around the peak value due to the large responses. The above effect is illustrated also by means of a realization of the vertical motion of the foil arranged in front of the bow of the ship for the above wave conditions and for a time duration of 25 peak periods as presented in Figure 4. The result is compared with the ship heaving motion shown in the same plot by using a dashed line, making clearer the enlargement of the foil vertical motion.

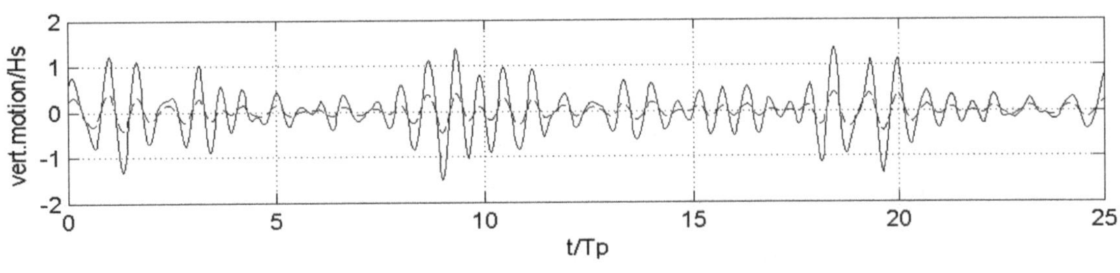

Figure 4. Realization of the vertical motion of the foil (solid line) at the bow of the ship in comparison with the ship heave motion (dashed line) for the wave conditions of Figure 3.

3. Flapping Thruster Performance in Irregular Waves

In Section 2, the augmentation of ship propulsion in waves using flapping foils is formulated and a mixed frequency–time domain method based on linearized ship dynamics coupled with unsteady lifting line theory is presented. The latter method is applied to obtain the responses of the ship-foil system in irregular waves corresponding to specific sea states. Using the responses of the system, the kinematics of the flapping thruster are obtained, permitting the detailed study of its performance in random head waves by cost-effective BEM and high fidelity RANSE solvers presented in more detail in this section.

3.1. BEM Model of the Flapping Thruster in Waves

In this subsection, the BEM model, developed by [10], is applied to obtain the performance of flapping wing thruster operating in random head waves. The problem is treated in the time domain and the flapping hydrofoil is denoted with a closed boundary $\partial D_B(t)$ moving with respect to an earth fixed reference frame. The model provides the two-dimensional lifting flow characteristics taking into account interaction with the free surface, and three-dimensional corrections can be approximately taken into account by means of coefficients based on aspect ratio, as discussed in the previous section (see, e.g., Equation (6)).

In the context of linear wave theory, the total wave potential $\Phi_T(x, y; t)$ is decomposed to the incident irregular wave potential $\Phi_I(x, y; t)$, and its disturbance is due to the operation of the foil $\Phi(x, y; t)$. The problem has been reformulated with respect to the disturbance potential satisfying on the solid boundaries ∂D_B the no entrance Neumann condition,

$$\frac{\partial \Phi(x, y; t)}{\partial n_B} = b\,,\ (x, y) \in \partial D_B, \text{ where } b = -\frac{\partial \Phi_I(x, y; t)}{\partial n_B} + \boldsymbol{V}_B \cdot \boldsymbol{n}_B \tag{17}$$

and the corresponding bottom boundary condition in the case of finite water depth. Moreover, the linearized free surface boundary conditions are applied on the mean free-surface level. In the above, $\boldsymbol{n}$ denotes the normal to the boundary unit vector, directed to the interior of the domain D. In order to treat lifting flow problems around bodies with sharp boundaries as the trailing edge of a hydrofoil, a vortex wake model must be used in order to model the generation of lift on the foil and the introduction of vorticity on the wake. In the present work, a curve of potential discontinuity is generated behind the foil and a simplified wake model is used. The trailing vorticity model as well as the implementation of a pressure-type Kutta condition are applied on the foil's trailing edge. More details are provided in [10].

Following a direct approach based on application of Green's theorem, the following set of boundary integral equations (on body boundary and the free surface) is obtained, with unknowns the potential on the body boundary Φ_B and on the free surface Φ_F, as well as their normal derivatives

$$\frac{1}{2}\Phi_{B/F}(\boldsymbol{x}_0; t) = \iint\limits_{\partial D_B} b(\boldsymbol{x}; t) G(\boldsymbol{x}_0|\boldsymbol{x}) - \Phi_B(\boldsymbol{x}; t)\frac{G(\boldsymbol{x}_0|\boldsymbol{x})}{\partial n}\, dS(\boldsymbol{x}) +$$

$$+\iint\limits_{\partial D_F} \frac{\partial \Phi_F(\boldsymbol{x}; t)}{\partial n} G(\boldsymbol{x}_0|\boldsymbol{x}) - \Phi_F(\boldsymbol{x}; t)\frac{\partial G(\boldsymbol{x}_0|\boldsymbol{x})}{\partial n}\, dS(\boldsymbol{x}) - \iint\limits_{\partial D_W} \mu_W(\boldsymbol{x}; t)\frac{\partial G(\boldsymbol{x}_0|\boldsymbol{x})}{\partial n}\, dS(\boldsymbol{x}) \tag{18}$$

where $G(\boldsymbol{x}_0|\boldsymbol{x})$ denotes the Green's function, which includes the mirror of the source singularity with respect to the bottom surface ($y = -H$):

$$G(\boldsymbol{x}_0|\boldsymbol{x}) = \frac{\ln r(\boldsymbol{x}_0|\boldsymbol{x})}{2\pi} + \frac{\ln r(\boldsymbol{x}_0|\boldsymbol{x}_m)}{2\pi},\ r(\boldsymbol{x}_0|\boldsymbol{x}) = |\boldsymbol{x}_0 - \boldsymbol{x}| \tag{19}$$

In the above equations, $\boldsymbol{x}_0 = (x_0, y_0)$ denotes a control point, $\boldsymbol{x}_0 = (x, y)$ is the integration point, and $\boldsymbol{x}_m = (x, -y - 2H)$ its image with respect the horizontal bottom.

Applying a low-order boundary element method, the body contour is approximated by a closed polygonal line of N_B straight line elements, and also the boundaries of the free surface and of the vortex wake are represented by N_F and $N_W(t)$ elements, respectively. It is noted here that the problem is treated in the time domain and thus the size vortex sheet emanating from the foil trailing edge increases in time, as it also happens with the number of wake elements. The various hydrodynamic quantities on the panels, such as the potential, its normal derivative, are assumed to be approximated by piecewise constant distributions. Moreover, a collocation scheme is used with control points at the center of each panel, and the following discretized form of Equation (18) is obtained.

$$\left[\begin{array}{c}\boldsymbol{\Phi}_{\mathrm{B}}\\ \frac{\partial\boldsymbol{\Phi}_{F}}{\partial n}\end{array}\right]=\boldsymbol{D}\cdot\left[\begin{array}{c}\boldsymbol{b}\\ \boldsymbol{\Phi}_{F}\end{array}\right]+\boldsymbol{P}(\boldsymbol{\mu}_{\mathrm{W}})+\boldsymbol{Z}\mu_{\mathrm{W1}} \tag{20}$$

where the vectors $\boldsymbol{\Phi}_B$, $\boldsymbol{\Phi}_F$, $\boldsymbol{b}$, $\frac{\partial\boldsymbol{\Phi}_F}{\partial n}$, and $\boldsymbol{\mu}_W$ consist of hydrodynamic quantities at collocation points of each boundary, $\boldsymbol{\mu}_W$ is the dipole strength in the foil's wake, and the matrixes $\boldsymbol{D}, \boldsymbol{P}, \boldsymbol{Z}$ are calculated as functions of panel integrals, that in the case of low order panel methods can be obtained analytically. The detailed expressions are provided in [10]. Equation (20) is the discrete Dirichlet to Neumann (*DtN*) operator that relates the potential with its derivative on the boundary, and also includes the unknown value of the potential jump or the dipole intensity μ_{W1} on the wake at the vicinity of the trailing edge. Applying the discrete *DtN* map, Equation (20), in the discrete form of the pressure-type Kutta condition and the free surface conditions, a system of ODEs is finally derived governing the evolution of the dynamical system in time, within the context of our approximation,

$$\frac{d\boldsymbol{U}}{dt}=f(\boldsymbol{U})\text{ where }\boldsymbol{U}=[\boldsymbol{\Phi}_F\quad\eta\quad\mu_{W1}]^T \tag{21}$$

More details concerning the analytic structure of $f(\boldsymbol{U})$ and the numerical method for the time integration, as well as for the numerical treatment of the horizontally infinite boundary can be found in [10].

After the solution has been obtained (at each time step), the pressure is calculated through the unsteady Bernoulli's theorem. Furthermore, forces and the moment are directly calculated through the integration of the instantaneous pressure acting on the solid boundary. The forces are corrected to include viscous effects according to the following empirical shear stress coefficient (see also [29]), as a function of Reynolds number $Re = Uc/v$ (where v is the kinematic viscosity) and the angle of attack a, as follows

$$c_{cor}=\frac{0.0858}{\left[\log_{10}Re-1.22\right]^2}+c_a(Re)\,a^2 \tag{22}$$

In the present study, where Re is of the order of 10^6, the parameter c_a is calibrated by RANSE simulation of flapping foil in random motion in unbounded domain as described in Section 4.3, and the corrections are used successfully in all the cases where the foil operates with the effect of the free surface in (Sections 4.4 and 4.5) and also exploited for a corrected prediction in Section 4.2 where also incident waves are considered.

3.2. *CFD model of the Flapping Thruster in Waves*

In this section, the theoretical background of the CFD methodology used in the present work is described. The incompressible equations are solved using the artificial compressibility formulation, while for the two phase flows, an additional equation of the volume fraction (VOF) [30], is added to the coupled system of equations; see also [31]. The governing equations are written in the following integral form:

$$\mathbf{\Gamma}\int_{\Omega}\frac{\partial\mathbf{Q}}{\partial\tau}d\Omega+\mathbf{\Gamma}_e\int_{\Omega}\frac{\partial\mathbf{Q}}{\partial t}d\Omega+\int_{\partial\Omega}(\mathbf{F}_c-\mathbf{F}_v)dS=\int_{\Omega}\mathbf{S}_q d\Omega \tag{23}$$

In the above system of equations, $\mathbf{Q}=[p,\mathbf{u},a]^T$ is the vector of the primitive variables, which contains the pressure, the velocity vector, and the volume fraction. The real and the fictitious time are denoted with t and τ, respectively. The inviscid fluxes are denoted with $\mathbf{F}_c$, while $\mathbf{F}_v$ are the viscous fluxes, and $\mathbf{S}_q$ are the source terms. The above system of equations is closed with the artificial compressibility parameter β, where it is assumed that there is a relation between density and pressure in pseudo-time $\beta=\left.\frac{\partial\rho}{\partial p}\right|_{\tau}$.

Although the system of equations is casted in primitive form, for time true computations, the conservative variables, $\mathbf{U} = [0, \rho\mathbf{u}, a]^T$, are used to advance the solution in time. For this purpose, the Jacobian transformation matrix $\mathbf{\Gamma}_e$ is introduced in Equation (23). Furthermore, in order to achieve better convergence characteristics, the preconditioner matrix $\mathbf{\Gamma}$ is used [32]. The preconditioner multiples the fictitious time derivatives and is also used in the inviscid flux evaluation to remove the dependency of the eigenvalues from the density, when multiphase flows are considered. The two matrices are given by

$$\mathbf{\Gamma} = \begin{bmatrix} \frac{1}{\rho_m} & 0 & 0 \\ 0 & \rho_m I & \mathbf{u}\Delta\rho \\ \frac{a_l}{\rho_m} & 0 & \beta \end{bmatrix}, \mathbf{\Gamma}_e = \begin{bmatrix} 0 & 0 & 0 \\ 0 & \rho_m I & \mathbf{u}\Delta\rho \\ 0 & 0 & \beta \end{bmatrix} \tag{24}$$

The governing equations are discretized using the finite volume method. The discretized form of the system Equation (23) for a given control volume Ω_I, is written as

$$\Gamma_e \frac{\partial(\mathbf{Q}\Omega_I)}{\partial t} + \Gamma\Omega_I \frac{\partial \mathbf{Q}}{\partial \tau} = -\mathbf{R}_{\Omega_I} \tag{25}$$

The residual $\mathbf{R}_{\Omega}$ is computed as a sum of the fluxes over the interfaces of the control volume with the addition of the source terms $\mathbf{S}_q$, which are considered to be constant in Ω_I.

$$\mathbf{R}_{\Omega_I} \simeq \sum_{j}^{N_f(\Omega_I)} (\mathbf{F}_c - \mathbf{F}_v)_j \Delta S_j + \Omega_I \mathbf{S}_q \tag{26}$$

The inviscid fluxes are evaluated by solving the preconditioned local Riemann problem between the neighbors of the face, using the approximate Riemann solver [33]. For the reconstruction of the velocity field, a piecewise linear interpolation scheme is used. For the pressure field, a similar methodology is followed, but due to the discontinuity of the pressure gradient, a density weighted interpolation scheme is adopted [34]. Furthermore, for the reconstruction of the VOF field, a high order interface capturing scheme is used [35], in order to reduce the numerical diffusion near the free surface.

For the turbulence modeling, the two equations model k-ω SST [36] is used, while a buoyancy term is added to the kinetic energy equation to suppress the turbulent viscosity at the free surface [37].

MaPFlow implements the technique of the deforming grids, which is described in [38], in order to account flows with objects that perform small amplitude motions.

For unsteady simulations, an implicit second order backwards difference scheme (BDF) is used [39], along with a dual time-stepping technique in order to facilitate convergence. Finally, for the inversion of the implicit operator, the Gauss–Seidel iterative method is used combined with the Reverse Cuthill–Mckee reordering scheme.

The above methodology was implemented as an extension in the compressible solver MaPFlow, see [40,41]. MaPFlow solves the unsteady Reynolds averaged Navier–Stokes equations on unstructured grids, in a multiprocessor environment utilizing the MPI protocol and is developed in the National Technical University of Athens.

3.3. Grid Generation

In this section, the basic numerical parameters for the CFD methodology are described. Figure 5 illustrates the basic numerical setup for the CFD simulations. All the CFD simulations employ the volume of fluid (VOF) approach to resolve both the air and the water phase. For the mesh generation, the BETA CAE ANSA preprocessor was used. The mesh is unstructured and extends 160 chords in the chordwise and 70 chords in the vertical direction.

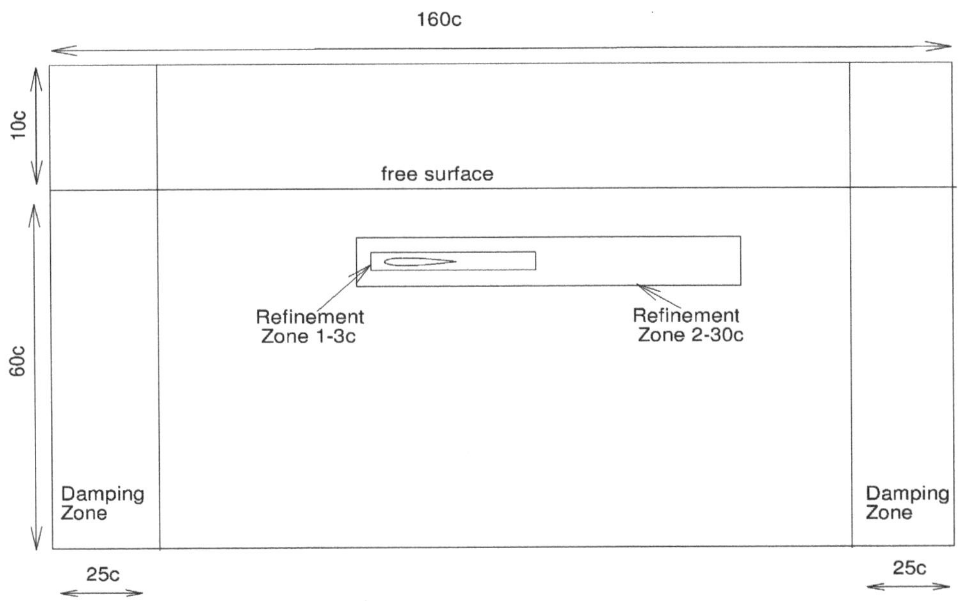

Figure 5. RANSE grid setup.

At the inlet and outlet boundaries, far field conditions are imposed while damping zones are employed to prevent numeral reflections from the outer boundaries reaching the domain of interest. Far-field conditions are also imposed on the top boundary (10 chords from the free surface) where only the air-phase is present, and similarly at the bottom is located 60 chords below the free surface to ensure a deep-water approximation. Finally, no-slip condition is employed for the NACA0012 hydrofoil. The body undergoes a combined heaving and pitching motion and consequently properly resolving the generated wake, which is essential to improve the accuracy of the numerical predictions. To that end, two refinement regions are defined in the hydrofoil wake. One refinement zone is at the near hydrofoil region spanning 3 chords (RF1) and another spanning 30 chords to resolve the far wake (RF2), as can be seen in Figure 6. The free surface is modeled though the VOF approach while the STACS scheme [35] is used for the reconstruction of the volume fraction. The computational mesh in the free surface region is structured (while the rest of the mesh remains unstructured) and the resolution was chosen to have 180 points per wavelength and 50 points per wave height (the maximum is considered). Regarding the boundary layer region, the first spacing is 1e-06 chords, ensuring a y+ below 1, while 600 nodes were placed around the hydrofoil. The boundary layer consists of 65 layers with a growth factor of 1.1. A snapshot of the computational mesh in the vicinity of the body is shown in Figure 7.

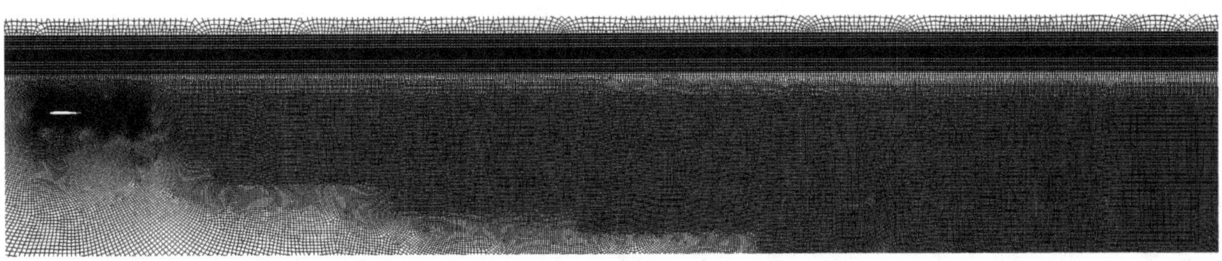

Figure 6. A snapshot of the finer computational mesh. The refinement region (RF2) behind the hydrofoil is evident as well as the structured region in the free surface vicinity.

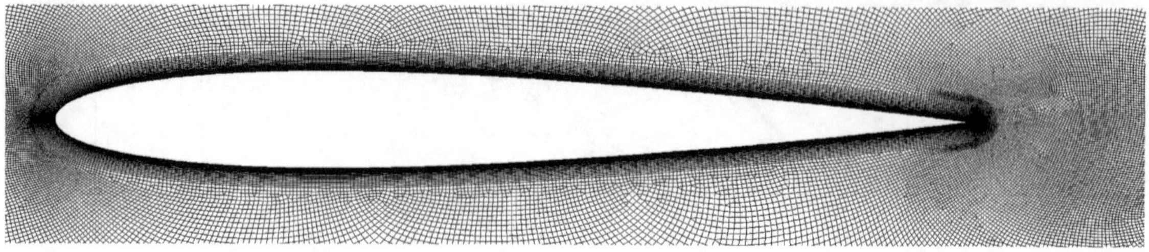

Figure 7. The computational grid near the hydrofoil.

In order to simulate the hydrofoil motion, the technique of deforming grids is utilized. In particular, the near grid follows the prescribed body motion, while the rest is adapted so that the displacement of the nodes decays as the distance from the moving boundary increases. At each timestep, the grid velocities are obtained by differences based on three snapshots.

Three successively refined grids are considered. Details are presented in Table 1. For the refinement zones (RF1, RF2) the cell sizes are presented with respect to the chord length. Grid-independence and convergence of numerical results is discussed in Section 4.5.

Table 1. Properties of computational meshes.

Grid	RF1 Cell Base Size	RF2 Cell Base Size	F.S. Nodes (Height)	F.S. Nodes (Length)	Grid Cells (Nc)
Grid-1	0.006	0.08c	20	120	3×10^5
Grid-2	0.004	0.04c	40	160	4.2×10^5
Grid-3	0.003	0.03c	50	180	5.8×10^5

4. Numerical Results and Discussion

In this section, a comparison between the BEM and the RANSE solvers is presented in a case of a steady foil beneath the free surface (Section 4.1). Then, the mixed frequency–time domain method presented in Section 2 is exploited to obtain the kinematics of the ship and wing in waves and study the performance of the flapping thruster using the BEM solver (Section 4.2). Results concerning the flapping foil at great submergence (without free-surface effects) are presented in Section 4.3, and close to the free surface in Section 4.4, as obtained by using the BEM and RANSE solvers. Finally, in Section 4.5 systematic investigation of free-surface, viscosity, and incident-wave effects is presented and discussed including convergence study of both solvers.

4.1. BEM and RANSE Calculations in Steady Conditions

Before we proceed to the unsteady calculations, we present a comparison between the BEM and the RANSE solvers in a case of a steady NACA0012 foil (Figure 8a) advancing below the free surface with velocity corresponding to foil Froude number $F_{foil} = U/\sqrt{gc} = 1.35$, with angle of attack $a = 5$ and submergence depth $d/c = 1.71$. The rotation axis is located on the chord at a distance $X_R/c = 0.33$ from the leading edge. This case corresponds to the wave resistance problem of a submerged hydrofoil steadily travelling near the calm free surface. The foil accelerates gradually until steady conditions are obtained, which is achieved after travelling a length of more than 100 wavelengths, which as predicted by linear theory in this case is $\lambda/c = 2\pi F_{foil}^2$. Comparison of simulations is presented in Figure 8, and it is observed that the calculated wavelength by both methods is in agreement with the prediction by the dispersion relation of linear theory. Calculated free surface elevation is comparatively shown in Figure 8b,c, where also the position of the foil is indicated at $x/c = 0,\ d/c = 1.71$ Furthermore, in Figure 8d,e, the calculated pressure coefficient on the foil $\left(c_p = p - p_{atm}/0.5\rho U^2\right)$ is shown, and the results by both methods indicate very good agreement.

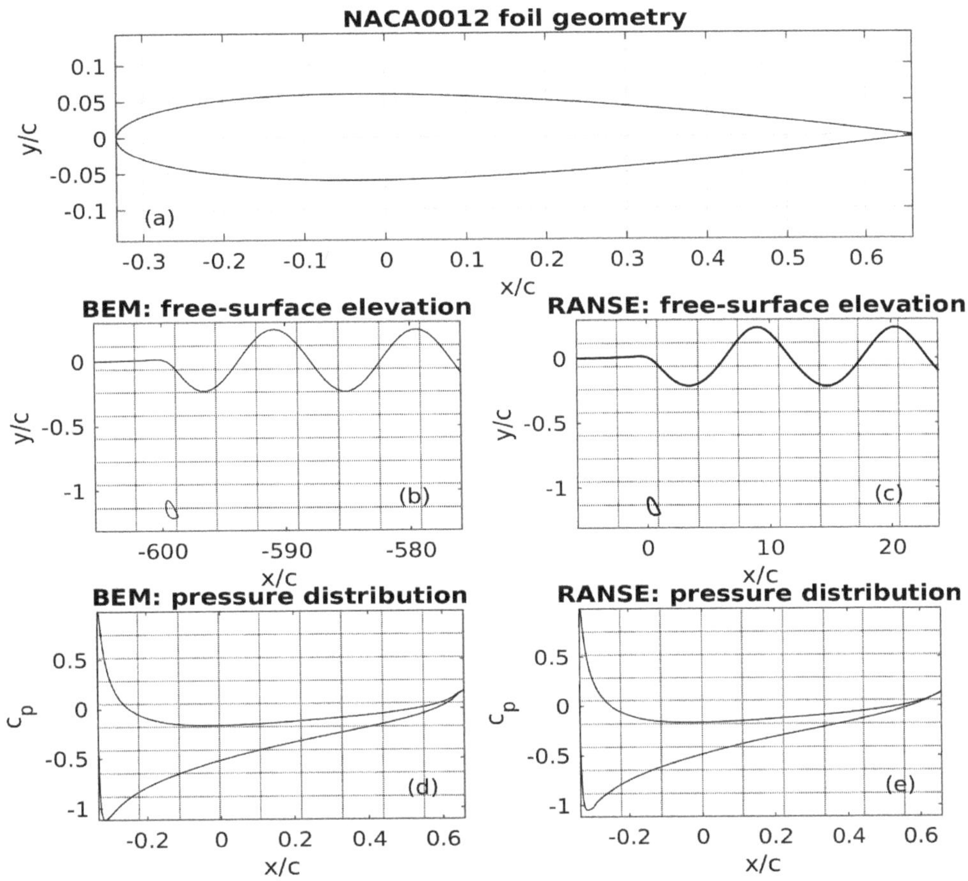

Figure 8. (**a**) NACA0012 hydrofoil considered for the wave resistance problem $F_{foil} = 1.35$ and submergence $d/c = 1.71$. Numerical results obtained from present BEM are shown in the left column subplots (**b**,**d**), while corresponding predictions by RANSE are shown in the right-hand side subplots (**c**,**e**). Calculated free surface elevation (**b**,**c**) and pressure coefficient $\left(c_p = p - p_{atm}/0.5\rho U^2\right)$ on the flapping foil (**d**,**e**).

4.2. Foil Motion and BEM Calculations in Waves

Next, unsteady numerical calculations are presented in Figure 9, considering the performance of the flapping thruster in the bow of the ship subjected to vertical oscillations due to ship responses in head waves and at the same time performing self-pitching oscillations about its pivot axis at $x_R/c = 0.33$ with controllable amplitude based on its vertical motion, as described in previous Sections 2 and 3.1. The flapping foil is located at the same mean submergence $d/c = 1.71$, advancing with the same as before velocity $F_{foil} = 1.35$ $\left(F_{ship} = 0.25\right)$, in irregular waves ($H_s/c = 0.86$ and Strouhal number $\omega_{p,en}H_s/(2\pi U) = 0.067$, where $\omega_{p,en}$ is the frequency of encounter that corresponds to the peak period of the spectrum; see Section 2.

Calculations have been done by using pitch control parameter $w = 0.5$. In particular, in Figure 9, the ship travelling in irregular waves is shown at five instants within a time interval corresponding to one modal period. The foil is located in front of the bow of the ship, at distance $x_{wing}/L = 0.6$ with respect to the midship section of the ship. In the same figure, the foil wake is plotted, with the calculated dipole intensity (potential jump) on the vortex sheet, which is illustrated by using arrows normal to the wake curve with length proportional to the local dipole strength. The latter result is associated with the memory effect of the generated lifting flow around the flapping foil operating in random incident waves.

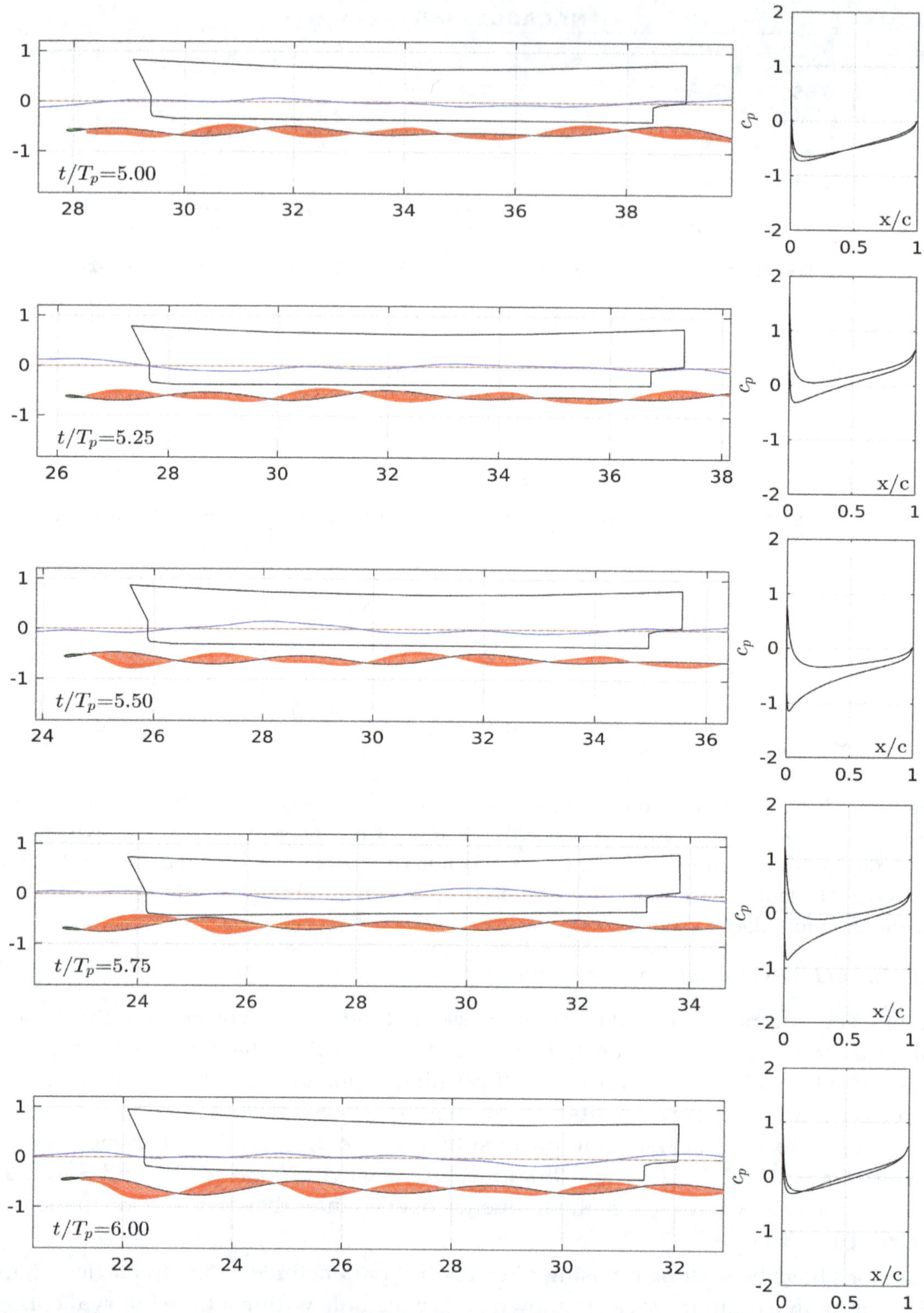

Figure 9. Boundary element method (BEM) simulation of the system ship and flapping thruster operating in head waves (for parameters see Figure 3). Results plotted at various instances in one modal period. In the right subplots, the distribution of the unsteady pressure coefficient on the flapping foil, located at the bow, is plotted. The unit length of both horizontal and vertical axes is equal to L/10.

Moreover, in the right subplots of Figure 9, the instantaneous distribution of the pressure coefficient on the foil, at the same time instants, as calculated by the present method. From the calculated pressure distributions, lift and thrust components are obtained at each time step by integration. In order to illustrate the relative magnitude of the incident and disturbance flow generated by the flapping motion of the foil in waves, for the same as before wave conditions and hydrofoil data, the calculated free-surface elevation normalized with respect to the significant wave height plotted in Figure 10 over the horizontal domain, at some instances during two modal periods.

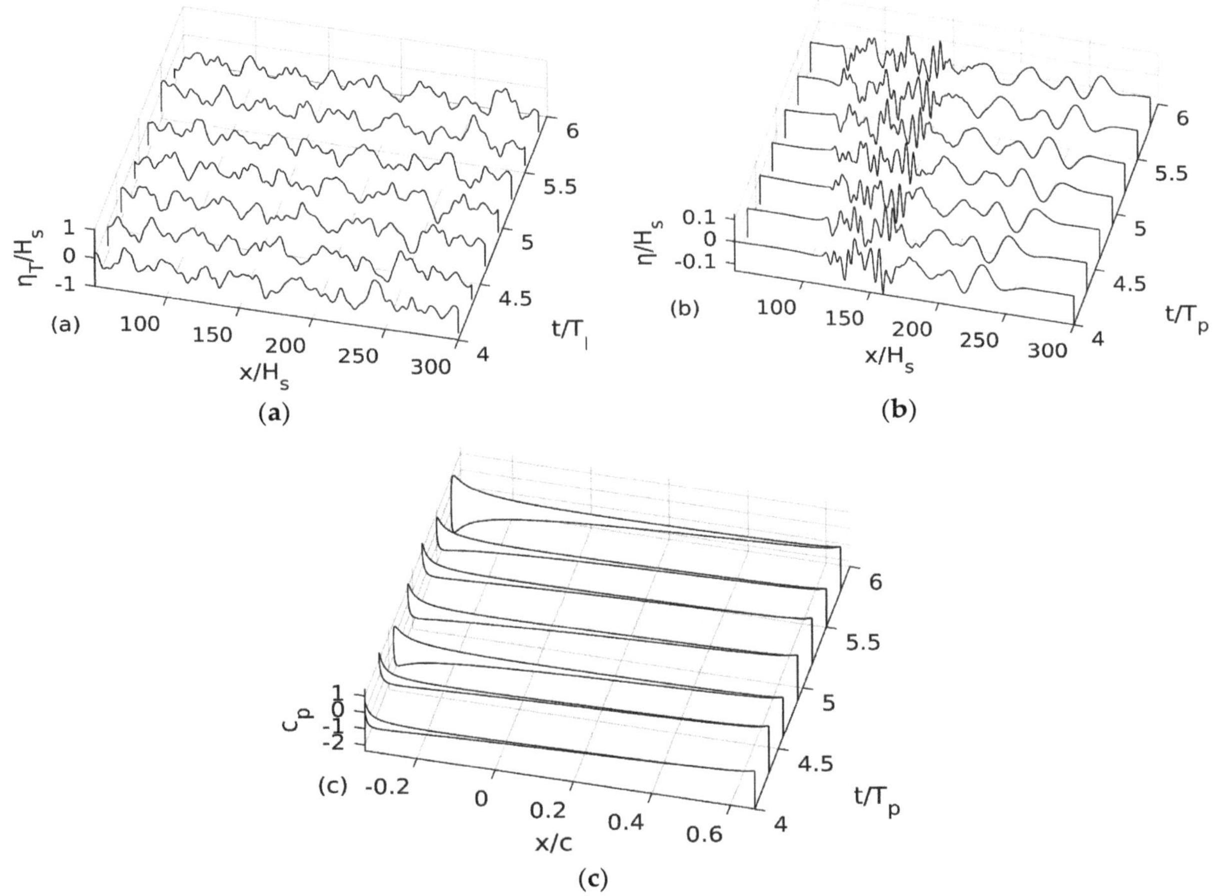

Figure 10. Time history of (**a**) the total free surface elevation and (**b**) the disturbance component due to the operation of the flapping foil, as well as (**c**) the unsteady pressure coefficient on the foil for a time interval of two peak periods, based on the present BEM simulation of Figure 9.

Furthermore, we present in Figure 11 the time series of several calculated quantities concerning thrust production by the examined system, as calculated by the present BEM for a time interval from 3 to 18 peak periods. The dynamic system is integrated starting from rest, and the transient effects have been died out in the first three periods. The responses of the examined ship and flapping foil system in waves are calculated by both the present BEM without (solid line) and with viscous corrections.

In particular, in Figure 11a, the time history of the angle of attack is presented excluding (a) and including (a_w) the effects of the incident wave field. The angle is calculated using Equation (5) and the effects of the free surface, the disturbance of the foil and the wake are approximately excluded from the calculation. The effect of the wave is either to increase or decrease the temporal value of the angle of attack by a small amount, resulting in a maximum value of $a_w = 15$ instead of $a = 12.9$ (at $t/T_p = 9.7$), and slightly reduce the corresponding rms value from 5.2 to 4.9. In general, the angle of attack is an indicator of the importance of viscous effects, which remain at moderate levels, rendering cost-effective potential methods capable to provide accurate predictions for the design of flapping foil thrusters as will be illustrated in the sequel through comparison against RANSE calculations in Sections 4.3–4.5.

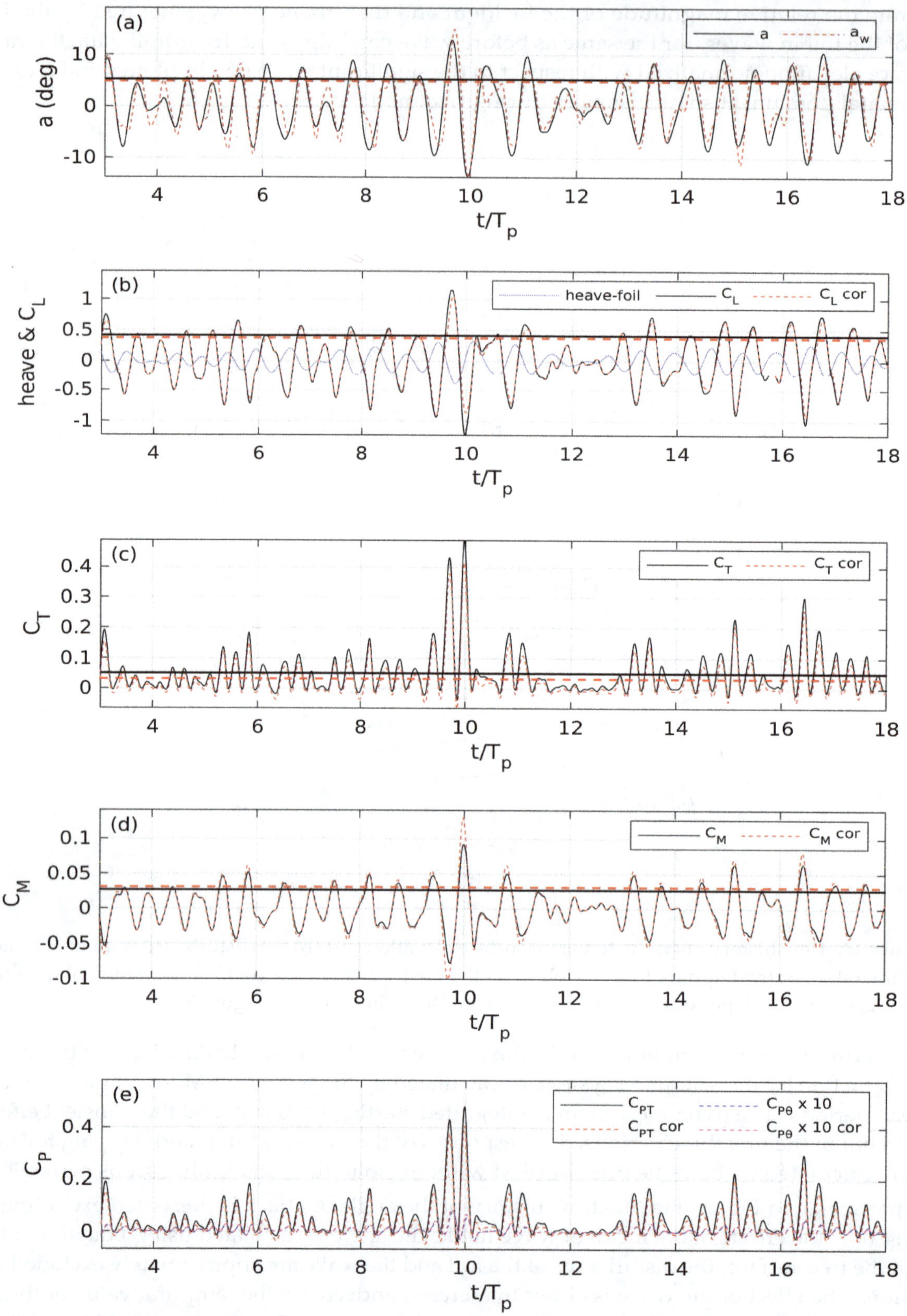

Figure 11. Evolution of motion and integrated hydrodynamic quantities for a NACA0012 flapping hydrofoil in random head waves, travelling at $F_{foil} = 1.35$, for time duration of 15 modal periods, for the simulation of Figure 9. Foil mean submergence $d/c = 1.71$ and control parameter $w = 0.5$. Plots of time history of (**a**) angle of attack, (**b**) foil heaving response and lift coefficient, (**c**) thrust coefficient, (**d**) moment coefficient, and (**e**) power coefficient.

In Figure 11b, we present the time history of foil's heaving motion, as produced by combined oscillatory heaving and pitching motion of the coupled ship-flapping foil system, together with the generated sectional lift coefficient $C_L = F_Y/0.5\rho U^2 c$. We observe that significant amplitudes of the lift force are produced. Moreover, the phase lag between foil's motion and lift force is approximately 180; Therefore, the generated lift acts as a restoring force, reducing the responses of a ship equipped with flapping hydrofoils. In Figure 11c, the dynamic evolution of sectional thrust coefficient $C_T = -F_x/0.5\rho U^2 c$ is shown, in the same time interval. We observe in this subplot that the thrust oscillations are in the interval $-0.06 \leq C_T \leq 0.49$, with an average value of 0.051 for the purely potential calculation, which is indicated by a horizontal black solid line and 0.033 for the corrected results indicated with a horizontal red dashed line. Although viscous effects cause almost a 35% decrease in mean value, the corresponding decrease is only 3.7% as concerns the peak value. In the examined case, the thrust production compared to the calm water resistance is found to reach levels of the order of 30%, which is considered to be important.

Additionally, the time history of moment required for the active pitching motion of the foil is presented in Figure 11d in terms of sectional moment coefficient $C_M = M/0.5\rho U^2 c^2$. As expected, viscous corrections increase the required torque input for the system to operate. Moreover, in Figure 11e, the power extracted by the examined system from the waves $C_{PT} = C_T$ is compared against the corresponding power that is necessary for the self-pitching motion of the foil $C_{P\theta} = P_\theta/0.5\rho U^3 c$, actually for tuning the instantaneous angle of attack in order to produce positive thrust. We observe that the latter quantity is practically insignificant and that the viscous effects modify slightly the predictions by reducing $C_{PT} = C_T$ and increasing $C_{P\theta}$. In the present section, the wave effects in the performance of the foil have been studied by means of a BEM and the effects of viscosity have been considered using empirical corrections. In the following subsections, the effects of viscosity and the free-surface boundary are also investigated by means of the CFD model and comparison between the BEM and RANSE are presented and discussed. First, in Section 4.3, the effects of the free-surface are neglected in order to derive calibration factors for the BEM model for viscous corrections. Subsequently, in Section 4.4, both models are applied and the predictions including the free-surface effects are compared.

4.3. BEM and RANSE Calculations of Flapping Foil in Unbounded Domain—Calibration of Viscous Corrections

In the present subsection, the effects of the wave and the free surface boundary are excluded from the modeling in order to derive estimates of the effects of viscosity for the flapping thruster. BEM and RANS calculations are compared to identify the range of applicability and the limitations of each scheme and to calibrate the correction formula (Equation (22)). To be more specific, in Figures 12–14, simulations with both numerical methods are presented for a flapping foil in heaving motion, pitching motion, and angle of attack as in Figure 11.

In Figure 14, we present the loads of the system in waves in unbounded domain, as calculated by the present BEM including corrections and RANSE. In subplots (a), (c), and (e), the plots of C_L, C_T, C_M histories are presented, respectively, whereas in subplots (b), (d), and (f), details during the 16th and 17th period are provided for comparison between the two methods.

The lift oscillations are in the interval $-1.24 \leq C_L \leq 1.23$, with an rms value of 0.45 for the BEM calculation and 0.39 for RANSE. Although viscous effects cause almost a 13% decrease in rms value of lift, the reduction is only 4.9% of the maximum peak value. The thrust oscillations are in the interval $-0.023 \leq C_T \leq 0.51$, with a mean value of 0.066 as obtained from BEM and 0.048 from RANSE. Although viscous effects cause almost a 27% decrease in mean value of thrust, the reduction is only 3.5% of the maximum peak value. The moment oscillations are in the interval $-0.13 \leq C_M \leq 0.13$ from BEM reaching the maximum value of 0.094, with mean value of 0.032. The corresponding value for RANSE is 0.038. Although viscous effects cause almost a 19% increase in rms value of foil self-pitching moment, that increase is only 6.4% of the maximum peak value. The corrected BEM results with appropriate selection of coefficient c_a provide compatible predictions with CFD model.

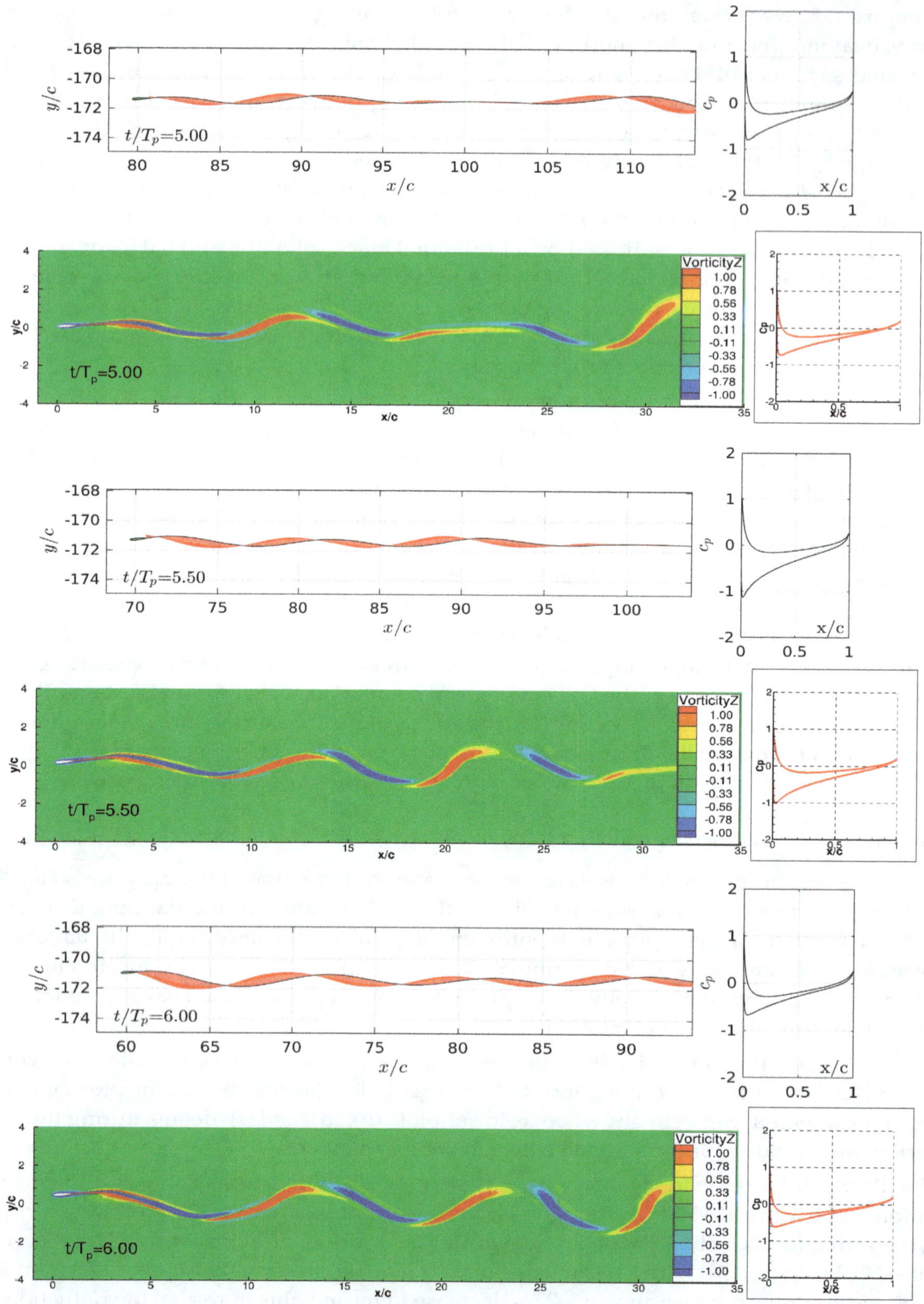

Figure 12. BEM and RANS simulation of the flapping thruster in vertical irregular motion and rotation about its pivot axis in unbounded domain without considering the free surface effects.

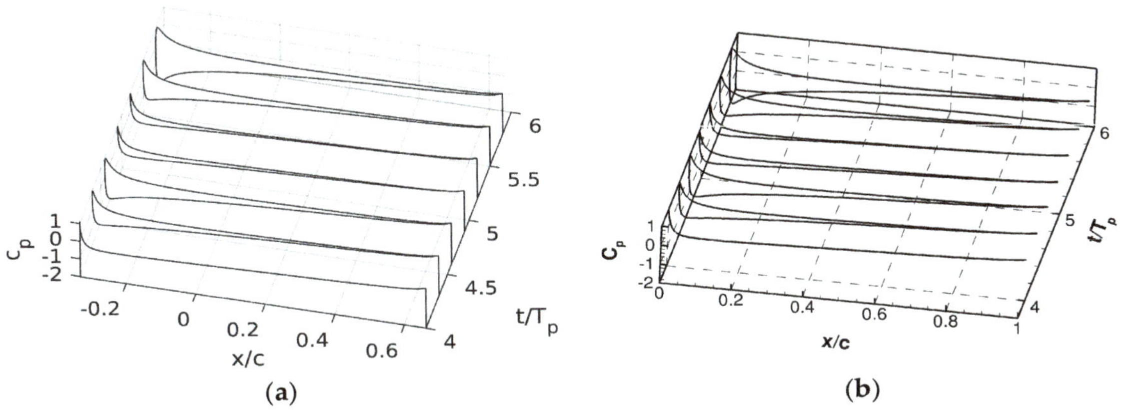

Figure 13. Time history of the pressure coefficient in unbounded domain during the 5th and 6th peak periods for the case of Figure 12. Comparison between BEM (**a**) and RANSE (**b**) calculations.

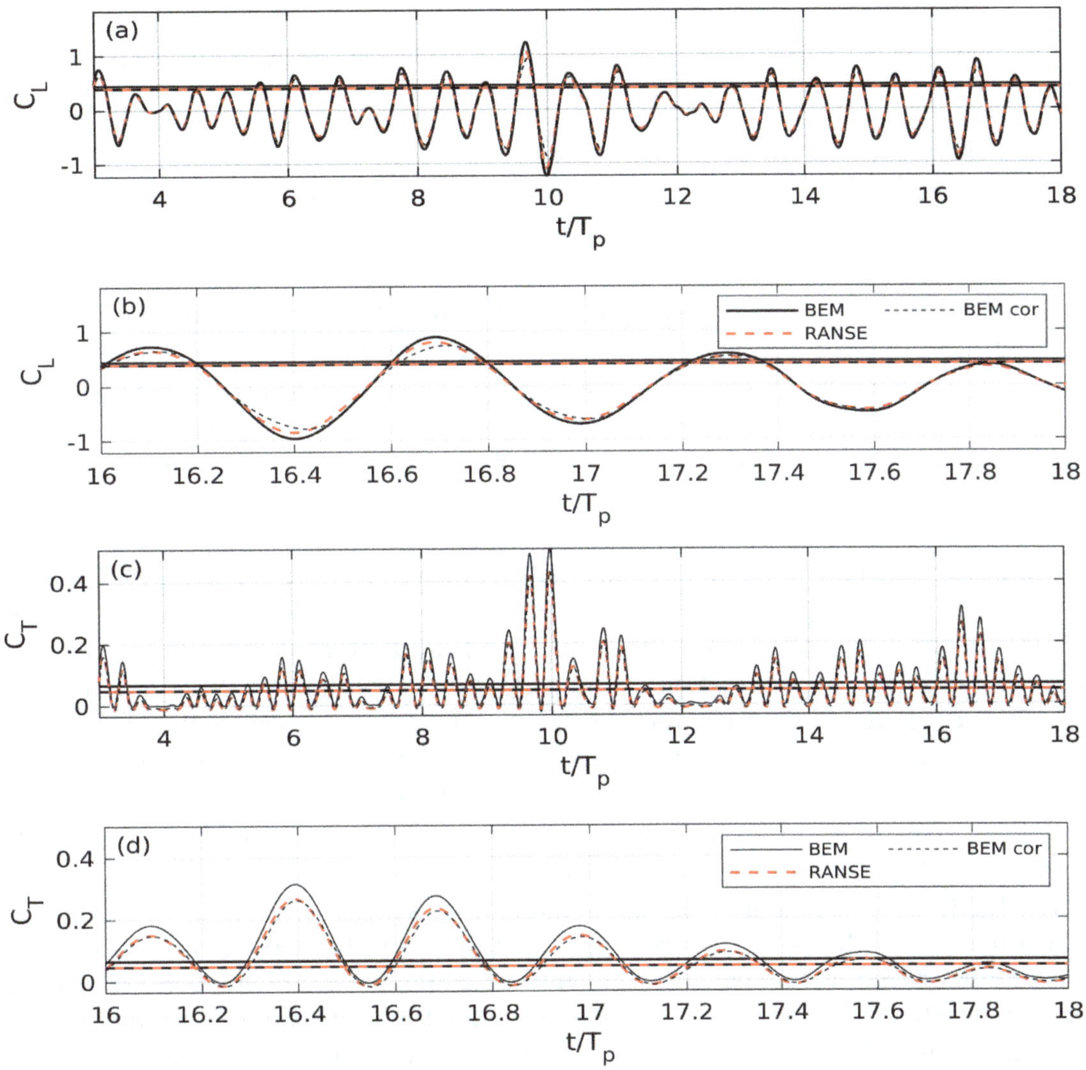

Figure 14. *Cont.*

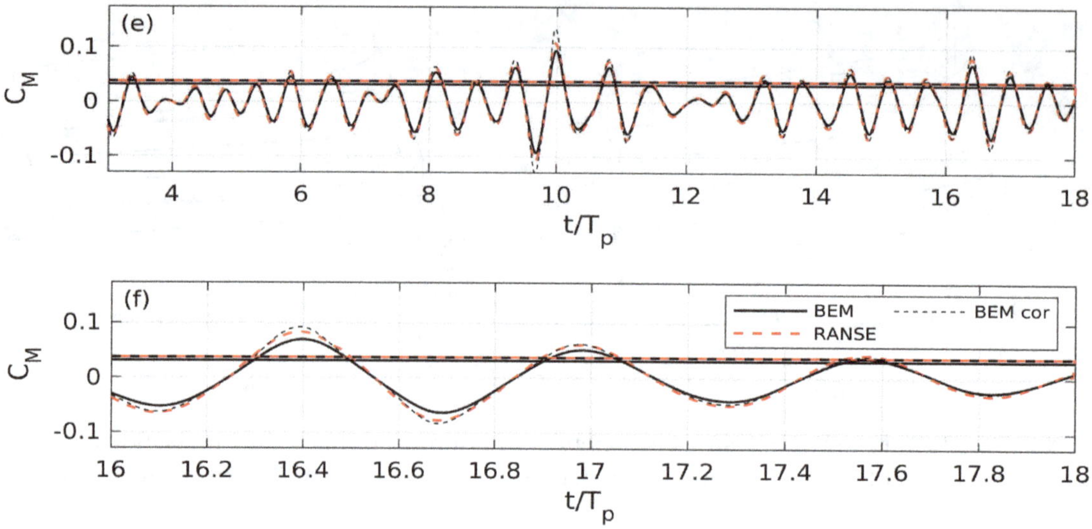

Figure 14. Time history of C_L, C_T, C_M on the flapping foil thruster in unbounded domain for the case of Figure 12. Subplots (**a**,**c**,**e**) show history over 18 peak periods and (**b**,**d**,**f**) details during the last two periods, showing in more detail the comparison between BEM, BEM corrected, and RANSE.

4.4. BEM and RANSE Calculation of Flapping Foil Beneath the Free Surface

Following the previous analysis, in this section the effect of the free surface is now taken into account. The flapping foil undergoes the same heaving and pitching motion as before, advancing with the same forward speed $F_{foil} = 1.35$. However, now it is considered completely submerged at a depth $d/c = 1.71$ as in Section 4.2. A comparison between BEM and RANS numerical predictions is presented in terms of foil loading as well as the foil-induced wave system.

In Figure 15, three snapshots of the flapping foil numerical predictions can be seen. BEM and RANSE simulations are presented during a time interval of one peak period. As in the previous subsection, BEM results concerning the dipole intensity in the vortex wake are shown, while for the CFD predictions vorticity contours are plotted.

Comparing the wake evolution between the two methods, it is clear that there is good qualitative agreement between the two solvers. Indeed, as the instantaneous pressure coefficients suggest (Figure 15), results are consistent between the two numerical methods. As stated in the previous section, apart from the absence of viscosity in the BEM predictions, the discrepancies between the two methods are due to the linearization of the wake vortex dynamics. When the free surfaces are considered, another source of error between the two methods emerges since the way the free surface is modeled is completely different. For the RANSE simulations the VOF approach is employed and thus the free surface is not explicitly defined. On the other hand, in the BEM approach, the free surface elevation is calculated as an explicit function of horizontal space variable x, also linearization of free-surface boundary conditions has been applied.

In Figure 16, the free surface elevation as well as the pressure distribution are plotted seven times between the 4th and the 6th period. It is observed that BEM and RANSE are in very good agreement both for the predicted free surface elevation as well as for the instantaneous pressure coefficient. It is noted that, contrary to the RANSE simulations (where the foil is stationary and the flow comes with $U_\infty = -U$), in the case BEM, the problem is solved in the earth-fixed frame of reference (the foil is travelling with forward speed U). However, the results in Figure 15 suggest that the predicted surface elevation in the near-foil regime is in very good agreement between the two methods. The same can be concluded by comparing the pressure coefficients presented in Figure 16c,d, from which it is evident that both solvers predict the very similar results for c_P.

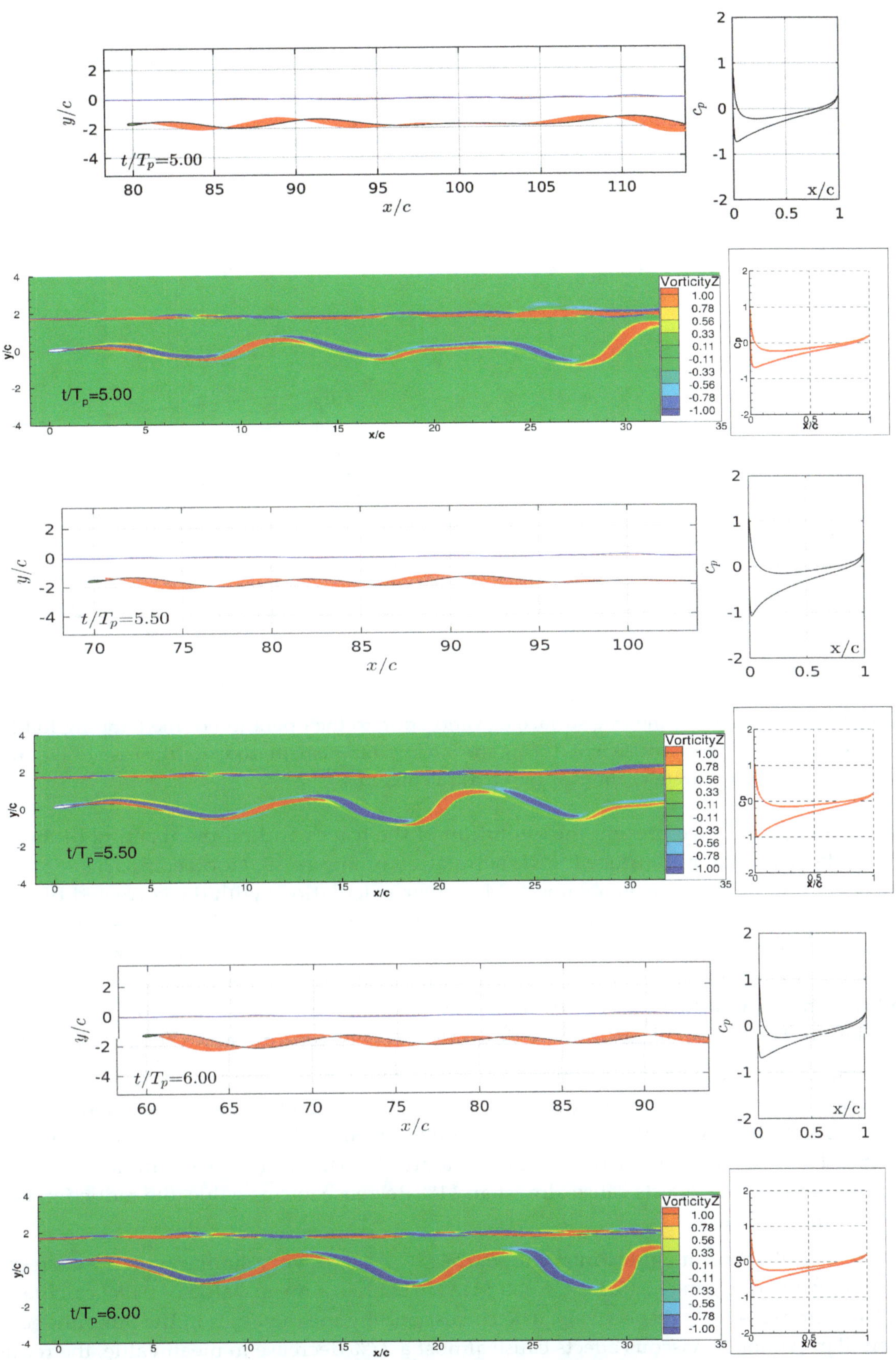

Figure 15. BEM and RANSE simulation of the system operating beneath the free surface in irregular vertical motion and rotation about its pivot axis without considering the incident wave effects.

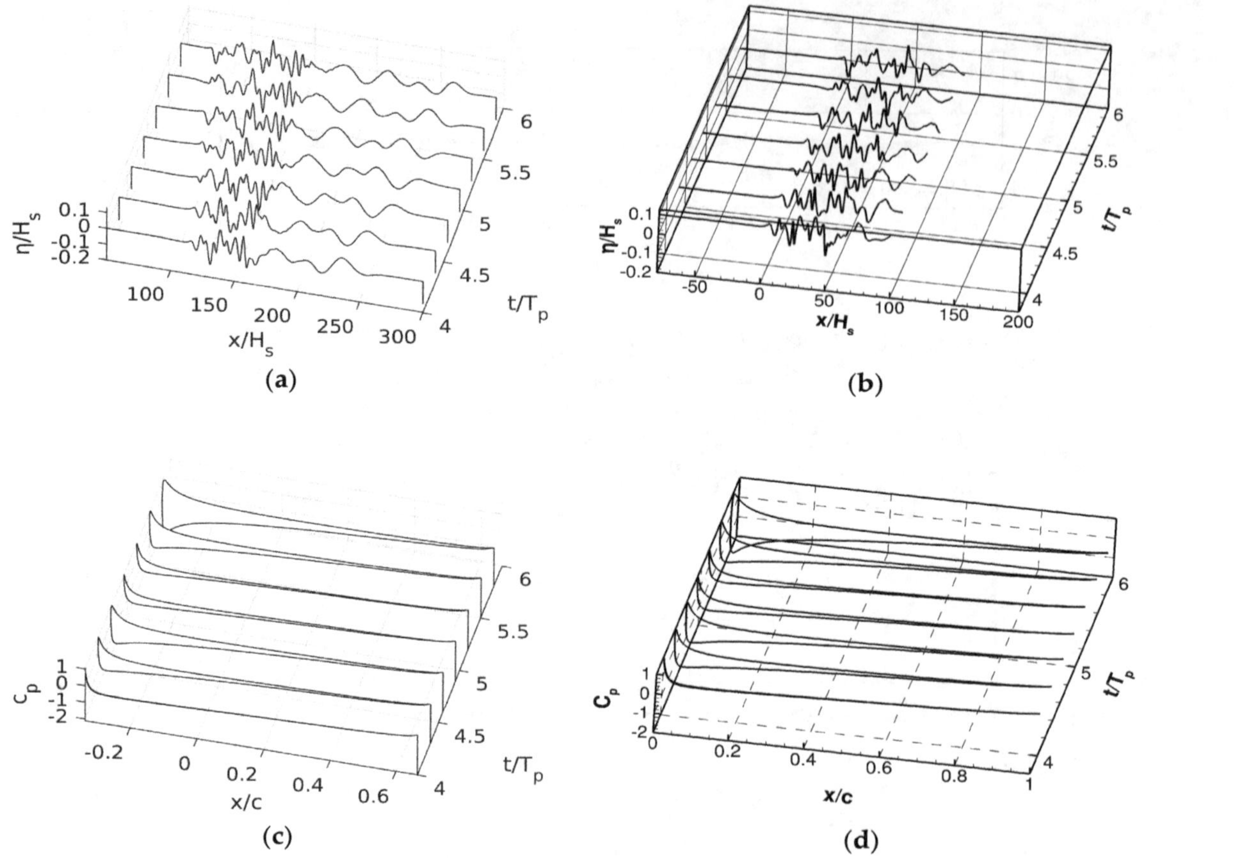

Figure 16. Time history of the free surface elevation due to the operation of the flapping foil (**a**,**b**) and the unsteady pressure coefficient during the 5th and 6th peak periods for the case of Figure 15. Comparison between BEM (**a**,**c**) and RANSE (**b**,**d**) calculations.

Finally, in Figure 17, we present the evolution of the forces and moment acting of the flapping foil operating beneath the free surface for the simulation of Figure 15. Good comparison between the present BEM and RANSE results is obtained. Moreover, calculations with the corrected BEM method are provided in the same figure. The viscous corrections are calibrated using the results of Section 4.3 for time integrated values (average and RMS). Again, simulation is over 18 peak periods, and the time-integrated quantities (averaged and RMS values calculated excluding the first 3 periods) are presented with horizontal lines. In Figure 17a,c,e, the time history of force coefficients is presented, whereas in Figure 17b,d,f, details during the time interval between 16th and 17th period are provided.

The sectional lift coefficient oscillations are found to be in the interval $-1.23 \leq C_L \leq 1.26$ that is slightly wider than the corresponding interval calculated in the previous section in unbounded domain ($-1.24 \leq C_L \leq 1.23$). The sectional lift coefficient obtained from BEM has an rms value of 0.45 and 0.40 from RANSE. The corresponding rms values in unbounded domain were of similar levels; i.e., 0.45 and 0.39. Although viscous effects cause almost an 11% decrease in rms value, the reduction is only 4% of the maximum peak value.

The thrust oscillations are in the interval $-0.026 \leq C_T \leq 0.51$, very similar with the corresponding result in unbounded domain ($-0.023 \leq C_T \leq 0.51$). Thrust has a mean value of 0.067 calculated from BEM and 0.049 from RANSE, respectively. The corresponding mean values in the unbounded domain are quite similar. Although viscous effects cause almost a 27% decrease in mean value, the reduction is only 3.5% of the maximum peak value.

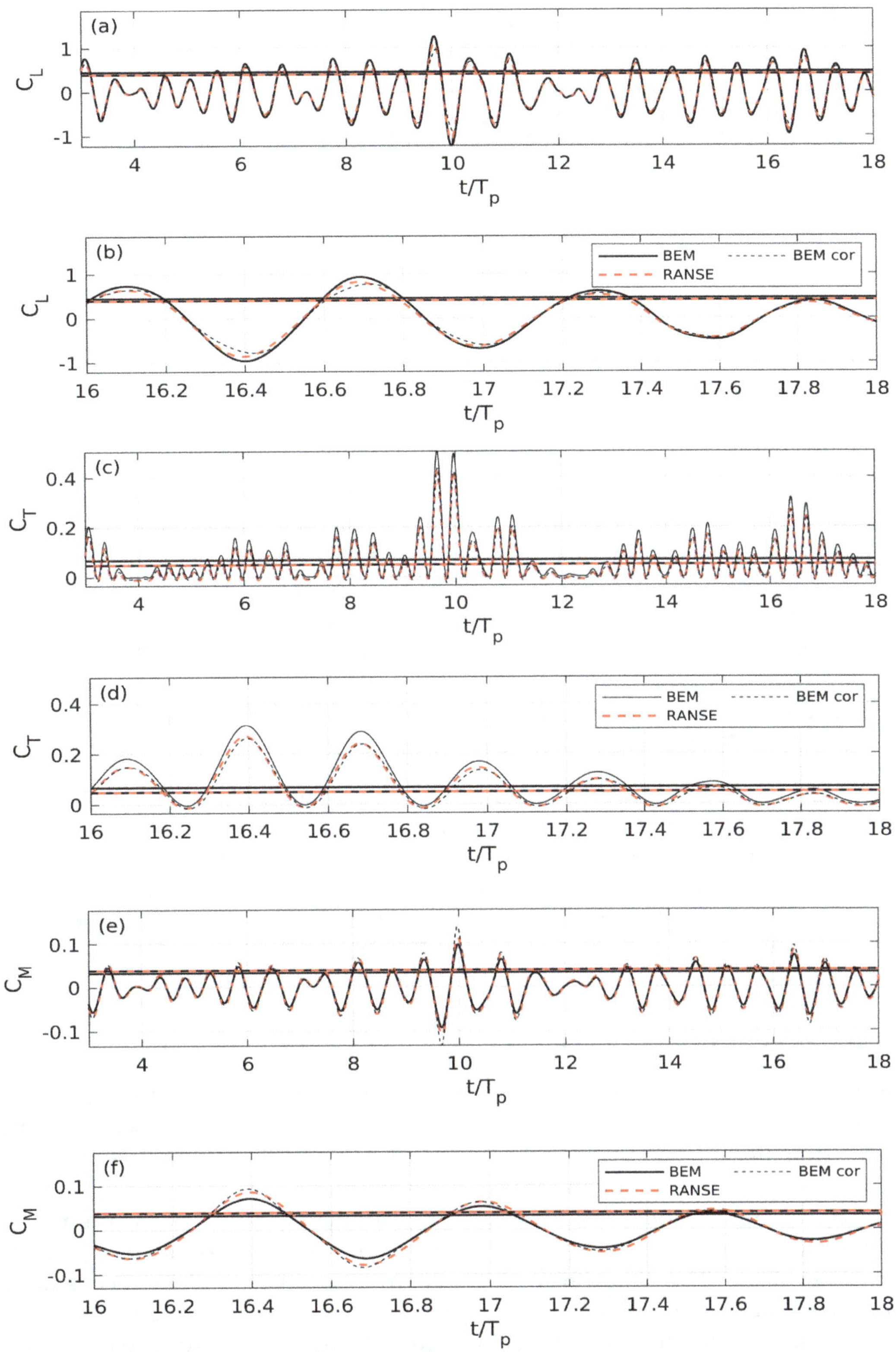

Figure 17. Time history of C_L, C_T, C_M on the flapping foil thruster beneath the free surface for the case of Figure 15. Subplots (**a**,**c**,**e**) show history over 18 peak periods and (**b**,**d**,**f**) details during the last two periods, showing in more detail the comparison between BEM, BEM corrected, and RANSE.

The self-pitching moment oscillations are in the interval $-0.13 \leq C_M \leq 0.14$ from BEM, reaching a maximum value of 0.097 and rms value 0.032, while the prediction from CFD analysis is 0.039. The corresponding mean values in unbounded domain were found to be very similar. Although viscous effects cause almost a 22% increase in rms value, the increase is only 7.2% of the maximum peak value. In the present case, where free-surface effects are considered, the corrected BEM predictions differ only at the third significant digit from the corresponding CFD. Moreover, the empirical corrections reduce the discrepancies at the peaks between BEM and RANSE calculations. Therefore, in the present case of Froude number $F_{foil} = 1.35$ and mean submergence $d/c = 1.71$, free-surface boundary does not affect significantly the calculations, and calibration using only one unbounded domain simulation is sufficient.

4.5. Effects of Waves, Free Surface, Viscosity, and Numerical Convergence

In this section, detailed investigation of viscosity, free-surface, and wave effects in a wide range of mean submergence of the flapping thruster is performed. Moreover, a detailed convergence study of the two numerical schemes is presented. In all cases, a NACA0012 heaving foil is considered in forward motion that corresponds to $F_{foil} = 1.35$ simultaneously pitching about a pivot axis located on the chord at distance $x_R/c = 0.33$ from the leading edge, resulting in angle of attack $a(t)$ and kinematics given in Figure 11.

The calculations in the unbounded domain are labeled as "inf" in Figure 18, while the calculations of the foil beneath the free surface without incident waves are denoted with "fs" and the calculations in irregular waves corresponding to $H_s/c = 0.86$ and Strouhal number $\omega_{p,en}H_s/(2\pi U) = 0.067$ are denoted with label "wave". The mean submergence parameter d/c varies from 1.14 to 18.29 and even in the "wave" calculations the same foil kinematics are used (ignoring the effect of the varying foil submergence to the ship-foil RAO). In all cases, BEM calculation with viscous corrections are also presented. The latter corrections are calibrated as described in Section 4.3 using data from unbounded domain BEM and RANSE simulations.

In Figure 18a–c, results concerning the lift, thrust, and moment coefficients are presented as functions of the mean submergence parameter d/c. Concerning viscosity effects, BEM and RANSE "fs" calculations present a similar trend with respect to mean submergence parameter d/c with approximately constant deviation −12%, −28%, and +18% for the mean values of C_L, C_T, and C_M, respectively. It is observed that the corrected BEM calculations are in very good qualitative and quantitative agreement with the RANSE calculations. The latter discrepancy is not significant compared to the total magnitude of the time signals; for more details see the discussion of Figures 11, 14 and 17 in previous sections. Numerical results indicate thrust production from the waves of the order of 20–30% of the calm water resistance in the examined cases.

As concerns the free surface effects, we observe that for foil submergence greater than 1–2 chord lengths they are not significant. To be more specific, the maximum increase with respect to the unbounded domain calculations is observed about $d/c = 1.71$ and is below 2% for all the load coefficients. Moreover, the maximum decrease is observed about $d/c = 4.57$ and is below 1.1% for the cases studied. However, it is noted that for the exact location of the maximum and minimum values more simulations with finer resolution of the d/c variation should be performed, and this is left as a subject for future work. The appearance of local extremes can be explained considering that the wave-making effects increase energy transfer from the foil to wave flow, reducing the load coefficients as the foil comes closer to the free surface. On the other hand, the free-surface boundary condition shows reverse effect as far as wave breaking does not occur.

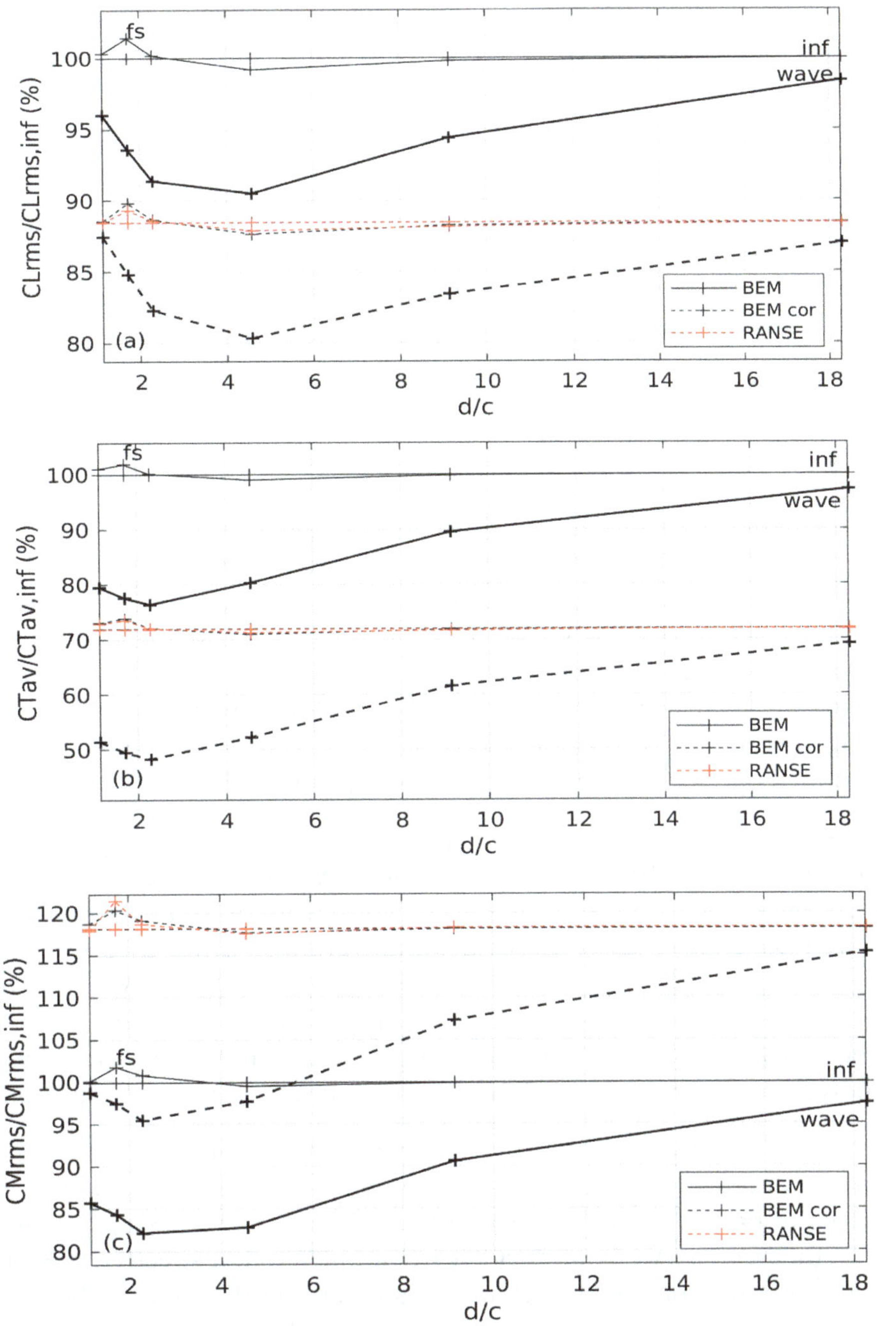

Figure 18. Free-surface, viscosity, and incident wave effects on the loads of a NACA 0012 flapping foil at $F_{foil} = 1.35$ performing heaving and pitching oscillations with kinematics as given in Figure 11. The calculations in unbounded domain are labeled as "inf", those of the foil beneath the free surface are denoted with "fs". The calculations in waves are denoted with "wave" using bold solid lines for the BEM calculations, and with bold dashed lines showing the corrected BEM results. BEM results are denoted with a solid line, BEM corrected with a black dashed line, and RANSE with red dashed line.

In addition, in the same figure results concerning load coefficients in presence of waves are plotted. With bold solid lines, the BEM calculations are indicated, and with bold dashed lines the corrected BEM ones. The comparison with RANSE calculations is left to be examined in future work. It is worth mentioning that calculations are in good agreement with the predictions in unbounded domain as the foil submergence increases. In general, the effect of the incident wave field in the present cases (ignoring the effect to the ship-foil RAO) is to reduce the load coefficients with maximum reduction 10%, 24%, and 18% for the mean C_L, C_T, and C_M, respectively. The difference is not significant compared to the total magnitude of the time signals; however, it should be considered in the modeling. This could be approximately calculated by considering wave-like background incident velocity field in the form of an unsteady gust.

Next, a convergence study has been performed for both numerical schemes in terms of space and time resolution for all load coefficients and the whole range of values of foil submergence. The error indicator used is $(C_F - C_{F,dense})/C_{F,\text{inf}}$, where the index F stands for lift, thrust, and moment, $C_{F,dense}$ denotes the value of the coefficient obtained using the finer grid, and $C_{F,\text{inf}}$ is the converged value in unbounded domain. All BEM calculations until now were implemented using the following numerical parameters, which were found enough for convergence of results: $N_B = 240$, $dt/T_P = 0.58\,\%$, $N_F = 814$ that corresponds to $dl/\lambda = 2\,\%$, where dl is the panel size on the free surface and λ the wavelength of the disturbance wave as predicted by the linear theory. Additionally, all RANSE calculations are performed using grid 3 (see Table 1) and $dt/T_P = 0.036\%$.

The most demanding goal was to achieve convergence with relative error significantly less than 1% in order to resolve the free-surface effects especially for the demanding RANSE simulations. This was achieved using the GRNET ARIS HPC infrastructure. In cases where the above goal is not achieved, the systematic study for the entire range of d/c and the good agreement between the corrected BEM and the RANSE calculations ensure that the trend of free-surface effects with respect to the submergence is correctly resolved. In BEM calculations, the number of panels in the body and the timestep are connected in order to ensure reasonable panel size ratio in the foil and wake panels near the trailing edge. Therefore, BEM convergence is presented only with respect to N_B. In RANSE simulations, convergence is examined by using three different timesteps and grid 3.

Turning into more details, we present in Figure 19 a convergence study for a flapping foil in unbounded domain for the case studied in Section 4.3. BEM convergence results with respect to the number of panels on the foil N_B and the size of the time step are plotted in Figure 19a–c. The relevant errors in all cases are reasonably small and mean sectional lift coefficient converges faster while the moment coefficient is found to be the most demanding. Moreover, a grid independence study of RANSE calculations is presented in Figure 19d–f using different grids (see Table 1). In this case, the level of error remains below 0.8% for all cases for grid 3. It was decided to present the finer grid predictions in the preceded analysis for the reasons explained before and since the wake is better resolved.

Furthermore, in Figure 20 we present a convergence study for a flapping foil beneath the free surface for the case studied in Section 4.4. BEM convergence results are plotted in Figure 20a–c with respect to the number of panels on the foil N_B and the size of the time step and with respect to the number of panels on the free-surface boundary N_F in Figure 20d–f, respectively. From the first case in Figure 20a–c, we conclude that convergence characteristics are similar to the unbounded domain study as expected with slightly faster convergence. From the second case in Figure 20d–f, it is seen that for the specific submergence ratio $d/c = 1.71$ the relative error reduces below 0.5% even for the coarser grids. However, we choose $N_F = 814$ to resolve adequately the most demanding case of $d/c = 1.14$. A grid independence study of RANSE calculations is presented in Figure 20g–i, using different grids according to Table 1. The relative errors are very small, but the convergence is found to be slightly more demanding in comparison with the unbounded domain configuration due to the additional requirements imposed by the free-surface.

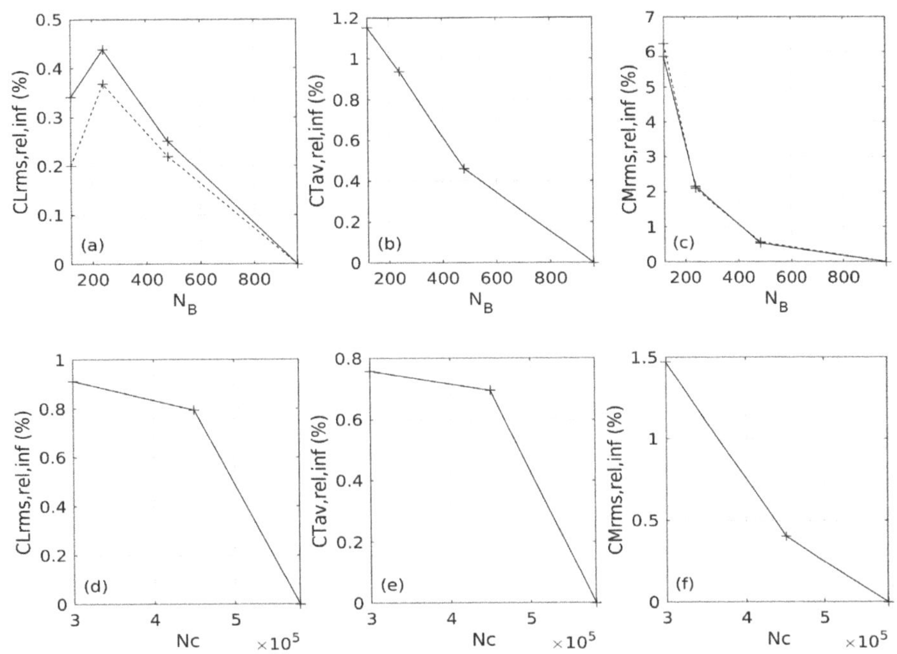

Figure 19. Convergence study for a flapping foil thruster in unbounded domain for the case studied in Section 4.3. BEM convergence with respect to the number of panels on the foil N_B and the time step (**a**–**c**). BEM results are shown using solid lines and BEM-corrected with dashed lines, respectively. Grid independence study of RANSE (**d**–**f**) with different grids (details are included in Table 1).

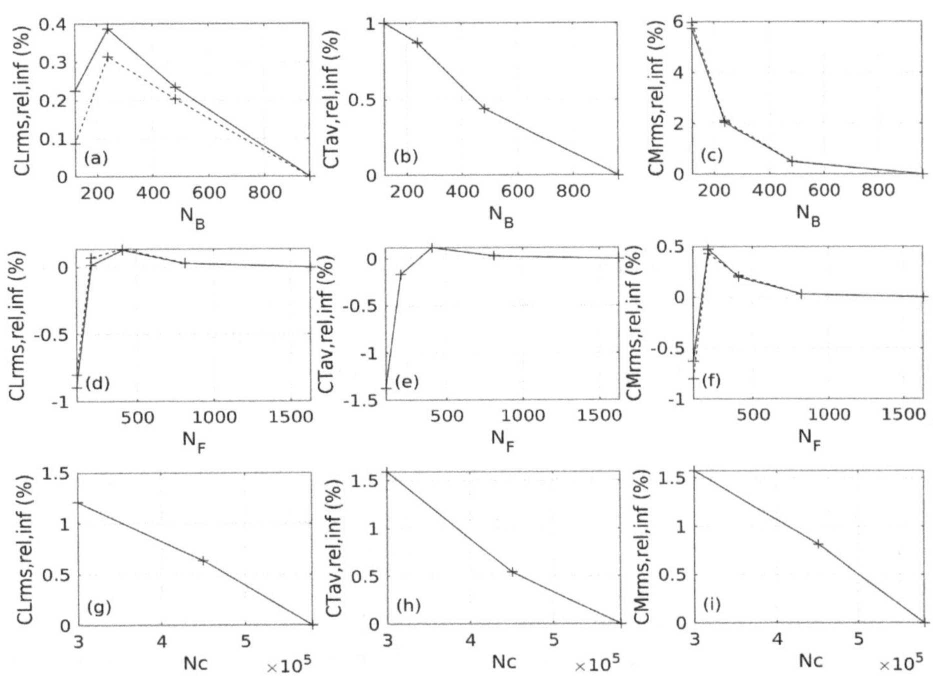

Figure 20. Convergence study for a flapping foil thruster beneath the free surface for the case studied in Section 4.4. (**a**–**c**) BEM convergence with respect to the number of panels on the foil N_B and the size of the time step, and (**d**–**f**) with respect to number of panels on the free-surface boundary N_F. BEM results are presented using solid lines and BEM corrected with dashed lines, respectively. Grid independence study of RANSE (**g**–**i**) with different grids (see Table 1).

Finally, in Figure 21, a convergence study for a flapping foil in waves for the case studied in Section 4.2 is presented. BEM convergence with respect to the number of panels on the foil N_B and the size of the time step are presented in Figure 21a–c and with respect to number of panels on the free-surface boundary N_F in Figure 21d–f. The findings are similar with those from the preceded figures; however, as explained in the beginning of the present section, the effect of waves causes a variation of reduction by 10%, 24%, and 18% for the mean sectional values of C_L, C_T, and C_M, respectively, and not below 2% as in the case of free-surface effects. Therefore, if the interest is in the design of the flapping foil device extracting energy from the waves in mean submergence larger than 1 chord length and not to the resolution of the details of the free-surface effects, a coarser spatial and temporal grid could be adequate. The enhancement of the present method with incorporation of nonlinear effects associated with the free surface dynamics and the leading edge flow separation, as well as full treatment of 3D effects, are important future contributions, supporting also the design of the considered systems; see, e.g., [20,43].

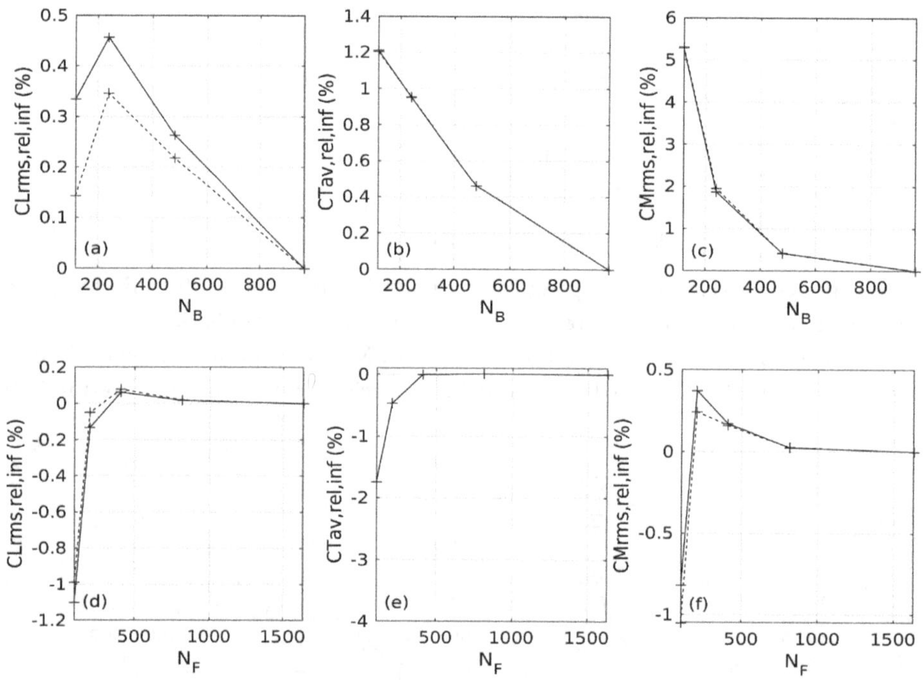

Figure 21. Convergence study for a flapping foil thruster in waves for the case studied in Section 4.2. (**a–c**) BEM convergence with respect to the number of panels on the foil N_B and the size of the time step, and (**d–f**) with respect to number of panels on the free-surface boundary N_F. BEM results are presented using solid lines and BEM corrected with dashed lines, respectively.

5. Conclusions

The concept of biomimetic flapping-foil thrusters for energy extraction and augmentation of ship prolusion has been further explored, focusing on the effects of free surface, incident wave, and viscosity. To this end, in-house GPU-BEM and HPC-RANSE solvers have been developed and results for cases of flapping foil in unbounded domain, beneath the free surface and in waves have been presented and discussed. For the determination of foil kinematics, a mixed frequency–time domain method based on linearized ship dynamics coupled with unsteady lifting line theory is applied, obtaining the response of the ship-foil system in irregular waves that correspond to specific sea states. After we have obtained the kinematics of the wing, an unsteady time domain boundary element method has been applied for the hydrodynamic analysis of flapping foils beneath the free surface in the presence of irregular incident waves. Numerical results concerning the thrust production have been presented, indicating significant energy extraction from the waves for the augmentation of ship's overall propulsion and enhancement of dynamic stability.

Free-surface, incident wave, and viscosity effects have been thoroughly studied indicating that in cases with foil submergence greater than a few chord lengths, the free-surface effects are minor. Moreover, BEM and RANSE calculations present quite similar trends with respect to mean submergence depth, presenting 12%, 28%, and 18% differences concerning the mean values of lift, thrust, and moment coefficients, respectively. From the simulations in unbounded domain, viscous corrections of the potential flow model have been derived, rendering the enhanced BEM solver cost-effective and useful for the systematic investigation and the optimization of the system operating relatively near the free surface in waves.

Future work is focused on the enrichment of the present models by the incorporation of nonlinear phenomena associated with the free surface dynamics and flow separation. Furthermore, full treatment of 3D effects including swept wings of low aspect ratio and effects of directional seas, as well as experimental verification of the examined ship flapping foil thruster are important directions for future research.

Author Contributions: This work is supervised by K.A.B. The BEM numerical scheme is developed and implemented in GPGPU by E.S.F. and the RANSE numerical scheme by G.P.P. The draft of the text is prepared by all authors and the numerical simulations were handled by E.S.F. and G.P.P. All authors have read and agreed to the published version of the manuscript.

Abbreviations

BEM	Boundary Element Method
RANSE	Reynolds Average Navier–Stokes Equations
GPGPU	General Purpose programming on Graphics Processing Units
HPC	High Performance Computing
RAO	Response Amplitude Operator
DtN	Dirichlet to Neumann map
ODEs	Ordinary Differential Equations
CFD	Computational Fluid Dynamics
VOF	Volume of Fluid
BDF	Backwards differentiation formula
STACS	Switching Technique for Advection and Capturing of Surfaces Scheme
RMS	Root Mean Square

Nomenclature

F_{foil}	Foil Froude number
F_n	Ship Froude number
x_R	Location of pivot axis from leading edge
η	Free-surface elevation
θ	Pitch angle
a	Angle of attack
a_w	Angle of attack with the effect of the wave
ω	Angular frequency of motion
C_T	Thrust coefficient
C_L	Lift coefficient
C_M	Moment coefficient
c_P	Pressure coefficient
d	Mean submergence
η	Froude efficiency

Φ	Potential field
N_B	Number of panels on the body
N_F	Number of panels on the free surface
N_c	Total number of panels of the CFD grid
H_s	Significant wave height
T_p	Peak period
c	Foil chord
L	Ship length
$C_{F,dense}$	Value of the load coefficient at the most refined case
$C_{F,\inf}$	Converged value of the load coefficient in unbounded domain
x_{wing}	Horizontal location of the foil
μ	Surface dipole intensity or potential jump

References

1. Triantafyllou, M.S.; Yue, D.K.P.; Triantafyllou, G.S. Hydrodynamics of Fishlike Swimming. *Annu. Rev. Fluid Mech.* **2000**, *32*, 33–53. [CrossRef]
2. Triantafyllou, M.; Techet, A.; Hover, F. Review of Experimental Work in Biomimetic Foils. *IEEE J. Ocean. Eng.* **2004**, *29*, 585–594. [CrossRef]
3. Taylor, G.K.; Triantafyllou, M.S.; Tropea, C. *Animal Locomotion*; Springer: Heidelberg, Germany, 2010; pp. 1–443.
4. Rozhdestvensky, K.V.; Ryzhov, V.A. Aero-hydrodynamics of flapping wing propulsors. *Prog. Aerosp. Sci.* **2003**, *39*, 585–633. [CrossRef]
5. Naito, S.; Isshiki, H. Effect of Bow Wings on Ship Propulsion and Motions. *Appl. Mech. Rev.* **2005**, *58*, 253–268. [CrossRef]
6. Belibassakis, K.; Politis, G.K. Hydrodynamic performance of flapping wings for augmenting ship propulsion in waves. *Ocean Eng.* **2013**, *72*, 227–240. [CrossRef]
7. Bøckmann, E.; Steen, S. Experiments with actively pitch-controlled and spring-loaded oscillating foils. *Appl. Ocean Res.* **2014**, *48*, 227–235. [CrossRef]
8. BIO-PROPSHIP: Augmenting Ship Propulsion in Rough Sea by Biomimetic-Wing System. Research Project Funded by National Strategic Reference Framework NSRF Greece 2007–2013. Available online: http://arion.naval.ntua.gr/~{}biopropship/index_en.html (accessed on 23 April 2020).
9. Belibassakis, K.; Filippas, E. Ship propulsion in waves by actively controlled flapping foils. *Appl. Ocean Res.* **2015**, *52*, 1–11. [CrossRef]
10. Filippas, E.; Belibassakis, K. Hydrodynamic analysis of flapping-foil thrusters operating beneath the free surface and in waves. *Eng. Anal. Bound. Elements* **2014**, *41*, 47–59. [CrossRef]
11. Bøckmann, E.; Steen, S. Model test and simulation of a ship with wavefoils. *Appl. Ocean Res.* **2016**, *57*, 8–18. [CrossRef]
12. Huang, S.W.; Wu, T.L.; Hsu, Y.T.; Guo, J.H.; Tsai, J.F.; Chiu, F.C. Effective energy-saving device of Eco-Ship by using wave propulsion. In Proceedings of the 2016 Techno-Ocean (Techno-Ocean), Kobe, Japan, 6–8 October 2016; pp. 566–570. [CrossRef]
13. Liu, P.; Su, Y.-M.; Liao, Y.-L. Numerical and experimental studies on the propulsion performance of a Wave Glide propulsor. *China Ocean Eng.* **2016**, *30*, 393–406. [CrossRef]
14. Bowker, J.A. Coupled Dynamics of a Flapping Foil Wave Powered Vessel. Ph.D Thesis, University of Southampton, Southampton, UK, 2018.
15. Bøckmann, E.; Yrke, A.; Steen, S. Fuel savings for a general cargo ship employing retractable bow foils. *Appl. Ocean Res.* **2018**, *76*, 1–10. [CrossRef]
16. Liu, P.; Liu, Y.; Huang, S.; Zhao, J.; Su, Y. Effects of Regular Waves on Propulsion Performance of Flexible Flapping Foil. *Appl. Sci.* **2018**, *8*, 934. [CrossRef]
17. Priovolos, A.; Filippas, E.; Belibassakis, K. A vortex-based method for improved flexible flapping-foil thruster performance. *Eng. Anal. Bound. Elements* **2018**, *95*, 69–84. [CrossRef]

18. Anevlavi, D.E.; Filippas, E.S.; Karperaki, A.; Belibassakis, K. A Non-Linear BEM–FEM Coupled Scheme for the Performance of Flexible Flapping-Foil Thrusters. *J. Mar. Sci. Eng.* **2020**, *8*, 56. [CrossRef]
19. Wu, X.; Zhang, X.; Tian, X.; Li, X.; Lu, W. A review on fluid dynamics of flapping foils. *Ocean Eng.* **2020**, *195*, 106712. [CrossRef]
20. Filippas, E. Hydrodynamic Analysis of Ship and Marine Biomimetic Systems in Waves Using GPGPU Programming. Ph.D. Thesis, NTUA, Athens, Greece, 2019.
21. Xiao, Q.; Zhu, Q. A review on flow energy harvesters based on flapping foils. *J. Fluids Struct.* **2014**, *46*, 174–191. [CrossRef]
22. Jeanmonod, G.; Olivier, M. Effects of chordwise flexibility on 2D flapping foils used as an energy extraction device. *J. Fluids Struct.* **2017**, *70*, 327–345. [CrossRef]
23. Koutsogiannakis, P.; Filippas, E.S.; Belibassakis, K. A Study of Multi-Component Oscillating-Foil Hydrokinetic Turbines with a GPU-Accelerated Boundary Element Method. *J. Mar. Sci. Eng.* **2019**, *7*, 424. [CrossRef]
24. Filippas, E.; Gerostathis, T.; Belibassakis, K. Semi-activated oscillating hydrofoil as a nearshore biomimetic energy device system in waves and currents. *J. Ocean Eng.* **2018**, *154*, 396–415. [CrossRef]
25. Lewandowski, E.M. *The Dynamics of Marine Craft - Maneuvering and Seakeeping*; World Scientific Pub Co Pte Lt.: Singapore, 2004; p. 22.
26. Sclavounos, P.; Borgen, H. Seakeeping analysis of a high-speed monohull with a motion control bow hydrofoil. *J. Ship Res.* **2004**, *48*, 77–117.
27. Politis, G.; Politis, K. Biomimetic propulsion under random heaving conditions, using active pitch control. *J. Fluids Struct.* **2014**, *47*, 139–149. [CrossRef]
28. Faltinsen, O.M. *Hydrodynamics of High-Speed Marine Vehicles*; Cambridge University Press (CUP): New York, NY, USA, 2006.
29. Blevins, R.D. *Applied Fluid Dynamics Handbook*; Van Nostrand Reinhold Co.: New York, NY, USA, 1984.
30. Hirt, C.; Nichols, B. Volume of fluid (VOF) method for the dynamics of free boundaries. *J. Comput. Phys.* **1981**, *39*, 201–225. [CrossRef]
31. Papadakis, G.; Filippas, E.; Ntouras, D.; Belibassakis, K. Effects of viscosity and nonlinearity on 3D flapping-foil thruster for marine applications. In Proceedings of the OCEANS 2019 - Marseille, Institute of Electrical and Electronics Engineers (IEEE), Marseille, France, 17–20 June 2019; pp. 1–10.
32. Kunz, R.F.; Boger, D.A.; Stinebring, D.R.; Chyczewski, T.S.; Lindau, J.W.; Gibeling, H.J.; Venkateswaran, S.; Govindan, T. A preconditioned Navier–Stokes method for two-phase flows with application to cavitation prediction. *Comput. Fluids* **2000**, *29*, 849–875. [CrossRef]
33. Roe, P. Approximate Riemann solvers, parameter vectors, and difference schemes. *J. Comput. Phys.* **1981**, *43*, 357–372. [CrossRef]
34. Queutey, P.; Visonneau, M. An interface capturing method for free-surface hydrodynamic flows. *Comput. Fluids* **2007**, *36*, 1481–1510. [CrossRef]
35. Darwish, M.; Moukalled, F. Convective Schemes for Capturing Interfaces of Free-Surface Flows on Unstructured Grids. *Numer. Heat Transfer Part B Fundam.* **2006**, *49*, 19–42. [CrossRef]
36. Menter, F.R. Two-equation eddy-viscosity turbulence models for engineering applications. *AIAA J.* **1994**, *32*, 1598–1605. [CrossRef]
37. Devolder, B.; Rauwoens, P.; Troch, P. Application of a buoyancy-modified k-ω SST turbulence model to simulate wave run-up around a monopile subjected to regular waves using OpenFOAM ®. *Coast. Eng.* **2017**, *125*, 81–94. [CrossRef]
38. Zhao, Y.; Tai, J.; Ahmed, F. Simulation of micro flows with moving boundaries using high-order upwind FV method on unstructured grids. *Comput. Mech.* **2002**, *28*, 66–75. [CrossRef]
39. Biedron, R.; Vatsa, V.; Atkins, H. Simulation of Unsteady Flows Using an Unstructured Navier-Stokes Solver on Moving and Stationary Grids. In Proceedings of the 23rd AIAA Applied Aerodynamics Conference; American Institute of Aeronautics and Astronautics (AIAA), Toronto, ON, Canada, 6–9 June 2005; 2005.
40. Papadakis, G.; Voutsinas, S. A strongly coupled Eulerian Lagrangian method verified in 2D external compressible flows. *Comput. Fluids* **2019**, *195*, 104325. [CrossRef]
41. Diakakis, K.; Papadakis, G.; Voutsinas, S. Assessment of transition modeling for high Reynolds flows. *Aerosp. Sci. Technol.* **2019**, *85*, 416–428. [CrossRef]

42. Politis, G. Unsteady Wake Rollup Modeling Using a Mollifier Based Filtering Technique. *Dev. Appl. Ocean. Eng.* **2016**, *5*, 1. [CrossRef]
43. Politis, G.K.; Tsarsitalidis, V. Flapping wing propulsor design: An approach based on systematic 3D-BEM simulations. *Ocean Eng.* **2014**, *84*, 98–123. [CrossRef]

3

Response of a Porous Seabed around an Immersed Tunnel under Wave Loading: Meshfree Model

Shuang Han [1,†], Dong-Sheng Jeng [1,*,†] and Chia-Cheng Tsai [2,3,4,†]

1 School of Engineering and Built Environment, Griffith University Gold Coast Campus, Queensland 4222, Australia; Shuang.han3@griffithuni.edu.au
2 Department of Marine Environmental Engineering, National Kaohsiung University of Science and Technology, Kaohsiung 80778, Taiwan; tsaichiacheng@gmail.com
3 Department of Marine Environment and Engineering, National Sun Yat-Sen University, Kaohsiung 80424, Taiwan
4 Center of Excellence for Ocean Engineering, National Taiwan Ocean University, Keelung 20224, Taiwan
* Correspondence: d.jeng@griffith.edu.au
† These authors contributed equally to this work.

Abstract: Seabed instability surrounding an immersed tunnel is a vital engineering issue regarding the design and maintenance for submarine tunnel projects. In this study, a numerical model based on the local radial basis function collocation method (LRBFCM) is developed to evaluate the seabed behaviour in a marine environment, in which the seabed is treated as the porous medium and governed by Biot's "$u - p$" approximation. As for the flow field above the seabed, the VARANS equations are used to simulate the fluid motion and properties. The present model is validated with analytical solutions and experimental data which show a good capacity of the integrated model. Both wave and current loading are considered in this study. Parametric studies are carried out to investigate the effects of wave characteristics and soil properties. Based on the numerical results, the maximum liquefaction depth around the immersed tunnel could be deeper under the wave loading with long wave period (T) and large wave height (H). Moreover, a seabed with lower permeability (K_s) and degree of saturation (S_r) is more likely to be liquefied.

Keywords: meshfree method; VARANS; Biot's "$u - p$" approximation; wave-current induced seabed response; immersed tunnel

1. Introduction

In recent years, to meet the continual improvement requirements in coastal transportation, the immersed tunnel has become one of the choices to fundamentally transform the transport in the region of oceans and rivers, replacing the conventional methods such as the ferry. The immersed tunnel has a history of about 100 years and it shows a good performance in reliability and applicability under complex natural dynamic loading; for example, the longest immersed tunnel in the world, the Hongkong–Zhuhai–Macao Bridge immersed tunnel. As an alternative to a bridge, the immersed tunnel has advantages in less environmental effects and no obstruction of navigation channels. Compared to bridges, the immersed tunnel is commonly constructed in a soft and loose seabed. Thus, the stability of the seabed soil surrounding the immersed tunnel in complex marine environments becomes one of the main concerns implicated in tunnel design and maintenance.

It has been recognized that the pore water pressures and stresses in seabeds are affected by the water pressures generated by the natural dynamic loading. If the pore water pressure reaches the limit value, the liquefaction could occur with the effective stress in seabed vanishing. To avoid seabed instability around the immersed tunnel, the study of seabed dynamic behaviour is necessary under the real hydrodynamic loading. Two mechanisms of wave-induced liquefaction has been figured out based on a mass of laboratory tests and field exploration [1–3], which are transient liquefaction and residual liquefaction. The transient liquefaction is motivated by the oscillatory excess pore water pressures under wave pressure vibration which usually happens with amplitude reduction and phase lag of pore pressure in seabed soil [4]. While the residual liquefaction is on the consequence of the excess pore water pressure build-up under cyclic wave loading [5]. Later, Jeng and Seymour [6] proposed a simplified approximation to predict the liquefaction process in a large seabed, and concluded that the residual mechanism is more essential under large waves while the transient mechanism is dominate in a seabed under small wave loading.

In the past 40 years, various analytical formulas have been developed and verified in regard to the seabed dynamic response [4,7,8]. However, the seabed dynamic response around the structure is difficult to be described by analytical methods. Thus, several numerical model was developed to simulate the soil behaviour around the offshore structures, such as breakwaters [9], and buried pipes [10]. However, the research of a wave-induced response around the undersea immersed tunnel is quite limited. For instance, a real case simulation of the Busan–Geoje fixed link in South Korea was conducted by Kasper et al. [11] under large waves (wave height up to 9.2 m) generated by typhoons. Nevertheless, this study did not consider the impact of the seabed liquefaction around the tunnel. Recently, [12,13] simulated the seabed transient and residual response around the immersed tunnel under wave loading based on Biot's consolidation equations neglecting the inertial terms for soil skeleton and fluid phase.

In natural ocean environments, current is another crucial component besides wave. For instance, a long-term mentoring data of the Lingding Bay, in which the Hong Kong–Zhuhai–Macau bridge tunnel located, shows the current co-exists with wave varying from bottom to surface as the consequence of the irregular semi-diurnal tide. The maximum velocity of the surface current reaches up to 1.5 m/s [14]. The interaction between current and wave is found to be able to affect the hydrodynamic properties directly and further impact the porous seabed dynamics. Ye and Jeng [15] investigated the transient response of a porous seabed under wave combined current loading firstly. Later, the current effects in the vicinity of a submarine pipeline were examined by Wen et al. [16] based on the commercial software ABAQUS. Lately, Liao et al. [17] simulated the residual seabed liquefaction under the flow field that wave and current generated simultaneously. To date, how current affect the liquefaction of seabed soil surrounding the immersed tunnel has not been examined yet to the author's knowledge.

The aforementioned numerical models mainly developed adopting the conventional methods such as the finite element method and finite difference method. In recent years, a new numerical technique has come up which uses a set of nodes instead of the conservative meshes to approximate the solution. In consequence of discretization of the partial differential equations directly on nodes instead of meshes, the meshfree method is more qualified in dealing with mesh entanglement problems and constructing the approximations with arbitrary order of continuity than the traditional mash-based numerical methods. The start point of the meshfree method was using the moving least-square (MLS) method to establish shape functions for a set of scatter nodes by Nayroles et al. [18]. After that, Belytschko et al. [19] used the Galerkin weak form to improve the diffuse element method, which formed a new element-free Galerkin method. The element-free Galerkin method has been widely used in soil mechanics problems. In addition, an innovative interpolation scheme to overcome the disadvantage of the MLS was proposed by Liu and Gu [20], which is called the point interpolation method (PIM). The original PIM method picks up the polynomial basis as the basis function which performs well for one dimensional problems. However, it is hard to determine the ranks when this method extended to multi-dimensional range, which is the result of basis function selection. In order to figure this problem, Wang et al. [21] proposed a point interpolation method based on the radial basis functions (RBF). This method maps the multi-dimensional space into one-dimensional space though a radial function, which makes choosing basis function easier. This method has been widely used on the geomechanics problems. For example, Wang and Liu [22] simulated the Biot's consolidation process by radial PIM method. The displacement and the pore water pressure were discretised by the same shape function in space, while the fully implicit integration scheme was adopted for time discretization to avoid the spurious ripple effect. In the literatures, two different type of the RBFs are commonly used, which are the multiquadrics (MQ) [23], and the Gaussian radial basis functions [24]. In this study, the MQ was used. These functions were first under intensive research in multivariate data interpolation and used to solve the partial differential equations by Kansa [25]. Thus, this method is usually called Kansa's method, and is known as the global RBF collocation method (GRBFCM). This method has been applied to computational fluid dynamics problems such as solutions of the Navier–Stokes equation [26], natural convections in porous media [27], numerical wave tanks [28], and solid–liquid phase change problems [29]. The GRBFCM has the obvious advantages in dealing with arbitrary and complex domain, which applied in many field, such as surface fitting, turbulence analysis, neural networks, meteorology. However, ill-conditions can occur when the resolution is high. For the purpose of overcoming the difficulties of ill-conditions as mentioned above, a localization procedure to transform the dense matrices into sparse matrices was proposed. On the basis of the multiquadric RBF [30], Lee et al. [31] proposed the local RBF collocation method (LRBFCM) for the first time. This method has been applied in various fields, such as the solutions of diffusion problem [32], Darcy flow in porous media [33], water wave scattering [34], macro-segregation phenomena [35] and so on. Recently, Wang et al. [36] adopted the LRBFCM based on the Biot's consolidation equation to simulate a wave-induced seabed response around a pipeline.

In this paper, an integrated numerical model is proposed to simulate the sandy seabed dynamic response and transient liquefaction in the vicinity of an immersed tunnel under natural complex loading. The flow model is developed based on IHFOAM [37], while the seabed model is established adopting the LRBFCM with Biot's "$u-p$" approximation which considered the inertial term of soil skeleton. The verification of the integrated model is carried out by comparisons with the analytical solution [7] and published experimental data [38]. In this study, the effects of the current on the seabed behaviour around the immersed tunnel are examined. Parametric studies are conducted in regard of the different wave characteristics and soil properties.

2. Theoretical Models

In this study, an integrated 2D numerical model is developed to simulate the fluid-structure seabed interaction, which is composed of two sub-models: the flow model and seabed model. The flow model is responsible for simulating the wave motion including the wave pressure and current velocity, while the seabed model evaluates the effective stress, displacement, and pore pressure in the seabed under the dynamic loading. The two sub-models adopt the one-way coupling algorithm which is connected by the continuous water pressure.

2.1. Flow Model

In this paper, the flow model is based on one of the solvers in OpenFOAM®, IHFOAM. Recently, some other open-source codes have been developed for modeling wave propagation and wave–structure interaction problems [39,40] as well. To simulate the coastal, offshore, and hydraulic engineering process, this model solves the three-dimensional Volume Averaged Reynolds Averaged Navier–Stokes (VARANS) equations with regard to two incompressible phases of air and water. The fluid model adopts the finite volume discretization and the volume of fluid (VOF) method [37]. Including continuity and momentum conservation equations; the VARANS equations as the governing mathematical expressions in this model can be expressed as:

$$\frac{\partial \langle u_i \rangle}{\partial x_i} = 0, \tag{1}$$

$$\frac{\partial \rho_f \langle u_i \rangle}{\partial t} + \frac{\partial}{\partial x_j}\left[\frac{1}{n}\rho_f \langle u_i \rangle \langle u_j \rangle\right] = -n\frac{\partial \langle p^* \rangle^f}{\partial x_i} + n\rho_f g_i + \frac{\partial}{\partial x_j}\left[\mu_{eff}\frac{\partial \langle u_i \rangle}{\partial x_j}\right] - [CT], \quad (2)$$

where $\langle\rangle$ is Darcy's volume averaging operator, while $\langle\rangle^f$ is the intrinsic averaging operator; ρ_f represents the density which is computed by $\rho_f = \alpha\rho_{water} + (1-\alpha)\rho_{air}$; α is the indicator function defined in (5); u_i is the velocity vector while n is the porosity; p^* is the pseudo-dynamic pressures; $\mathbf{g}$, g_i is the gravitational acceleration; μ_{eff} is the efficient dynamic viscosity, defined as $\mu_{eff} = \mu + \rho_f \nu_{turb}$, in which μ is the molecular dynamic viscosity and ν_{turb} is the turbulent kinetic viscosity, given by the chosen turbulence model. The $k-\epsilon$ turbulence model is adopted in this study; the last term in (2) corresponds to the resistance of porous media, which is shown as:

$$[CT] = A\langle u_i \rangle + B|\langle u \rangle|\langle u_i \rangle + C\frac{\partial \langle u_i \rangle}{\partial t}, \quad (3)$$

compared to factors A and B, factor C is less significant. Moreover, 0.34 (kg/m^3) is often adopted for the value of C by default [41].

In present fluid model, the values of A and B are derived by Engelund [42]'s formulae, which also employed in Burcharth and Andersen [43].

$$A = E_1\frac{(1-n)^3}{n^2}\frac{\mu}{d_{50}^2}, \text{ and } B = E_2(1+\frac{7.5}{KC})\frac{1-n}{n^2}\frac{\rho_f}{d_{50}}, \quad (4)$$

in which d_{50} is the medium grain diameter of the materials; KC is the Keulegan Carpenter number, which is defined as $KC = u_m T_0/(nd_{50})$; u_m is the maximum oscillating velocity while T_0 is the period of the oscillation. E_1 and E_2 are parameters that define the linear and non-linear friction terms. The default values of these parameters ($E_1 = 50$ and $E_2 = 1.2$) are used in this study [44].

A two-phased fluid mixture containing air and water is taken into consideration in each cell, which can be controlled by an indicator function α. In the present model, α is defined as the quantity of water per unit of volume, which varies from 0 (air) to 1 (water):

$$\alpha = \begin{cases} 1, & \text{water} \\ 0, & \text{air} \\ 0 < \alpha < 1, & \text{free surface} \end{cases} \quad (5)$$

As an phase function, α can represent any variation of fluid properties considering the mixture properties, such as density and viscosity:

$$\Phi = \alpha\Phi_{water} + (1-\alpha)\Phi_{air}, \quad (6)$$

where Φ_{water} and Φ_{air} stand for the properties of water and air, separately, such as density of the fluid.

The fluid movement could be traced by solving the advection equation as [44]:

$$\frac{\partial \alpha}{\partial t} + \frac{1}{n}\frac{\partial \langle u_i \rangle \alpha}{\partial x_i} + \frac{1}{n}\frac{\partial \langle u_{ci} \rangle \alpha(1-\alpha)}{\partial x_i} = 0, \quad (7)$$

where $|u_c| = min\,[c_\alpha|u|, max(|u|)]$, in which the default value of c_α is 1, however, the user can specify a greater value to enhance the compression of the interface, or zero to eliminate it.

The solving algorithm used in flow model is PIMPLE, which is a combination of PISO (Pressure Implicit with Splitting of Operators) and SIMPLE (Semi-Implicit Method for Pressure-Linked Equations) algorithms. In this study, the $\kappa-\epsilon$ RAS (Reynolds-averaged simulation) turbulence model is adopted to model the turbulent viscosity ν_{turb} as:

$$\nu_{turb} = \frac{C_\mu \kappa^2}{\epsilon} \quad (8)$$

where C_μ is an empirical constant; κ is the turbulence kinetic energy while ϵ is the turbulence energy dissipation rate, separately.

The IHFOAM implements the wave generation and active wave absorption in the fluid domain, which introduce several boundary conditions: (i) the inlet boundary condition allows the generation of wave according to different wave theories as well as adding different steady current flow; (ii) the outlet boundary condition applies an active wave absorption theory to prevent the re-reflection of incoming wave; (iii) the slip boundary condition (zero-gradient) is applied on the bottom of the fluid domain and the lateral boundary of the numerical wave flume; (iv) the top boundary condition is set as the atmospheric pressure. The details of IHFOAM could be found in Higuera et al. [37].

2.2. Seabed Model

It is widely known that saturated soil considered as a multi-phase material is formed by soil particles and the void of the skeleton. The pores of the solid phase are filled with the water and trapped air distributed through the body. Therefore, in order to simulate the interaction between the soil skeleton and pore water in a porous seabed, a seabed model is established based on the partially dynamic Biot's equation (also known as "$u-p$" approximation) [45] with consideration of acceleration inertia term of flow and soil particles.

With the assumptions of a homogeneous and isotropic seabed and compressible pore fluid, the mass conservation equation of pore fluid can be expressed as [46]:

$$K_s\nabla^2 p_s - \gamma_w n_s \beta_s \frac{\partial p_s}{\partial t} + K_s \rho_f \frac{\partial^2 \epsilon_s}{\partial t^2} = \gamma_w \frac{\partial \epsilon_s}{\partial t}, \tag{9}$$

where p_s is pore water pressure, γ_w is the unit weight of water, K_s is the soil permeability, n_s is soil porosity; β_s is the compressibility of pore fluid while ϵ_s is volume strain, which can be expressed as:

$$\beta_s = \frac{1}{K_w} + \frac{1-S_r}{P_{wo}}, \quad \text{and} \quad \epsilon_s = \frac{\partial u_s}{\partial x} + \frac{\partial w_s}{\partial z}, \tag{10}$$

where u_s and w_s are the soil displacements in the x- and z- direction, respectively; K_w is the bulk modulus of pore fluid ($K_w = 1.95 \times 10^9 \mathrm{N/m^2}$ [4]); S_r is the degree of saturation and P_{wo} is absolute static water pressure, which defined as $P_{wo} = \gamma_w d$, in which d is the water depth.

Based on Newton's second law, the force equilibrium equation in a poro-elastic medium in horizontal and vertical directions can be given as:

$$\frac{\partial \sigma_x'}{\partial x} + \frac{\partial \tau_{xz}}{\partial z} = -\frac{\partial p_s}{\partial x} + \rho \frac{\partial^2 u_s}{\partial t^2}, \tag{11}$$

$$\frac{\partial \tau_{xz}}{\partial x} + \frac{\partial \sigma_z'}{\partial z} = -\frac{\partial p_s}{\partial z} + \rho \frac{\partial^2 w_s}{\partial t^2}, \tag{12}$$

where σ_x' and σ_z' are the effective normal stresses in horizontal and vertical direction respectively; τ_{xz} is shear stress component; ρ is the average density of a porous seabed and can be obtained by $\rho = \rho_f n_s + \rho_s(1-n_s)$, in which ρ_f is the fluid density while ρ_s is the solid density.

Based on the pore-elastic theory, the effective normal stresses and shear stress can be expressed in term of soil displacements:

$$\sigma_x' = 2G\left[\frac{\partial u_s}{\partial x} + \frac{\nu_s}{1-2\nu_s}\epsilon_s\right], \tag{13}$$

$$\sigma_z' = 2G\left[\frac{\partial w_s}{\partial z} + \frac{\nu_s}{1-2\nu_s}\epsilon_s\right], \tag{14}$$

$$\tau_{xz} = G\left[\frac{\partial u_s}{\partial z} + \frac{\partial w_s}{\partial x}\right], \tag{15}$$

where the shear modulus G is defined with Young's modulus (E) and the Poisson's ratio (ν_s) in the form of $E/2(1+\nu_s)$.

Substituting (13)–(15) into (11)–(12), we have the governing equations for force balance as

$$G\nabla^2 u_s + \frac{G}{1-2\nu_s}\frac{\partial}{\partial x}(\frac{\partial u_s}{\partial x} + \frac{\partial w_s}{\partial z}) = \frac{\partial p_s}{\partial x} + \rho_s \frac{\partial^2 u_s}{\partial t^2}, \tag{16}$$

$$G\nabla^2 w_s + \frac{G}{1-2\nu_s}\frac{\partial}{\partial z}(\frac{\partial u_s}{\partial x} + \frac{\partial w_s}{\partial z}) = \frac{\partial p_s}{\partial z} + \rho_s \frac{\partial^2 w_s}{\partial t^2}, \tag{17}$$

2.3. Boundary Conditions

At seabed surface ($z=0$), the vertical effective stress and shear stress is assumed to be zero and the water pressure is directly acting on it.

$$\sigma_z' = \tau_{xz} = 0, \text{ and } p_s = P_b, \tag{18}$$

where P_b is the dynamic wave pressure at the seabed surface, which is obtained by the wave model.

For the soil resting on the seabed bottom and lateral boundaries, zero displacements and impermeable are considered at the seabed bottom ($z=-h$), i.e.,

$$u_s = w_s = 0, \quad \text{and} \quad \frac{\partial p_s}{\partial z} = 0, \tag{19}$$

In addition, the boundary condition of tunnel surface is assumed to be impermeable with zero displacements, i.e.,

$$u_s = w_s = 0, \quad \text{and} \quad \frac{\partial p_s}{\partial n_v} = 0, \tag{20}$$

where n_v denotes normal vector of the tunnel surface.

2.4. Meshfree Model for the Seabed Domain

The LRBFCM is adopted to solve the governing equations listed above. First, an approximation function $\Phi(y_i)$ which can either stand for displacement or pore pressure in the "$u-p$" formulation is considered in the computational geometry. This function is composed of an arbitrarily distributed points series $y_j(j=1,2,\cdots,n)$ located both in the computational domain and on its boundary. Approximate $\Phi=\Phi(x)$ around y_i by the RBF $\chi(r_m)$ to construct a linear equation for each node y_n as,

$$\Phi(x) \approx \sum_{m=1}^{K} \alpha_m \chi(r_m), \tag{21}$$

where α_m is the undetermined coefficient for $\chi(r_m)$, and $\chi(r_m)$ is the MQ defined by,

$$\chi(r_m) = \sqrt{{r_m}^2 + c^2}, \tag{22}$$

with r_m, c, and x_m being the Euclidean distance from x to x_m, the shape parameter [30], and the positions of the K nearest neighbour nodes around the prescribed centre $x_1=y_n$, respectively. An algorithm based on the kd-tree is adopted to search the K nearest neighbour nodes [47].

Then, (21) is collocated on the K nearest neighbour nodes, which can be expressed as:

$$[\Phi]_{K\times 1} = [\chi]_{K\times K}\,[\alpha]_{K\times 1} \tag{23}$$

$$[\Phi]_{K\times 1} = [\Phi(x_1), \Phi(x_2), \cdots, \Phi(x_K)]^T, \tag{24}$$

$$[\chi]_{K\times K} = \begin{bmatrix} \chi(\| \mathbf{x}_1 - \mathbf{x}_1 \|) & \chi(\| \mathbf{x}_1 - \mathbf{x}_2 \|) & \cdots & \chi(\| \mathbf{x}_1 - \mathbf{x}_K \|) \\ \chi(\| \mathbf{x}_2 - \mathbf{x}_1 \|) & \chi(\| \mathbf{x}_2 - \mathbf{x}_2 \|) & \cdots & \chi(\| \mathbf{x}_2 - \mathbf{x}_K \|) \\ \vdots & \vdots & \ddots & \vdots \\ \chi(\| \mathbf{x}_K - \mathbf{x}_1 \|) & \chi(\| \mathbf{x}_K - \mathbf{x}_2 \|) & \cdots & \chi(\| \mathbf{x}_K - \mathbf{x}_K \|) \end{bmatrix}, \tag{25}$$

$$[\alpha]_{K\times 1} = \begin{bmatrix} \alpha_1 \\ \alpha_2 \\ \vdots \\ \alpha_K \end{bmatrix}, \tag{26}$$

Inversion (23) gives,

$$[\alpha]_{K\times 1} = [\chi]_{K\times K}{}^{-1} [\Phi]_{K\times 1} \tag{27}$$

Here, we assume a linear differential operator of both governing equation and the boundary condition which represented by L. If collocated the $L\Phi(x)$ on $x_1 = y_n$, (23) can be written as:

$$L\Phi(y_n) = \sum_{m=1}^{K} \alpha_m L\chi(r_m)\,|_{x=x_1} \tag{28}$$

which can be expressed as the vector form:

$$L\Phi(y_n) = [L\chi]_{1\times K} [\alpha]_{K\times 1} \tag{29}$$

In (29) mentioned above, the $L\chi(r_m)$ is on behalf of the results of the differential operator act on the the RBF $\chi(r_m)$. Then, by combining (29) and (27), it can be obtained as:

$$L\Phi(y_n) = [C]_{1\times K} [\phi]_{K\times 1} \tag{30}$$

$$[C]_{1\times K} = [L\chi]_{1\times K} [\chi]_{K\times K}^{-1} \tag{31}$$

$$[L\chi]_{1\times K} = \left[L\chi(r_1)|_{x=x_1} \quad L\chi(r_2)|_{x=x_2} \quad \cdots \quad L\chi(r_K)|_{x=x_K} \right] \tag{32}$$

Obviously, the value of the row vector $[C]_{1\times K}$ can be obtained if all variables L, χ and x_j are known.

For all y_n in computational domain, the linear equations can be obtained by the above-mentioned localization procedure on either the governing equation or boundary conditions. Next, these equations can be assembled into the system matrix, as:

$$[A]_{N\times N} [\Phi]_{N\times 1} = [B]_{N\times 1} \tag{33}$$

where

$$\begin{aligned}[\Phi]_{N\times 1} = [&u_1(y_1), u_1(y_2), \cdots, u_1(y_N), u_2(y_1), u_2(y_2), \cdots, u_2(y_N),\\ &p(y_1), p(y_2), \cdots, p(y_N)]^T,\end{aligned} \tag{34}$$

which, $[\phi]_{N\times 1}$ is the sought solution, while $[B]_{N\times 1}$ is a column vector contributed from the external loadings. It is noticeable that (33) is a sparse system matrix which is similar to both finite difference method as well as the finite element method. In the present numerical model, the direct solver of SuperLU [48] is adopted to solve the sparse system, (33). The procedure of the LRBFCM is finished here.

In numerical strategy, it is necessary to integrate the governing equations in the time domain if the boundary conditions or the extrernal loadings are time-dependent. In the present model, the single-step time integration method of the Newmark method [49] is adopted, which could handle each time step independently when the first- and second-order time derivatives exist at the same time. More details can be found in some previous studies [50,51].

2.5. Integration Procedure of Flow Model and Seabed Model

For the integration procedure in this study, the one-way coupling algorithm is adopted for pore pressure delivery from the wave model to the seabed model. As demonstrated in Figure 1, the flow model is responsible to simulate the wave–current interaction which includes wave generation and fluid propagating according to the given wave/current characteristics. The water pressure will be extracted from the results after solving the VARANS equations, then used as the boundary conditions of seabed surface as the external loading in the seabed model, while the soil model can carry out the seabed behaviour under the external wave loading combined with the input parameters of seabed model. The displacement, the pore water pressure and the effective stresses could be determined by solving the Biot's "$u - p$" approximation.

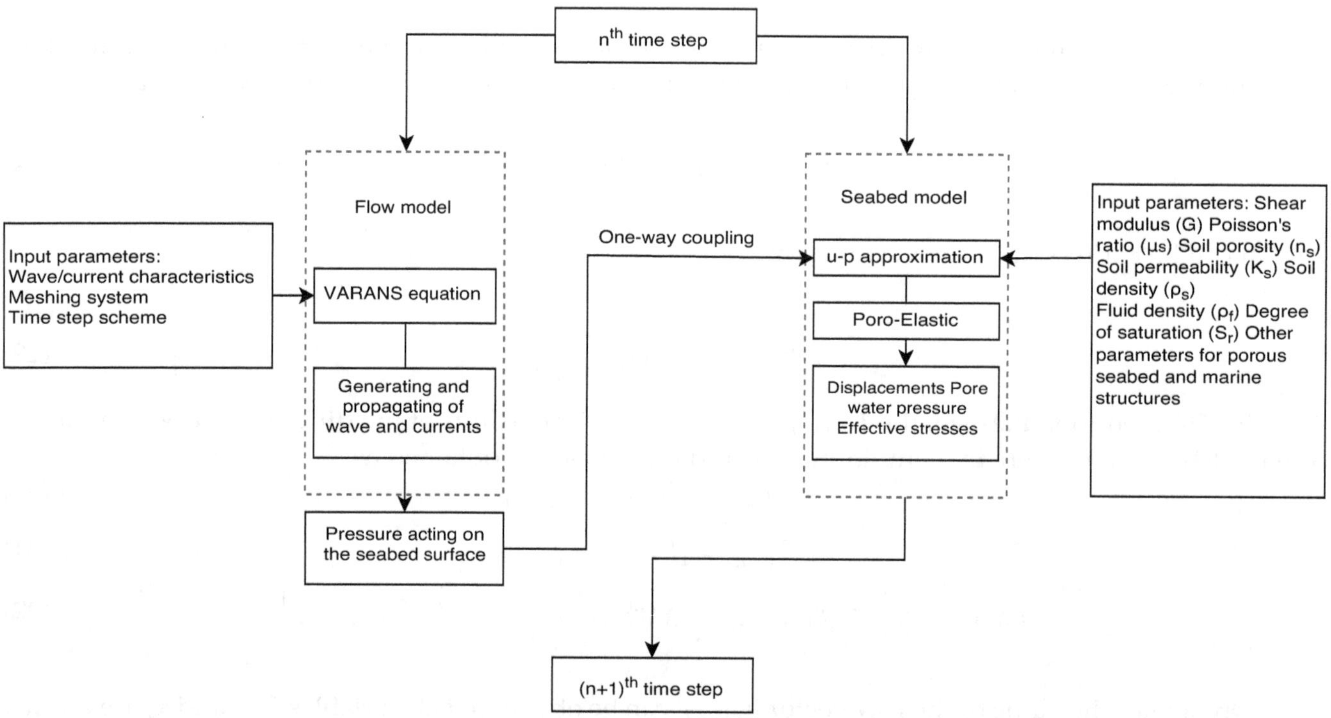

Figure 1. The coupling process of the integrated model.

2.6. Convergence Tests

Before conducting parametric studies of the wave-induced dynamic response in a porous seabed under wave loading, it is necessary to check the convergence of a newly proposed numerical model. The convergence tests are carried out in regards of the nodes distance size (Δx), shape factor (c) and the node number of the local region (K), which could have an influence on the numerical accuracy and computational efficiency.

Firstly, the small node distance size makes the results more accurate, however, it will result in enormous computational cost. As shown in Figure 2, the Δx is equal to $L/50$, $L/100$, and $L/200$ respectively (L is the wavelength). The non-dimensional pore water pressures (p_s/p_0) are depicted, $_0$ represents the amplitude of linear wave pressure at the seabed surface. From the figure, the result for the case of Δx equal to $L/50$ is slightly difference from the others, while the results are almost the same for Δx equal to $L/100$ and $L/200$, which indicate the model is convergent with a node distance that is smaller than $L/100$.

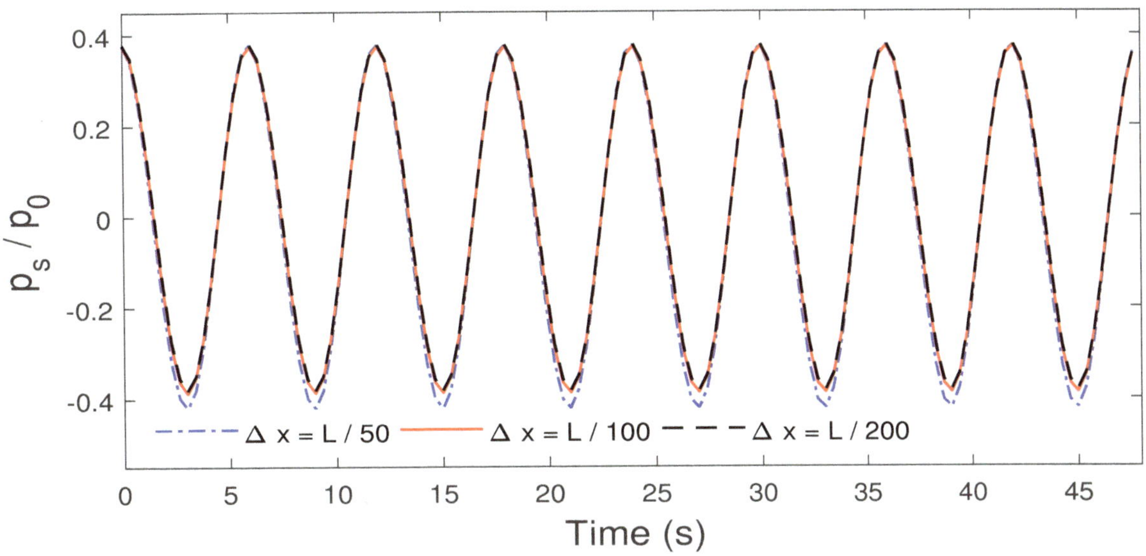

Figure 2. Time variation of dynamic pore water pressure in porous seabed under wave loading for different mesh conditions.

Next, the convergence test of shape factor is conducted.The shape factor is generally equal to 15–60 times the maximum Euclidean distance between two adjacent nodes by convention. The c adopts 3.975 ($15 \times \Delta x$), 7.95 ($30 \times \Delta x$), and 15.9 ($60 \times \Delta x$) separately. From the Figure 3, it can be conclude that the numerical results are not sensitive to the affect the shape factor on the consequence of the almost the same results obtained for this set of shape factors.

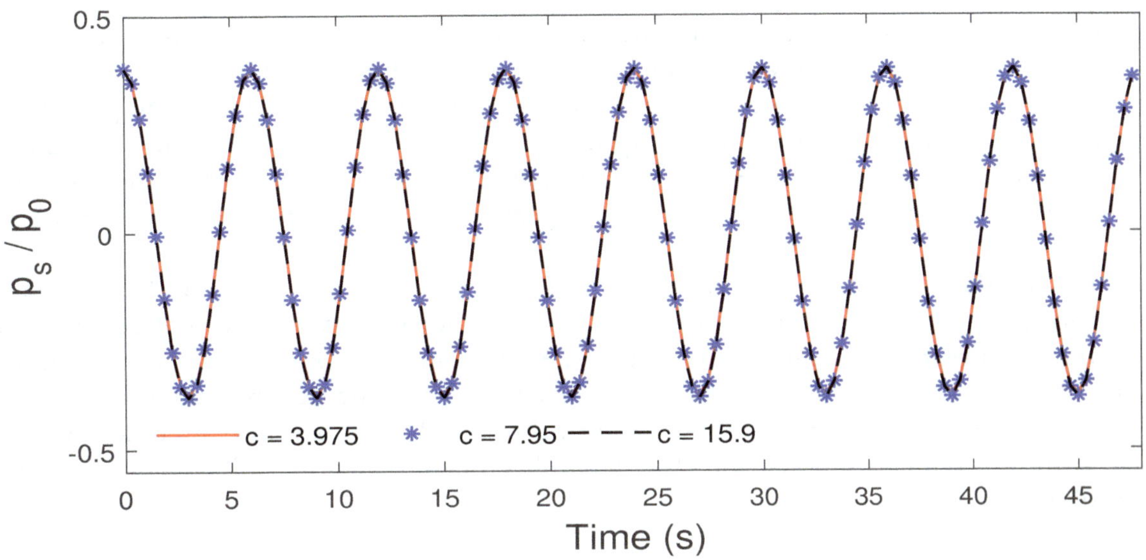

Figure 3. Time variation of dynamic pore water pressure in porous seabed under wave loading for different shape factors.

Moreover, the effect of the number of nearest neighbour nodes in a local region is examined. The dynamic pore water responses in a seabed are depicted in Figure 4 regards for the different numbers of neighbour nodes in the local region, which are 5, 9 and 13, respectively. From the figure, the result shows a good tendency during the whole process when $K = 9$. It also can be seen that for $K = 5$, the amplitude of the result is slightly different with the result of $K = 9$. When $K = 13$, the value of $|p_s|/p_0$ is even beyond 1 after 2 s, which is obviously wrong. This condition might be the ill-condition in this case. Thus, the number of the nodes located in the local region is 9 in this study.

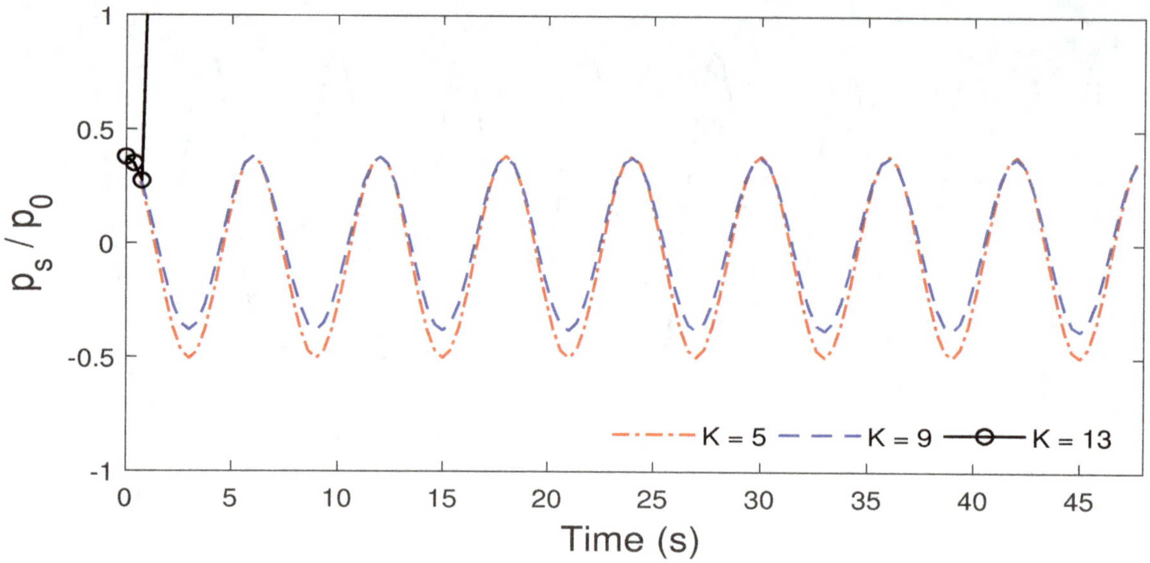

Figure 4. Time variations of dynamic pore water pressure in porous seabed under wave loading for different local region nodes number.

3. Verification of the Proposed Model

Before conducting the parametric study, it is essential to verify the capability of the newly developed meshfree model. In this section, the proposed model is verified by comparison with analytical solutions and a set of published laboratory experiments from a previous study.

3.1. *Comparison of the Present Model and the Analytical Solution for Wave-Seabed Interactions*

To verify the proposed meshfree model under the dynamic condition, the present model is used to simulate the saturated isotropic seabed under linear wave loading based on a partially dynamic model. The result of the numerical solution are compared with the analytical solution presented by Hsu and Jeng [7]. The linear wave loading is applied to the seabed surface. The parameters for the comparison are given as: wave period $T = 10$ s; water depth $d = 20$ m; degree of saturation $S_r = 0.975$; Poisson's ratio $\nu_s = 0.4$; soil porosity $n_s = 0.35$; soil permeability $K_s = 10^{-2}$ m/s; shear modulus $G = 5 \times 10^6$ Pa.

The vertical distributions of the wave-induced pore pressure ($|p_s|/p_0$), effective stresses ($|\sigma_z'|/p_0$) and shear stress ($|\tau_{xz}|/p_0$) versus the soil depth (z/H) in a porous seabed are plotted in Figure 5. As depicted in the figure, the blue lines represent the results obtained from analytical solution, while red circles denote soil behaviour simulated from the present "$u - p$" time-dependent model. The figure clearly shows that the result obtained by the numerical method are in a good agreement with that of analytical solution, which demonstrates the capability of meshfree model for seabed dynamic response simulation.

3.2. *Comparison of the Present Model with the Laboratory Experiments and Fem Results for Combined Waves and Current Loading*

It is necessary to verify the performance of the integration model including both fluid and soil models under the circumstance of complex nature loading. There are numerous laboratory experiments related to the seabed response around the marine structures to date. However, it is quite limited in terms of the experimental data available for the case of immersed tunnel. Thus, the verifications of the integration model are carried out by comparison with the laboratory experiments and the FEM (Finite Element Method) results from DIANA-SWANDYNE II [52] for the seabed without the structure instead in this section. Qi and Gao [38] conducted a series of flume tests considering wave and wave combined currents as dynamic loading, respectively.

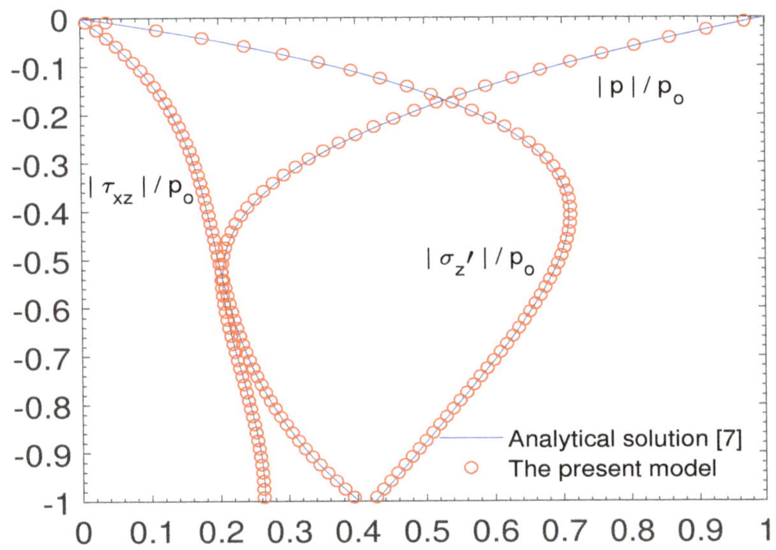

Figure 5. Comparison of vertical distribution of maximum oscillatory pore pressure, effective normal stress and shear stress with the analytical solution [7].

The first validation of this section is compared to the laboratory experiments conducted under wave loading only [38]. The input data for the first validation are: wave height $H = 0.12$ m, water depth $d = 0.5$ m, wave period $T = 1.4$ s, seabed thickness $h = 1.2$ m, degree of saturation $S_r = 1.0$, shear modulus $G = 10^7$ N/m^2, Poisson's ratio $\nu_s = 0.3$, permeability $K_s = 1.88 \times 10^{-4}$ m/s. Figure 6 depicts the wave patterns with corresponding dynamic pore water pressure of the seabed, which are predicted by the present model and obtained from the experiment and FEM model separately. It can be seen that the result obtained from both the wave model and seabed model are in good agreement with the test data, which indicate that the present model is capable for simulating the wave motion in the fluid domain as well as the corresponding soil response of a sandy seabed.

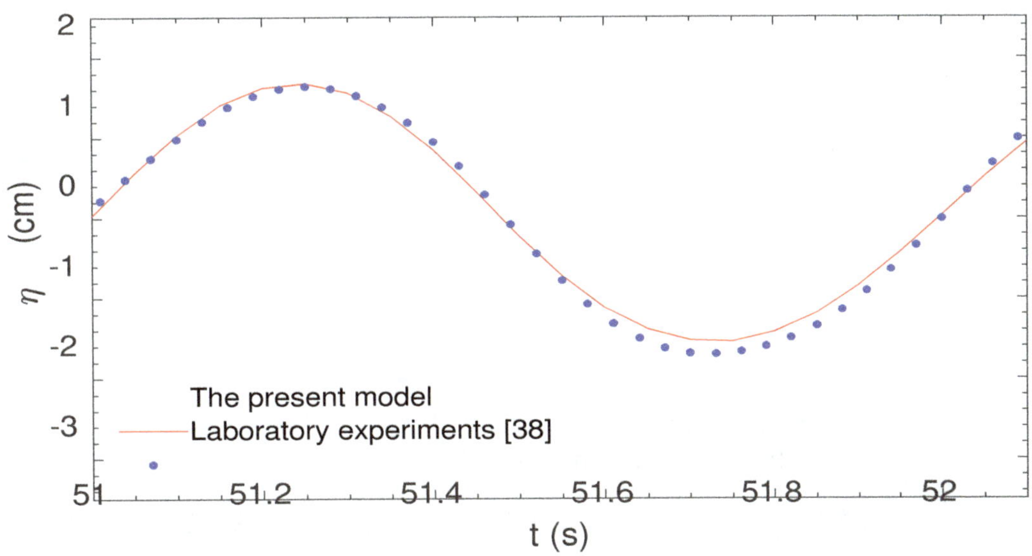

(**a**) Water surface elevation (η)

Figure 6. *Cont.*

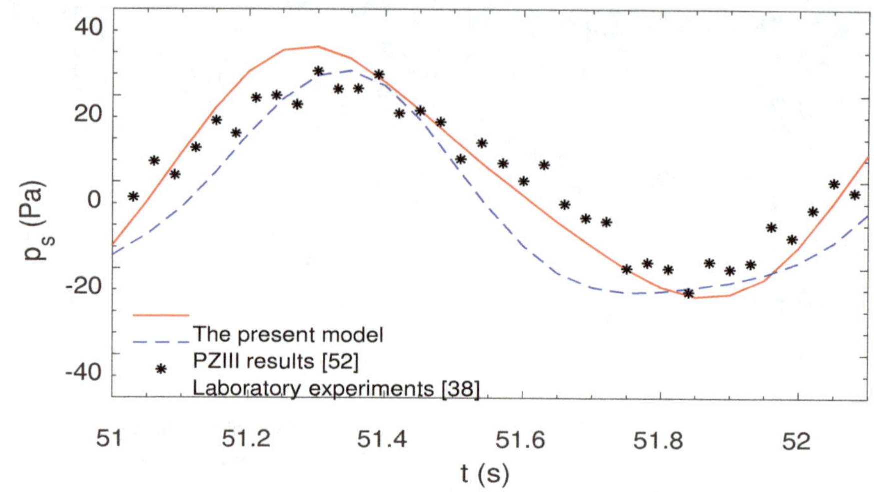

(**b**) Wave-induced pore pressure in seabed (p_s)

Figure 6. Validation of the (**a**) Water surface elevation (η) and (**b**) Wave-induced pore pressure in seabed (p_s) under wave loading at $z = -0.1$ m against the experimental data (which was from [38]) and FEM result data (caculated from DIANA-SWANDYNE II [52]).

The second validation in this section is to compare with the previous laboratory experiments conducted by Qi and Gao [38]. Unlike the previous case, this test simulates the seabed dynamic response under wave and current, which are generated synchronously. The following current with velocity of 0.05 m/s is adopted. The wave parameters and the soil properties are the same as above. As shown in Figure 7, the fluid pattern tracked by the fluid model matches well with the experiment data, while the pore water pressure simulated by the present model in correspondence with that obtained from the experiment and the FEM model. Thus, the current model performs well for simulating a more realistic marine dynamic elastic behaviour including both the fluid and soil parts.

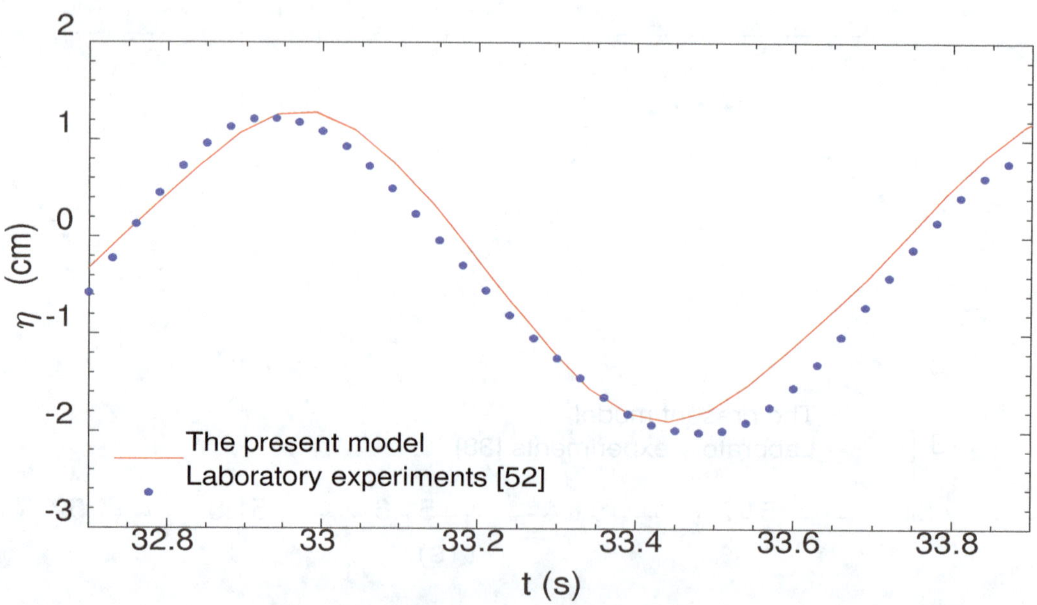

(**a**) Water surface elevation (η)

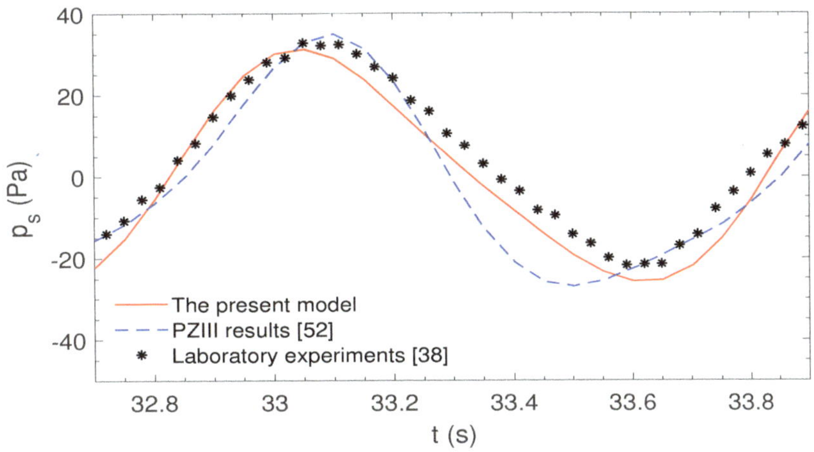

(**b**) Wave-induced pore pressure in seabed (p_s)

Figure 7. Validation of the (**a**) Water surface elevation (η) and (**b**) Wave-induced pore pressure in seabed (p_s) under wave combined current loading at $z = -0.1$ m against the experimental data (which was from [38]) and FEM result (calculated from DIANA-SWANDYNE II [52]).

4. Dynamic Response of the Seabed

In this study, a new meshfree model is developed, based on Biot's "$u - p$" approximation to simulate the dynamic sandy seabed behaviour around an immersed tunnel under complex natural loading including wave and current. As shown in Figure 8, the external loadings are assumed to be propagating over the porous seabed in which the immersed tunnel is buried. The buried depth is defined as b adopting 0.5 m below the seabed surface in this case. The sandy seabed foundation is treated as an elastic two-phase medium above a rigid impermeable bottom with 200 m for seabed length (Lx) and 40 m for seabed thickness (h). The seawater depth is specified as d. The immersed tunnel is assumed to be placed on the trench dredged on the middle of the seabed of the computational domain. The trench is back-filled by the same type of loose sand with the seabed soil. The tunnel geometry, wave profile and seabed profile in this case are roughly the same as the actual conditions of the Hong Kong–Zhuhai–Macao bridge tunnel, which could provide a reference of the sandy seabed dynamic for such a large immersed tunnel under wave loading.

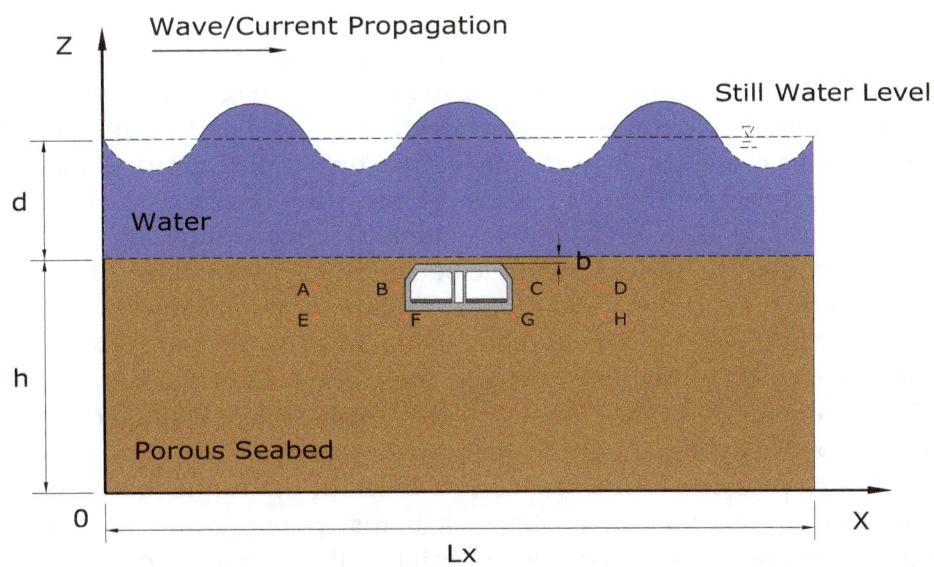

Figure 8. The sketch of the computational domain of wave-seabed-tunnel interaction problem.

The detailed dimensions of the immersed tunnel are given in the Figure 9. As shown in figure, the immersed tunnel in this study is considered as an elastic material comprising two traffic tubes of 30 m long and 9 m high (cross section). The boundary condition of the immersed tunnel is treated as impermeable with zero pore pressure gradient. No relative displacement is assumed between the seabed soil and the tunnel frame on the consequence of the high fraction exists between the concrete tunnel surface and seabed soil.

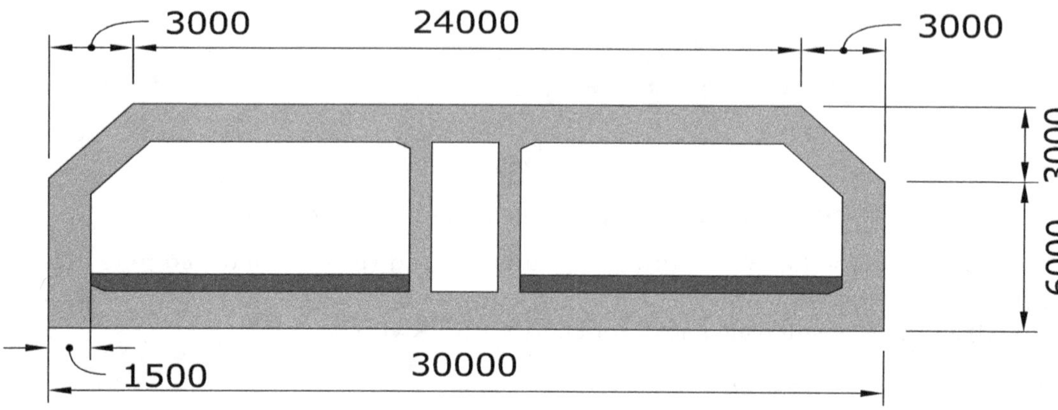

Figure 9. Cross section of an immersed tunnel element (Unit: mm).

The configuration of the fluid domain and seabed domain can be found in Table 1, as well as the wave characteristics, seabed soil properties and modelling parameters. The seabed foundation is assumed to be composed of relative dense deposited sand [53]. Figure 10 shows the applicability range of the different wave theories [54]. The wave characteristics adopted in this study are in the range of Stokes second-order wave ($H = 4$ m, $T = 10$ s, $d = 30$ m, shown as a red star in Figure 10), which is generated and simulated by fluid model. The node number of the local region (K) for local RBF method is 9, while a positive constant c which is known as the shape factor equals 6. The total number of nodes in this case is 771,140. The convergence test has been done to check the stability of the present parameter configuration, which is quite enough to obtain an accuracy and detailed result. The time step Δt set in this case is 0.5 s, while 20 time steps are contained in one wave period.

The aim of this section is tracking the dynamic soil response during the wave propagating over the seabed around the immersed tunnel. During the wave propagation from the left to the right of the computational domain continuously, the effective stresses and pore water pressures show a correlation trend with the change of the water pressure acting on the surface of the seabed. As shown in Figure 11, the oscillatory wave-induced pore water pressures, horizontal displacements (u_s) and vertical displacement (w_s) for the computational domain of the seabed at $t = 13$ s are presented, respectively. It is found that the seabed dynamic behaviour with immersed tunnel is not periodic symmetry any more under the cyclic wave loading. The figure shows that the existence of the immersed tunnel has an obvious influence on the dynamic behaviour of the sandy seabed soil nearby by comparing the region located on the leftward and the rightward of the tunnel. It can be seen in Figure 11 that the placement of a tunnel weakens the displacement change of local area in a way, while the fluctuation of the dynamic pore water pressure decrease around as well. In Figure 11c, the dynamic pore water pressure of the seabed soil beneath the tunnel bottom shows a different tendency from the surrounding that the positive oscillatory pore pressure occurs on the left corner while the negative occurs on the right corner. In order to figure out a more detailed dynamic soil response in the vicinity of the immersed tunnel, the results of dynamic pore water pressure of some typical locations are analysed.

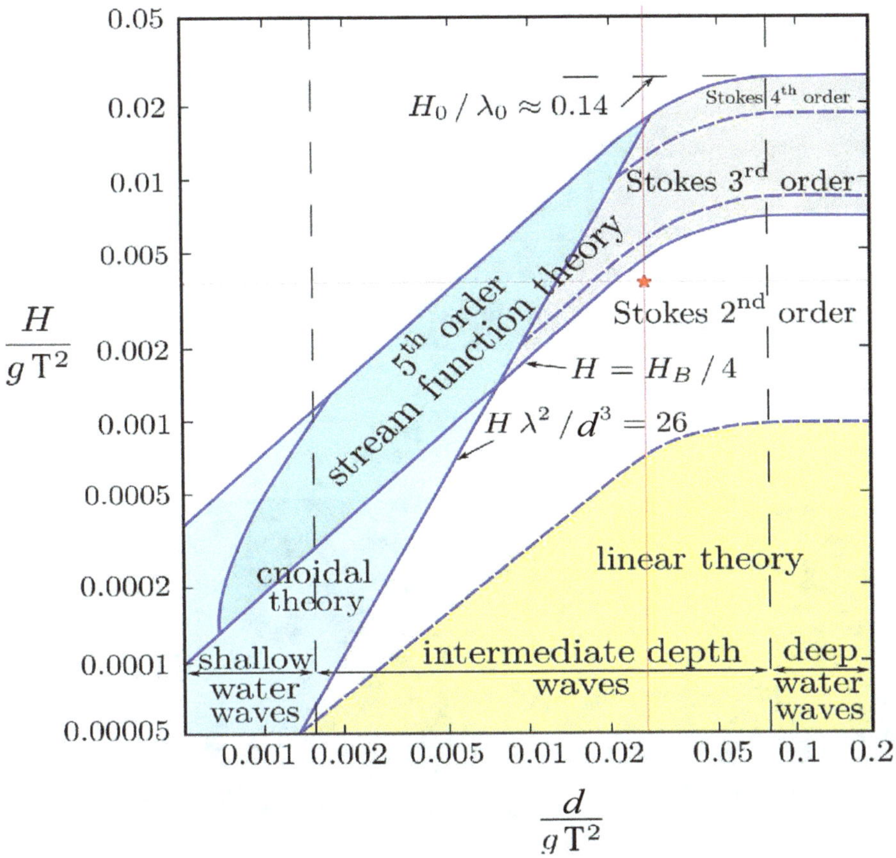

Figure 10. Wave theories range of applicability [54]. The red star represents the wave characteristics used in this study.

Table 1. Parameters used for standard case in parametric study.

Wave Characteristics	**Value**	**Unit**
Wave period (T)	10	s
Wave height (H)	4	m
Wavelength (L)	142.355	m
Water depth (d)	30	m
Soil Properties	**Value**	**Unit**
Poisson's ratio (ν_s)	0.20	-
Permeability (K_s)	1.0×10^{-5}	m/s
Porosity (n_s)	0.42	-
Degree of saturation (S_r)	0.95	-
Shear modulus (G)	5.0×10^6	Pa
Density of soil particles (ρ_s)	2650	kg/m^3
Tunnel buried depth (b)	0.5	m
Modelling Parameters	**Value**	**Unit**
Number of KNN (K)	9	-
Shape factor (c)	6	-
Nodes number (x direction)	2000	-
Nodes number (z direction)	400	-
Δt	0.5	s
Time steps in a period (Ndt)	20	-

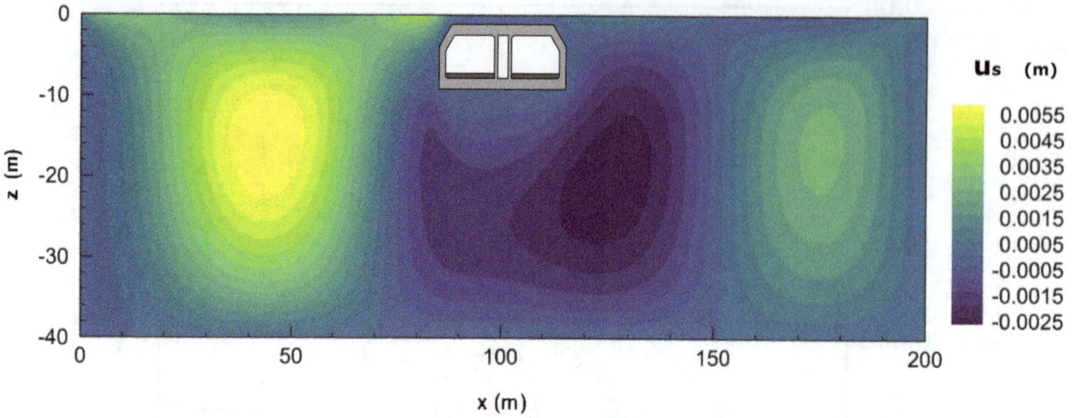

(**a**) Horizontal displacement

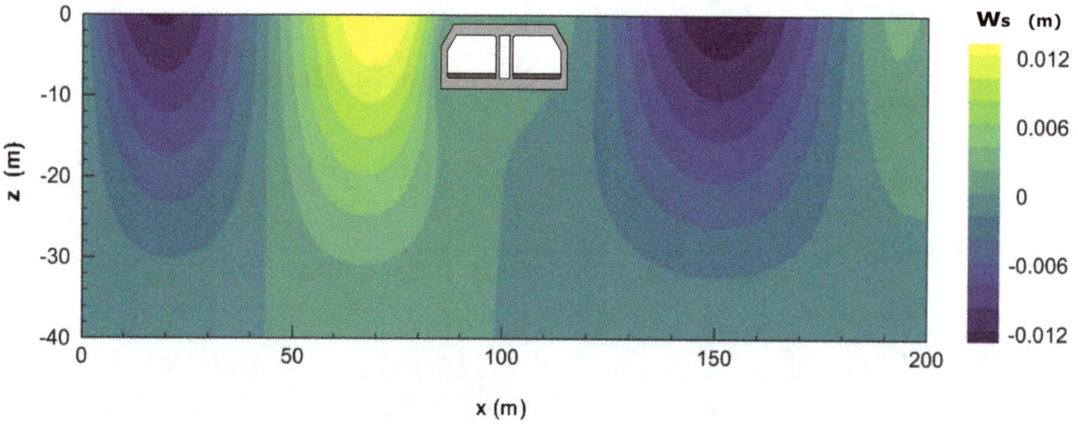

(**b**) Vertical displacement

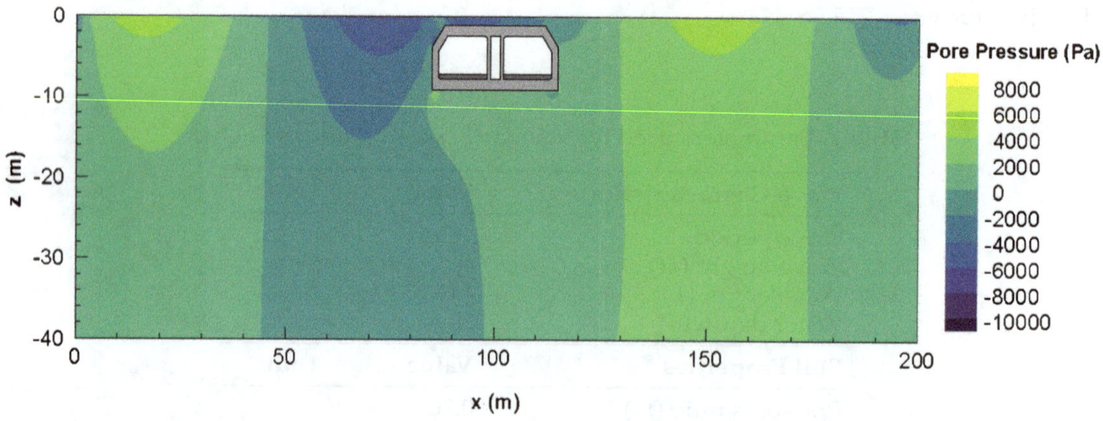

(**c**) Dynamic pore water pressure

Figure 11. The dynamic distributions of the seabed around tunnel under the wave loading when $t = 13$ s including (**a**) horizontal displacement; (**b**) vertical displacement, (**c**) pore water pressure.

As shown in Figure 8, points A to D are a set of symmetrical nodes about the tunnel at the depth of -5 m in the seabed ($x = 60, 82, 118, 140$ m, respectively), while points E to F located at the depth of -10 m ($x = 60, 85, 115, 140$ m respectively). Points F and G are set on 0.5 m below the two base corners of the immersed tunnel. Figure 12 depicts the time series of the dynamic pore water pressure generated by the two point sets. From the figure, it can be seen that the vibration of the pore water pressure is more violent on the remote seabed from the tunnel, which is consistent with the conclusion mentioned above. Furthermore, the amplitude reduced for point F and G below the tunnel are slightly less than the points B and C on two sides of the tunnel. The pore water pressure vibration of the soil on points F and G is even larger than on points B and C, which indicates that the seabed foundation beneath the immersed tunnel is more likely to be unstable due to transient liquefaction. In addition, a phenomenon occurred which is that there is a phase difference of dynamic pore pressure brought out under the base corners of the tunnel (points F and G), as shown in Figure 12b.

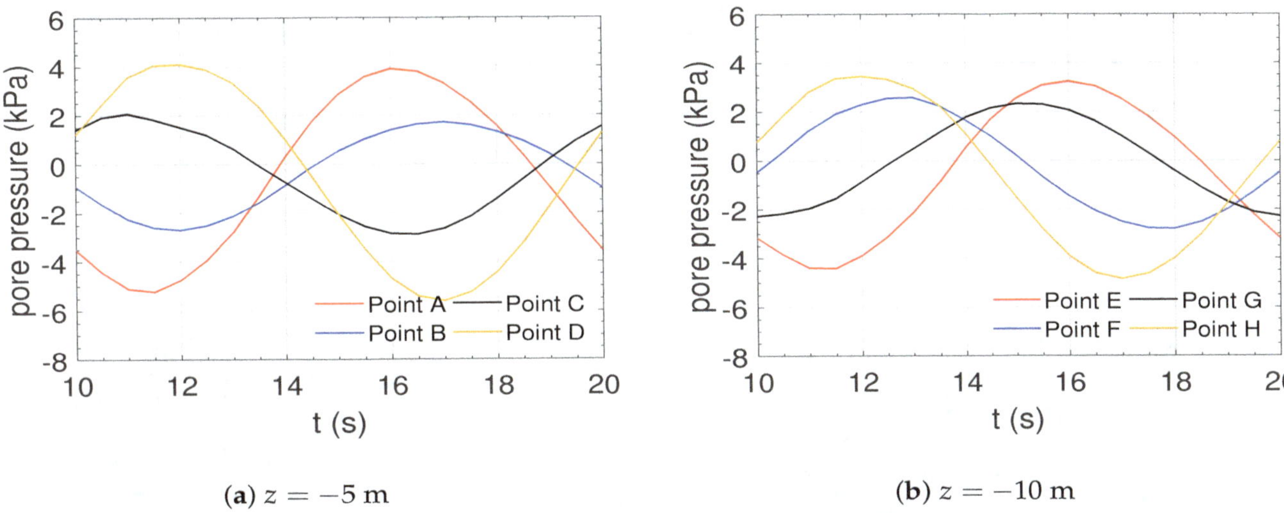

Figure 12. Time series of wave-induced pore pressure in seabed foundation at **(a)** $z = -5$ m; **(b)** $z = -10$ m.

5. Wave-Induced Liquefaction

The stability of the coastal structures and its seabed foundation is one of the main concerns for engineering design procedure. The wave-induced liquefaction in a porous seabed is one of the most significant unstable factors. Zen and Yamazaki [55] pointed out that the liquefaction of the porous seabed is responding to the variation of the ocean wave, which is actually caused by the periodic upward seepage force. Thus, we proposed to estimate the liquefaction state in two-dimensions, i.e.,

$$\sigma_0' + (P_b - p_s) \leq 0 \tag{35}$$

where σ_0' represents the initial effective stress, P_b is the wave pressure acting on seabed, while p_s is the wave-induced transient pore pressure. The value of $P_b - p_s$ is equal to the excess pore pressure generated by the wave loading ($|u_e|$).

Figure 13 shows the wave-induced transient liquefaction area around the immersed tunnel at three typical times ($t = 12$ s, $t = 14$ s and $t = 15.5$ s) separately. As shown in the figure, the transient liquefaction area moves along the direction of the wave propagation. The previously liquefied area is able to recover as the wave trough go away. This process is repeated periodically under the cyclic wave loading. The maximum liquefaction depth in this case is 0.8 m below the seabed surface, as seen in Figure 13a, while the soil that covered the tunnel is fully liquefied during one wave period which illustrated in Figure 13b. Thus, the back filling soil above the tunnel can not protect the immersed tunnel any more in this circumstance. Moreover, the maximum liquefaction depth of the rightward seabed of the tunnel is 0.6 m, which is slightly shallow than that in leftwards of 0.8 m.

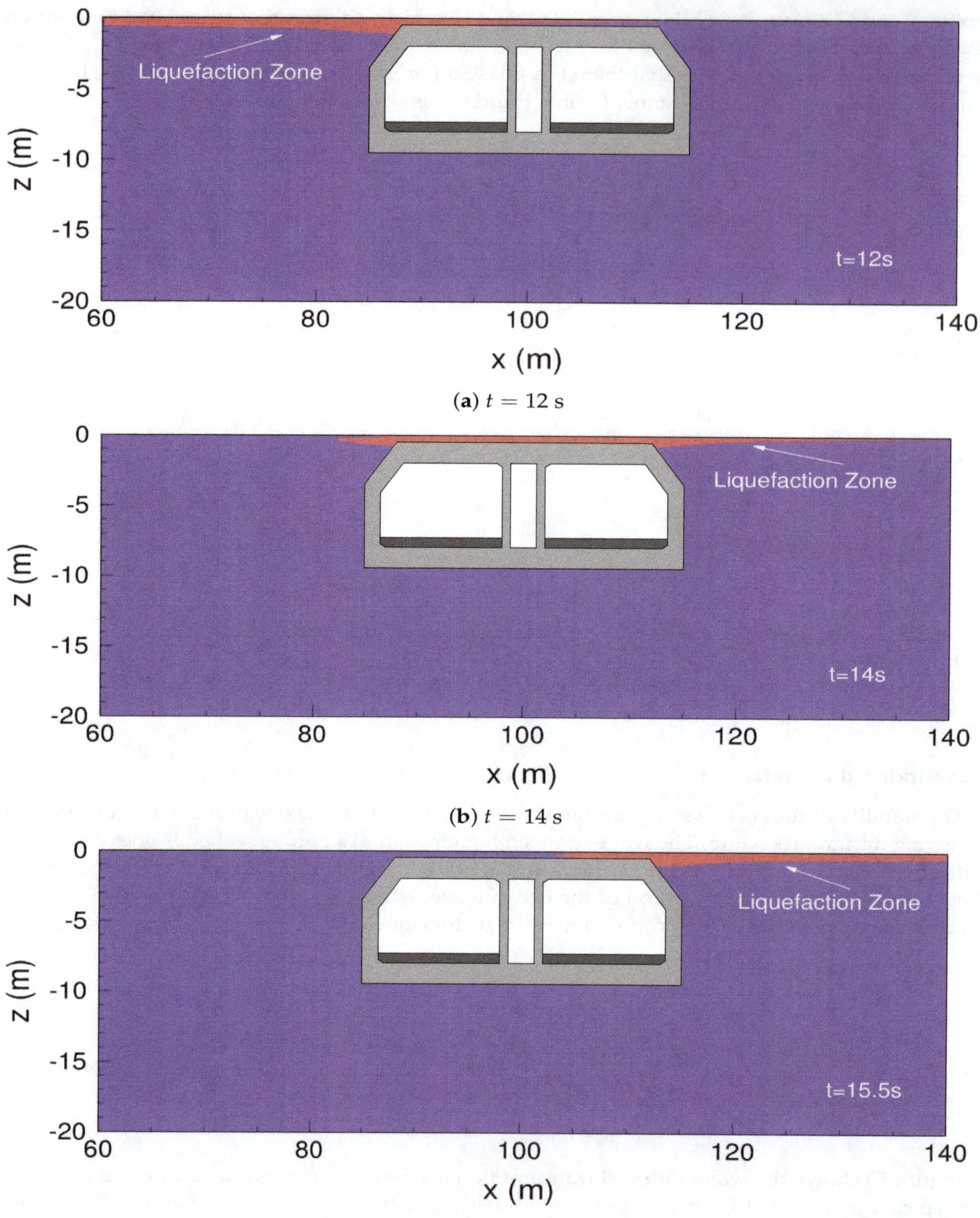

(**a**) $t = 12$ s

(**b**) $t = 14$ s

(**c**) $t = 15.5$ s

Figure 13. The liquefaction area surrounding immersed tunnel for (**a**) $t = 12$ s; (**b**) $t = 14$ s and (**c**) $t = 15.5$ s respectively.

6. Parametric Study

In this section, parametric studies of the seabed liquefaction around the immersed tunnel are carried out, which include the wave characteristics, soil properties and the effects of current specifically.

6.1. Effects of Wave Characteristics

Basically, the wave-induced oscillatory liquefaction phenomenon of the porous seabed is highly relative to the water pressure propagating on it. As a consequence, the relation of the wave characteristics and the liquefaction depth in the vicinity of the tunnel is discussed in detail with respect to the wave period (T), wave height (H), and still water depth (d), separately. Table 2 lists the input data involving the wave variables and other parameters. Figure 14 characterizes the maximum liquefaction depth of the vertical section at $x = 80$ m which is on the left side of the tunnel in regard of the different wave conditions.

Table 2. Parameters used for in parametric study of wave characteristics.

Wave Characteristics	Value	Unit
Wave period (T)	8 or various	s
Wave height (H)	2 or various	m
Water depth (d)	30 or various	m
Soil Properties	**Value**	**Unit**
Poisson's ratio (ν_s)	0.20	-
Permeability (K_s)	1.0×10^{-5}	m/s
Porosity (n_s)	0.42	-
Degree of saturation (S_r)	0.95	-
Shear modulus (G)	5.0×10^{6}	Pa
Tunnel buried depth (b)	0.5	m

It is well known that the wave height is in positive correlation to the value of the water pressure acting on the seabed precisely. Furthermore, the water pressure will affect the wave-induced pore water pressure in seabed deposits. As illustrated in Figure 14a, the maximum liquefaction depth grows deeper with an increase of the wave height, which shows the positive relationship between the wave height and the potential of liquefaction.

The wave period is another key wave parameter that influences the wave length directly, which will further affect the seabed transient liquefy process. In this study, the wave periods are from 8 to 10 s. From Figure 14b, the maximum liquefaction depth is progressively increase with the rise of wave periods. It can be concluded that the liquefaction degree is more intense in the case of a long wave period.

Lastly, the influence of still water depth is discussed. The water depth ranges from 30 to 40 m. Unlike wave height and period, the maximum oscillatory liquefaction depth decreases with the raising of still water depth, as shown in Figure 14c, which indicates that the wave-induced transient liquefaction is more likely to occur in the shallow water areas.

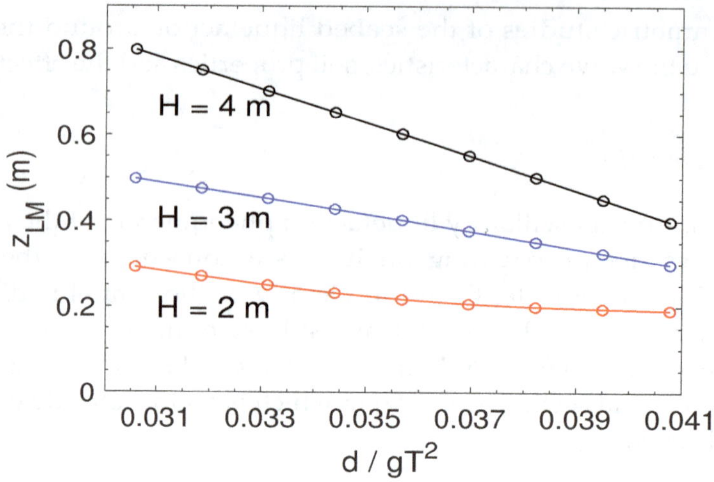

(**a**) Effects of wave heights

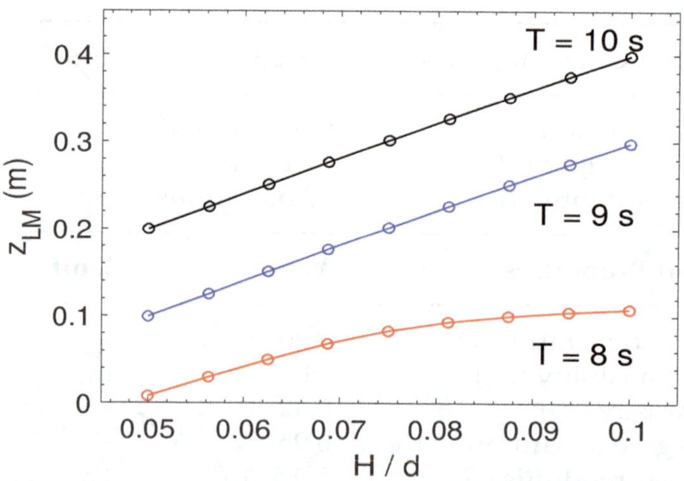

(**b**) Effects of wave periods

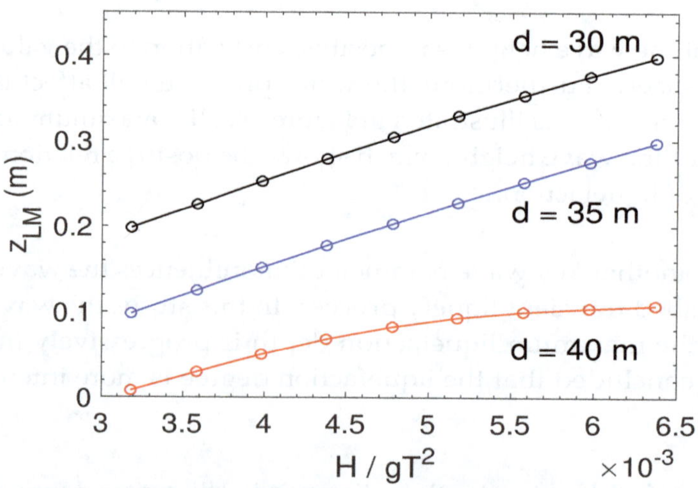

(**c**) Effects of water depths

Figure 14. The maximum liquefaction depth of seabed surrounding the tunnel with various (**a**) wave heights; (**b**) wave periods; (**c**) water depths.

6.2. Effects of Soil Properties

Besides the wave characteristics, the deposit conditions are found to be essential in a wave-seabed structure interactions. The effects of two parameters of soil properties are discussed individually, i.e., degree of saturation (S_r), and soil permeability (K_s). A series of topical values are selected in this section, which are from 1.0×10^{-5} to 1.0×10^{-3} for soil permeability, and 0.925–0.975 for degree of saturation, respectively. The wave condition and other parameters used in this section can be found in Table 3.

Table 3. Parameters used for in parametric study of soil properties.

Wave Characteristics	Value	Unit
Wave period (T)	10	s
Wave height (H)	4	m
Water depth (d)	30	m
Soil Properties	**Value**	**Unit**
Poisson's ratio (ν_s)	0.20	-
Permeability (K_s)	1.0×10^{-5} or various	m/s
Porosity (n_s)	0.42	-
Degree of saturation (S_r)	0.925 or various	-
Shear modulus (G)	5.0×10^6	Pa
Tunnel buried depth (b)	0.5	m

To examine how the different soil parameters affect the seabed liquefaction around the immersed tunnel, Figures 15 and 16 are depicted to show the vertical distribution of the excess pore pressure ($|u_e|/\sigma_0'$) and the maximum liquefaction depth of the vertical section at $x = 80$ m in seabed with various soil properties. Figure 15b illustrates that the maximum liquefaction depth is larger in the seabed with relative low degree of saturation. While the same conclusion can be draw for the various of soil permeability in Figure 16b, i.e., the smaller the soil permeability adopted, the deeper the liquefaction occurs in the vicinity of the tunnel.

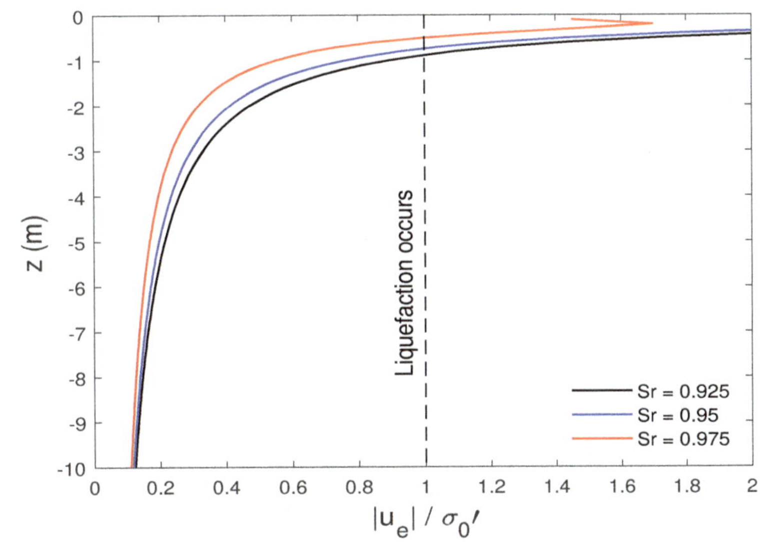

(**a**) distribution of the excess pore pressure ($|u_e|/\sigma_0'$) under wave trough

Figure 15. *Cont.*

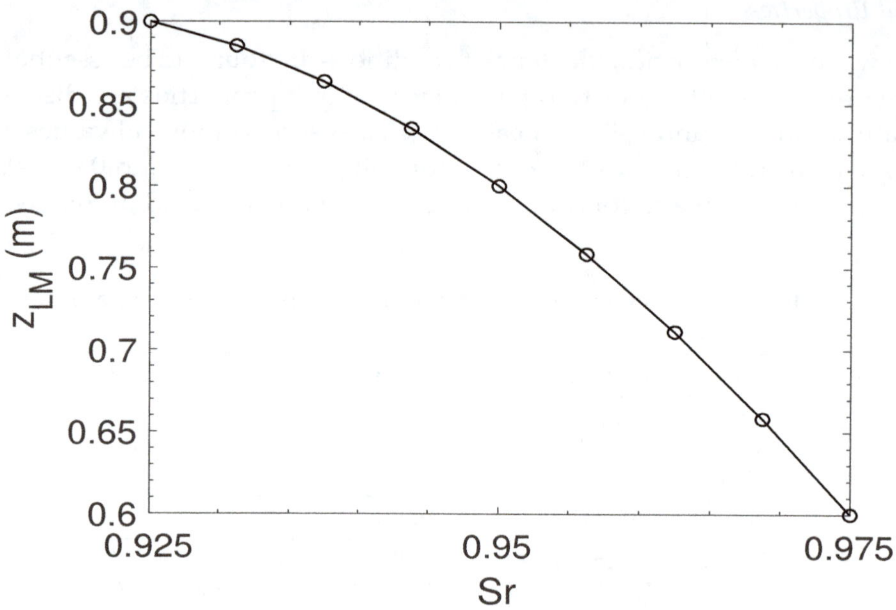

(**b**) maximum liquefaction depth of sandy seabed

Figure 15. (**a**) The vertical distribution of the excess pore pressure ($|u_e|/\sigma_0'$) under wave trough and (**b**) the maximum liquefaction depth of the sandy seabed surrounding the tunnel with various degree of saturation.

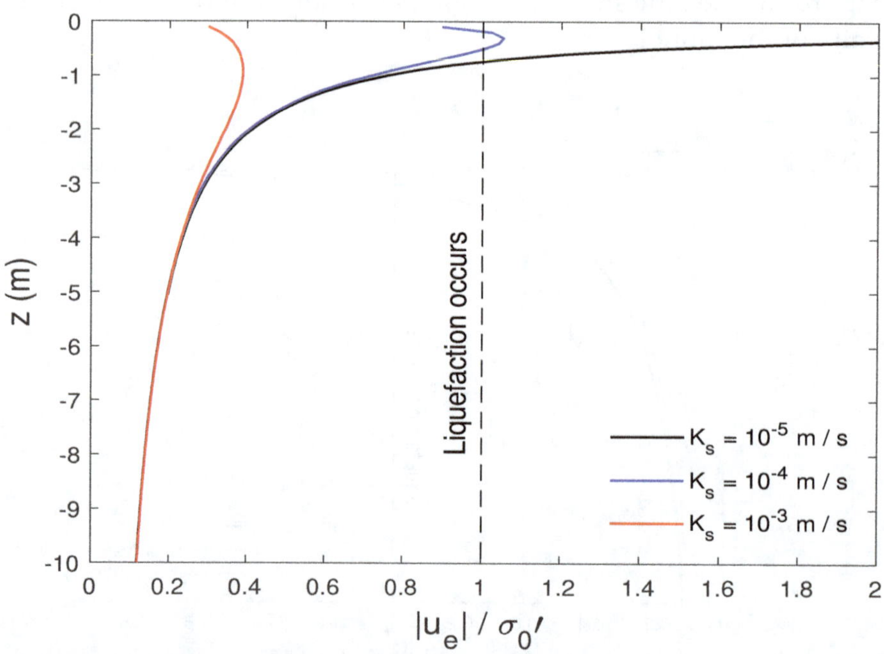

(**a**) distribution of the excess pore pressure ($|u_e|/\sigma_0'$) under wave trough

Figure 16. *Cont.*

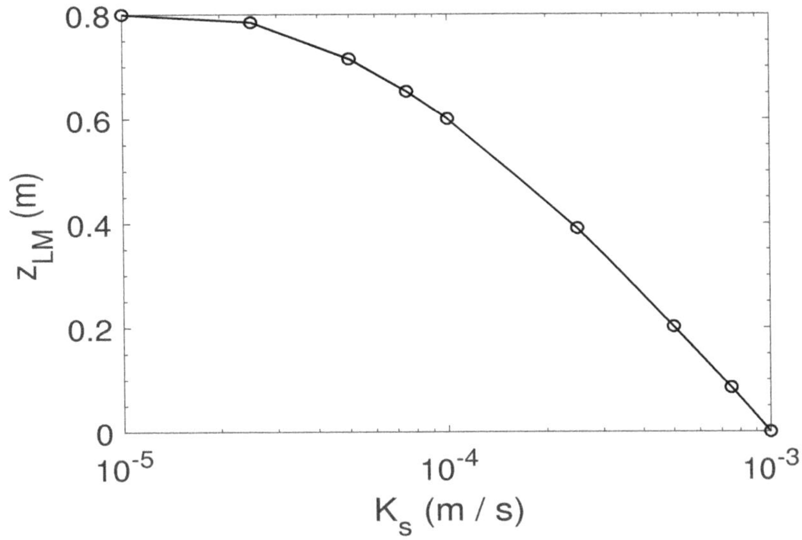

(**b**) maximum liquefaction depth of sandy seabed

Figure 16. (**a**) The vertical distribution of the excess pore pressure ($|u_e|/\sigma_0'$) under wave trough and (**b**) the maximum liquefaction depth of the sandy seabed surrounding the tunnel with various soil permeability.

6.3. *Effects of Current*

In reality, the fluid circumstance above the seabed are quite complex, and need to considered in the wave–current interaction. The motion of the current is able to influence the wave propagation, which further affects the seabed liquefaction process. This section aims to investigate the relationship between seabed liquefaction and the current around the immersed tunnel. The seabed response under a second-order Stokes wave ($T = 10$ s, $H = 4$ m and $d = 35$ m) with a series of following currents ($U_0 = 0.5, 1.0, 1.5$ m/s) and opposing currents ($U_0 = -0.5, -1.0, -1.5$ m/s) are compared with the no current case. Other parameters used in this study are listed in Table 4.

Table 4. Parameters used in study of effects of current.

Wave Characteristics	Value	Unit
Wave period (T)	10	s
Wave height (H)	4	m
Water depth (d)	35	m
Current velocity (U_0)	-1.5 or various	m/s
Soil Properties	**Value**	**Unit**
Poisson's ratio (ν_s)	0.20	-
Permeability (K_s)	1.0×10^{-5}	m/s
Porosity (n_s)	0.42	-
Degree of saturation (S_r)	0.95	-
Shear modulus (G)	5.0×10^6	Pa
Tunnel buried depth (b)	0.5	m

As seen in Figure 17, the maximum liquefied depth in seabed around the tunnel ($x = 80$ m) is 0.4 m, 0.6 m and 0.7 m when the current velocity U_0 takes on -1.5 m/s, 0 m/s and 1.5 m/s, which indicate the following current could deeper the liquefaction zone while the opposing current could decrease the liquefaction depth. Moreover, Figure 18 shows the liquefaction area around the immersed tunnel (when wave trough travels above the cross section $x = 80$ m) triggered by the wave combined different currents velocities. It can be concluded that not only the liquefied depth, but the liquefaction zone changes around the tunnel are also positively relative to the velocity of the current. Thus, oscillatory liquefaction is more likely to occur under the following current and the opposing current is able to decrease the potential of liquefaction.

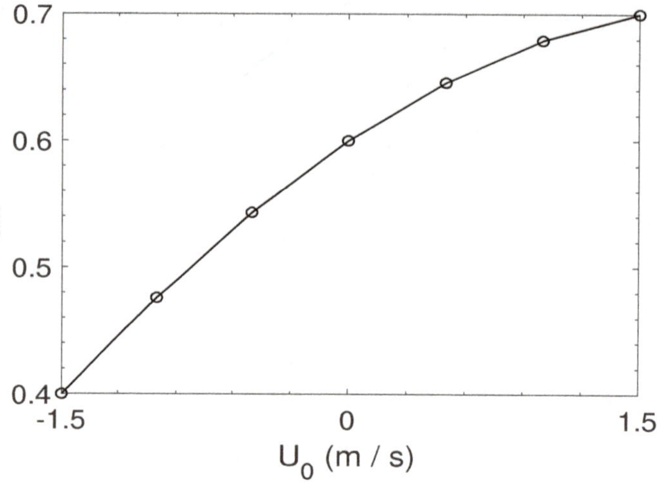

Figure 17. The liquefaction conditions around the tunnel with different wave height.

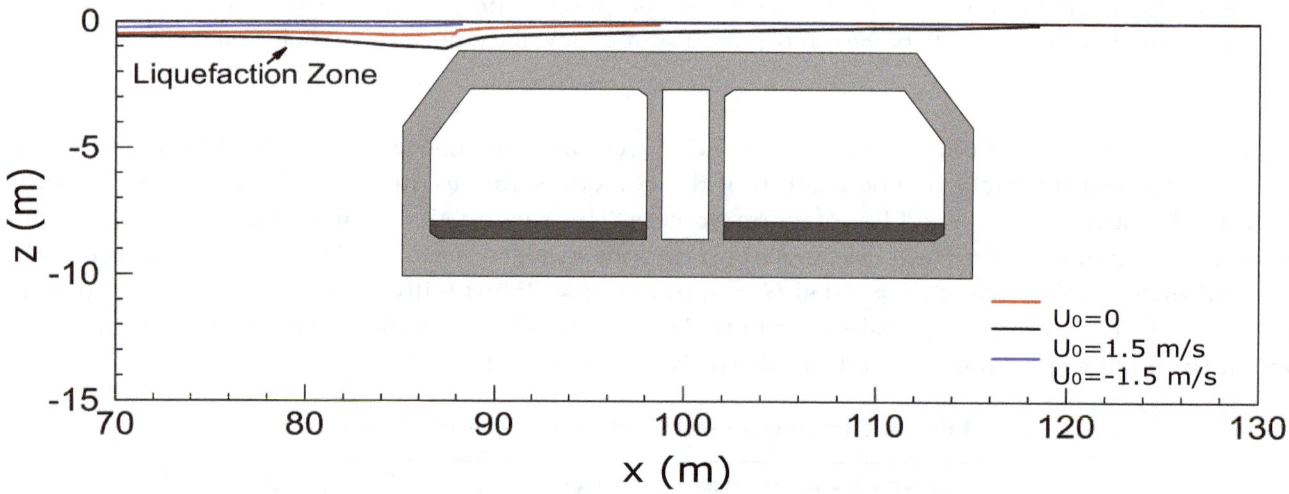

Figure 18. Effect of currents U_0 on maximum liquefaction depth around the immersed tunnel.

7. Conclusions

In this study, an integrated numerical model including a LRBFCM seabed model based on Biot "$u - p$" approximation is proposed to investigate the flow field dynamics and corresponding seabed behaviour around an immersed tunnel under second-order stokes wave combined current with various velocities. The oscillatory liquefaction under the different wave characteristics, soil properties, and current velocities are discussed in detail. From the numerical result, the following conclusions can be drawn:

(1) The newly developed meshfree model is well validated by comparison with the analytical solution, the laboratory experiments and previous numerical results. The LRBFCM is examined to be reliable in simulation of wave-induced oscillatory liquefaction behaviour of a seabed. Moreover, on the consequence of mesh-independence, the present model is more efficient than the conventional model based on FEM (Finite element method) or FVM (Finite volume method).
(2) The existence of the immersed tunnel affects surrounding seabed dynamic behaviours, which are able to weaken the displacement and dynamic pore pressure change nearby. Furthermore, the maximum liquefied depth on the right of the tunnel is smaller than that on the left.

(3) The wave-induced transient liquefaction around the tunnel is highly relative to the wave characteristics. It is found that the seabed liquefaction is more likely to occur under a shallow water area with the waves of large wave height and long period.

(4) The parametric studies show that the soil properties around tunnel have a significant impact on the liquefaction behaviour as well. The smaller the soil permeability and degree of saturation adopted, the deeper the liquefaction occurs in the vicinity of the tunnel.

(5) Based on the numerical result, the occurrence of currents could obviously affect wave-induced liquefaction. The following current can aggravate the seabed liquefaction while the opposing current can decrease the liquefied risk around the tunnel.

Author Contributions: Conceptualization, D.-S.J.; methodology, C.-C.T. and S.H.; validation, S.H.; formal analysis, S.H.; writing—original draft preparation, S.H.; writing—review and editing, D.-S.J. and C.-C.T.; supervision, D.-S.J. and C.-C.T.

Acknowledgments: The first author is grateful for the support of High Performance Computing Cluster "Gowanda" to complete this research, and the supporting of scholarship from Griffith University. The third author is grateful for the support of the Ministry of Science and Technology of Taiwan under the grant no. MOST 106-2918-I-022-001.

Abbreviations

p_0	The amplitude of linear wave pressure at the seabed surface
P_b	The dynamic wave pressure at the seabed surface
γ_ω	Unit weight of water
n_s	Soil porosity
β_s	The compressibility of pore fluid
K_f	The bulk modulus of pore fluid
$\langle\rangle$	Darcy's volume-averaging operator
$\langle\rangle^f$	The intrinsic averaging operator
ρ_{water}, ρ_{air}	Density of water and air
ρ_f, ρ_s	Density of fluid and solid
ρ	Average density of a porous seabed
α	The indicator function
u_i	The velocity vector
g_i	The gravitational acceleration
μ_{eff}	The efficient dynamic viscosity
μ	The molecular dynamic viscosity
ν_{turb}	The turbulent kinetic viscosity
d_{50}	The medium grain diameter of the material
KC	The Keulegan-Carpenter number
T_0	The period of the oscillatory
Φ_{water}, Φ_{air}	Water and air properties
σ'_x, σ'_z	The effective normal stresses in the x- and z-directions
σ'_0	The initial effective stress
τ_{xz}	The shear stress
u_s, w_s	The soil displacement in the x- and z-directions
b	Tunnel buried depth
K_w	The true bulk modulus of the elasticity of water
S_r	The degree of saturation
P_{wo}	The absolute water pressure
p_s	The pore water pressure

u_e	The excess pore pressure geberated by the wave loading
α_m	The undetermined coefficient related to RBF of a linear equation
p^*	The psedo-dynamic pressure
G	The shear modulus
K	The number of nearest neighbour nodes
H	Wave height
d	Water depth
h	Seabed thickness
T	Wave period
U_0	Current velocity
ν_s	Poisson's ratio
K_s	Soil permeability
η	Wave profile
E	Young's modulus
L	Wavelength
L_x	Seabed length
z	Soil depth
κ	The turbulence kinetic energy
ϵ	The turbulence energy dissipation rate
ϵ_s	The volume strain
C_μ	An empirical constant relate to the turbulent viscosity
c	Shape factor
N_{dt}	Time steps in a period
FEM	Finite Element Method
FVM	Finite Volume Method
VARANS	Volume-Average Reynolds-Average Navier-Stokes
VOF	Volume of fluid
2D	Two-dimensional
3D	Three-dimensional
LRBFCM	Local radial basis function collocation method
MLS	Moving least-square method
PIM	Point interpolation method
RBF	Radial basis function
MQ	Multiquadrics
GRBFCM	Global radial basis function collocation method

References

1. Zen, K.; Yamazaki, H. Mechanism of wave-induced liquefaction and densification in seabed. *Soils Found.* **1990**, *30*, 90–104. [CrossRef]
2. Sumer, B.M.; Fredsøe, J. *The Mechanics of Scour in the Marine Environment*; World Scientific Publishing Co. Pte. Ltd.: Singapore, 2002.
3. Jeng, D.S. *Mechanics of Wave-Seabed-Structure Interactions: Modelling, Processes and Applications*; Cambridge University Press: Cambridge, MA, USA, 2018.
4. Yamamoto, T.; Koning, H.; Sellmeijer, H.; Hijum, E.V. On the response of a poro-elastic bed to water waves. *J. Fluid Mech.* **1978**, *87*, 193–206. [CrossRef]
5. Seed, H.B.; Rahman, M.S. Wave-induced pore pressure in relation to ocean floor stability of cohesionless soils. *Mar. Geotechnol.* **1978**, *3*, 123–150. [CrossRef]
6. Jeng, D.S.; Seymour, B.R. A simplified analytical approximation for pore-water pressure build-up in a porous seabed. *J. Waterw. Port Coast. Ocean Eng. ASCE* **2007**, *133*, 309–312. [CrossRef]
7. Hsu, J.R.C.; Jeng, D.S. Wave-induced soil response in an unsaturated anisotropic seabed of finite thickness. *Int. J. Numer. Anal. Methods Geomech.* **1994**, *18*, 785–807. [CrossRef]
8. Jeng, D.S. *Porous Models for Wave-Seabed Interactions*; Springer Science & Business Media: Berlin/Heidelberg, Germany, 2013.

9. Jeng, D.S.; Ye, J.H.; Zhang, J.S.; Liu, P.L.F. An integrated model for the wave-induced seabed response around marine structures: Model verifications and applications. *Coast. Eng.* **2013**, *72*, 1–19. [CrossRef]
10. Zhao, H.Y.; Jeng, D.S. Accumulation of pore pressures around a submarine pipeline buried in a trench layer with partially backfills. *J. Eng. Mech. ASCE* **2016**, *142*, 04016042. [CrossRef]
11. Kasper, T.; Steenfelt, J.; Pedersen, L.; Jackson, P.; Heijmans, R. Stability of an immersed tunnel in offshore conditions under deep water wave impact. *Coast. Eng.* **2008**, *55*, 753–760. [CrossRef]
12. Chen, W.; Fang, D.; Chen, G.; Jeng, D.; Zhu, J.; Zhao, H. A simplified quasi-static analysis of wave-induced residual liquefaction of seabed around an immersed tunnel. *Ocean Eng.* **2018**, *148*, 574–587. [CrossRef]
13. Chen, W.Y.; Liu, C.L.; Duan, L.L.; Qiu, H.M.; Wang, Z.H. 2D Numerical study of the stability of trench under wave action in the immersing process of tunnel element. *J. Mar. Sci. Eng.* **2019**, *7*, 57. [CrossRef]
14. CCCC Second Flight Engineering Survey & Design Institute Co., Ltd. (CCCC SFES & DI). *Geological Investigation Report on Immersed Tunnel of Hong Kong-Zhuhai-Macao Bridge in Construction Documents Design Phase*; CCCC Second Flight Engineering Survey & Design Institute Co., Ltd.: Guangzhou, China, 2009. (In Chinese).
15. Ye, J.; Jeng, D.S. Response of seabed to natural loading-waves and currents. *J. Eng. Mech. ASCE* **2012**, *138*, 601–613. [CrossRef]
16. Wen, F.; Jeng, D.S.; Wang, J.H. Numerical modeling of response of a saturated porous seabed around an offshore pipeline considering non-linear wave and current interactions. *Appl. Ocean Res.* **2012**, *35*, 25–37. [CrossRef]
17. Liao, C.C.; D-S.; Lin, Z.; Guo, Y.; Zhang, Q. Wave (current)-induced pore pressure in offshore deposits: A coupled finite element model. *J. Mar. Sci. Eng.* **2019**, *6*, 83. [CrossRef]
18. Nayroles, B.; Touzot, G.; Villon, P. Generalizing the finite element method: Diffuse approximation and diffuse elements. *Comput. Mech.* **1992**, *10*, 307–318. [CrossRef]
19. Belytschko, T.; Lu, Y.Y.; Gu, L. Element-free Galerkin methods. *Int. J. Numer. Methods Eng.* **1994**, *37*, 229–256. [CrossRef]
20. Liu, G.R.; Gu, Y. A point interpolation method for two-dimensional solids. *Int. J. Numer. Methods Eng.* **2001**, *50*, 937–951. [CrossRef]
21. Wang, J.; Liu, G.; Lin, P. Numerical analysis of biot's consolidation process by radial point interpolation method. *Int. J. Solids Struct.* **2002**, *39*, 1557–1573. [CrossRef]
22. Wang, J.; Liu, G. On the optimal shape parameters of radial basis functions used for 2-D meshless methods. *Comput. Methods Appl. Mech. Eng.* **2002**, *191*, 2611–2630. [CrossRef]
23. Hardy, R.L. Theory and applications of the multiquadric-biharmonic method 20 years of discovery 1968–1988. *Comput. Math. Appl.* **1990**, *19*, 163–208. [CrossRef]
24. Powell, M. The theory of radial basis function approximation in 1990, Advances in Numerical Analysis II: Wavelets, Subdivision, and Radial Functions (WA Light, ed.). *Oxf. Univ. Press. Oxf.* **1992**, *105*, 210.
25. Kansa, E.J. Multiquadrics—A scattered data approximation scheme with applications to computational fluid-dynamics—II solutions to parabolic, hyperbolic and elliptic partial differential equations. *Comput. Math. Appl.* **1990**, *19*, 147–161. [CrossRef]
26. Mai-Duy, N.; Tran-Cong, T. Indirect RBFN method with thin plate splines for numerical solution of differential equations. *CMES Comput. Model. Eng. Sci.* **2003**, *4*, 85–102.
27. Šarler, B.; Perko, J.; Chen, C.S. Radial basis function collocation method solution of natural convection in porous media. *Int. J. Numer. Methods Heat Fluid Flow* **2004**, *14*, 187–212. [CrossRef]
28. Wu, N.J.; Chang, K.A. Simulation of free-surface waves in liquid sloshing using a domain-type meshless method. *Int. J. Numer. Methods Fluids* **2011**, *67*, 269–288. [CrossRef]
29. Kovacevic, I.; Poredos, A.; Sarler, B. Solving the Stefan problem with the radial basis function collocation method. *Numer. Heat Transf. Part B Fundam.* **2003**, *44*, 575–598. [CrossRef]
30. Hardy, R.L. Multiquadric equations of topography and other irregular surfaces. *J. Geophys. Res.* **1971**, *76*, 1905–1915. [CrossRef]
31. Lee, C.K.; Liu, X.; Fan, S.C. Local multiquadric approximation for solving boundary value problems. *Comput. Mech.* **2003**, *30*, 396–409. [CrossRef]
32. Šarler, B.; Vertnik, R. Meshfree explicit local radial basis function collocation method for diffusion problems. *Comput. Math. Appl.* **2006**, *51*, 1269–1282. [CrossRef]
33. Kosec, G.; Sarler, B. Local RBF collocation method for Darcy flow. *Comput. Model. Eng. Sci.* **2008**, *25*, 197–208.

34. Tsai, C.C.; Lin, Z.H.; Hsu, T.W. Using a local radial basis function collocation method to approximate radiation boundary conditions. *Ocean Eng.* **2015**, *105*, 231–241. [CrossRef]
35. Kosec, G.; Založnik, M.; Šarler, B.; Combeau, H. A meshless approach towards solution of macrosegregation phenomena. *Comput. Mater. Contin.* **2011**, *22*, 169–195.
36. Wang, X.X.; Jeng, D.S.; Tsai, C.C. Meshfree model for wave-seabed interactions around offshore pipelines. *J. Mar. Sci. Eng.* **2019**, *7*, 87. [CrossRef]
37. Higuera, P.; Lara, J.L.; Losada, I.J. Realistic wave generation and active wave absorption for Navier-Stokes models: Application to OpenFOAM. *Coast. Eng.* **2013**, *71*, 102–118. [CrossRef]
38. Qi, W.G.; Gao, F.P. Physical modelling of local scour development around a large-diameter monopile in combined waves and current. *Coast. Eng.* **2014**, *83*, 72–81. [CrossRef]
39. Ahmad, N.; Bihs, H.; Chella, M.A.; Arntsen, Ø.A.; Aggarwal, A. Numerical modelling of arctic coastal erosion due to breaking waves impact using REEF3D. In Proceedings of the 27th International Ocean and Polar Engineering Conference, International Society of Offshore and Polar Engineers, San Francisco, CA, USA, 25–30 June 2017.
40. Nangia, N.; Patankar, N.A.; Bhalla, A.P.S. A DLM immersed boundary method based wave-structure interaction solver for high density ratio multiphase flows. *arXiv* **2019**, arXiv:1901.07892.
41. de Jesus, M.; Lara, J.L.; Losada, I.J. Three-dimensional interaction of waves and porous coastal structures: Part I: Numerical model formulation. *Coast. Eng.* **2012**, *64*, 57–72. [CrossRef]
42. Engelund, F. *On the Laminar and Turbulent Flows of Ground Water through Homogeneous Sand*; Akademiet for de Tekniske Videnskaber: København, Denmark, 1953.
43. Burcharth, H.F.; Andersen, O.K. On the one-dimensional steady and unsteady porous flow equations. *Coast. Eng.* **1995**, *24*, 233–257. [CrossRef]
44. Higuera, P. Application of Computational Fluid Dynamics to Wave Action on Structures. Ph.D. Thesis, Universidade de Cantabria, Cantabria, Spain, 2015.
45. Zienkiewicz, O.C.; Chang, C.T.; Bettess, P. Drained, undrained, consolidating and dynamic behaviour assumptions in soils. *Géotechnique* **1980**, *30*, 385–395. [CrossRef]
46. Biot, M.A. General theory of three-dimensional consolidation. *J. Appl. Phys.* **1941**, *26*, 155–164. [CrossRef]
47. Bentley, J.L. Multidimensional binary search trees used for associative searchings. *Commun. ACM* **19752**, *18*, 509–517. [CrossRef]
48. Li, X.S. An overview of SuperLU: Algorithms, implementation, and user interface. *ACM Trans. Math. Softw. (TOMS)* **2005**, *31*, 302–325. [CrossRef]
49. Newmark, N.M. A method of computation for structural dynamics. *J. Eng. Mech. Div. ASCE* **1959**, *85*, 67–94.
50. Ye, J.H.; Jeng, D.S.; Wang, R.; Zhu, C. Validation of a 2-D semi-coupled numerical model for fluid-structure-seabed interaction. *J. Fluids Struct.* **2013**, *42*, 333–357. [CrossRef]
51. Zienkiewicz, O.C.; Chan, A.H.C.; Pastor, M.; Schrefler, B.A.; Shiomi, T. *Computational Geomechanics with Special Reference to Earthquake Engineering*; John Wiley and Sons: Hoboken, NJ, USA, 1999.
52. Li, Z.; Jeng, D.S.; Zhu, J.F.; Zhao, H. Effects of principal stress rotation on the fluid-induced soil response in a porous seabed. *J. Mar. Sci. Eng.* **2019**, *7*, 123. [CrossRef]
53. Hicher, P.Y. Elastic properties of soils. *J. Geotech. Eng.* **1996**, *122*, 641–648. [CrossRef]
54. Le Mehaute, B. *An Introduction to Hydrodynamics and Water Waves*; Springer Science & Business Media: Berlin/Heidelberg, Germany, 2013.
55. Zen, K.; Yamazaki, H. Oscillatory pore pressure and liquefaction in seabed induced by ocean waves. *Soils Found.* **1990**, *30*, 147–161. [CrossRef]

4

Wave-Induced Seafloor Instability in the Yellow River Delta: Flume Experiments

Xiuhai Wang [1,2,3], Chaoqi Zhu [1,2,3,4,*] and Hongjun Liu [1,2,3,*]

1 College of Environmental Science and Engineering, Ocean University of China, Qingdao 266000, China; showseas@ouc.edu.cn
2 Shandong Provincial Key Laboratory of Marine Environment and Geological Engineering, Qingdao 266000, China
3 Laboratory for Marine Geology, Qingdao National Laboratory for Marine Science and Technology, Qingdao 266000, China
4 Shandong Provincial Key Laboratory of Marine Ecology and Environment & Disaster Prevention and Mitigation, Qingdao 266000, China
* Correspondence: george-zhu@foxmail.com (C.Z.); hongjun@ouc.edu.cn (H.L.)

Abstract: Geological disasters of seabed instability are widely distributed in the Yellow River Delta, posing a serious threat to the safety of offshore oil platforms and submarine pipelines. Waves act as one of the main factors causing the frequent occurrence of instabilities in the region. In order to explore the soil failure mode and the law for pore pressure response of the subaqueous Yellow River Delta under wave actions, in-lab flume tank experiments were conducted in this paper. In the experiments, wave loads were applied with a duration of 1 hour each day for 7 consecutive days; pore water pressure data of the soil under wave action were acquired, and penetration strength data of the sediments were determined after wave action. The results showed that the fine-grained seabed presented an arc-shaped oscillation failure form under wave action. In addition, the sliding surface firstly became deeper and then shallower with the wave action. Interestingly, the distribution of pores substantially coincided with that of sliding surfaces. For the first time, gas holes were identified along with their positioning and angle with respect to the sediments. The presence of gas may serve as a primer for submarine slope failures. The wave process can lead to an increase in the excess pore pressure, while the anti-liquefaction capacity of the sediments was improved, causing a decrease in the excess pore pressure resulting from the next wave process. Without new depositional sediments, the existing surface sediments can form high-strength formation under wave actions. The test results may provide a reference for numerical simulations and engineering practice.

Keywords: wave; seafloor instability; pore pressure; slide surface; gas distribution

---

1. Introduction

The Yellow River Delta is one of largest-scale deltas with the fastest progradation rate in the world. The Yellow River brings abundant high-concentration sediments from the Loess Plateau, which are rapidly deposited in the subaqueous delta of the estuary, forming a high-moisture under-consolidated silty seabed. It is estimated that the average annual sediment transport volume exceeds 800 million tons from 1950 to 2005 [1], and over 70–90% of the sediments were deposited in the sea area within 30 km from the estuary [2], gradually forming the modern subaqueous Yellow River Delta. Such newly formed seabeds under the influence of waves and storm surges are prone to instability, thereby forming unstable seabed landforms of subaqueous delta in the estuary of the Yellow River.

Previous studies reported that various ocean dynamic processes can affect the delta of the Yellow River [3–5]. The yearly winter storms coming from the northwest have the most powerful effect from

October to March. During the winter season, the prevailing northwest winter winds becomes stronger than grade 8 on average 6.4 times each year. The wave height generated by winter storms can reach 7 m [3]. In addition, the weak southeastern winds prevail from July to August. Occasionally, hurricanes or typhoons occur in a given year. For example, Typhoon Lekima meandered over the Yellow Sea and Bohai Sea in 2019. Available survey results showed that geological disasters such as liquefaction, landslide, subsidence and depression, gully, scarp, disturbed strata, and erosion are common across the subaqueous Yellow River Delta [6–9]. The Yellow River Delta is rich in oil and gas resources and is the location of Shengli Oilfield, currently the largest offshore oil field in China. Seabed instability disasters seriously threatens the safety and stability of the offshore oil platform, submarine pipeline, wharf, and embankment [10–12]. For example, on December 3, 1998, the CB6A-5 oil production platform of Chengdao Oilfield collapsed, and the casing pipes at the bottom of the oil well ruptured, causing a material oil spill lasting half a year [13]. In November 2003, a submarine landslide occurred under the action of ocean power near the oil production platform CB12B, resulting in the interruption of two submarine cables [14]. In 2010, Operation Platform No. 3 of Shengli Oilfield was overturned due to the instability of the seabed stratum during a storm surge, leading to the death of two people and a direct economic loss of RMB 5.92 million [15].

There are two main types of seabed instability under wave action. In one, the shear stress is greater than the shear strength of the soil, causing the seabed to be unstable; in the other, it is the variation in the pore water pressure in the seabed under wave action that causes the seabed to undergo liquefaction, resulting in the loss of effective stress in the soil, thereby leading to instability. With respect to the studies on the instability of the seabed under wave action, especially in the subaqueous Yellow River Delta, plenty of field surveys, numerical analyses, laboratory tests, and in situ observation studies have been carried out [16–21]. However, few wave flume experiments have comprehensively investigated landforms, failure mode, gas distribution, pore pressure response, and sediment strength. In addition, failure of the seabed has been understudied, even though some interesting phenomena have been identified. For example, previous studies have confirmed fine particle migration due to wave action. In order to deeply explore the soil failure mode, gas distribution, and pore pressure response law under wave action, in-lab wave flume experiments were designed and carried out in this study. For the first time, gas holes were identified along with their positioning and angle with respect to the sediments. The presence of gas may serve as a primer for submarine slope failure.

2. Materials and Methods

2.1. Experiment Process

The flume used in the experiment was 120 cm × 50 cm × 120 cm. In order to observe the variation in soil mass, the outer wall of the flume was made of transparent glass. In the flume experiment, the sediment thickness was set to be 45 cm, and the depth of overlying water was set to be 40 cm. The sediments used in the experiment were taken from the Yellow River Delta, and its median particle size was 0.058 mm. The particle-size distribution curve of these sediment samples is shown in Figure 1.

Prior to each experiment, powdery sediments were taken, and their contents were controlled at around 40%, we then added water to have a water content of 40%, and the powdery sediment was mixed with the stirrer. The deposited sediments were then transferred to the flume until the thickness of the sediment hit 45 cm; and the pore pressure sensors were then buried with a burial depth of 0 cm, 25 cm, 35 cm, and 40 cm, respectively, as shown in Figure 2. Afterward, water was filled into the flume, avoiding the formation of erosion pits on the surface of the soil mass during the water-filling process due to high water-filling speed, and we stop adding water when the height of the water filling reached 40 cm. We kept the model flume still for 24 hours after its preparation was finished in order to complete drainage consolidation. We then performed manual wave generation with the duration of 1 hour each day and proceeded for 7 consecutive days. Simultaneously, we measured the pore water pressure and observed the experimental phenomenon. We discharged the overlying water body when the experiment ended, and determined the penetration strength of the soil mass with a micro penetrometer at different depths.

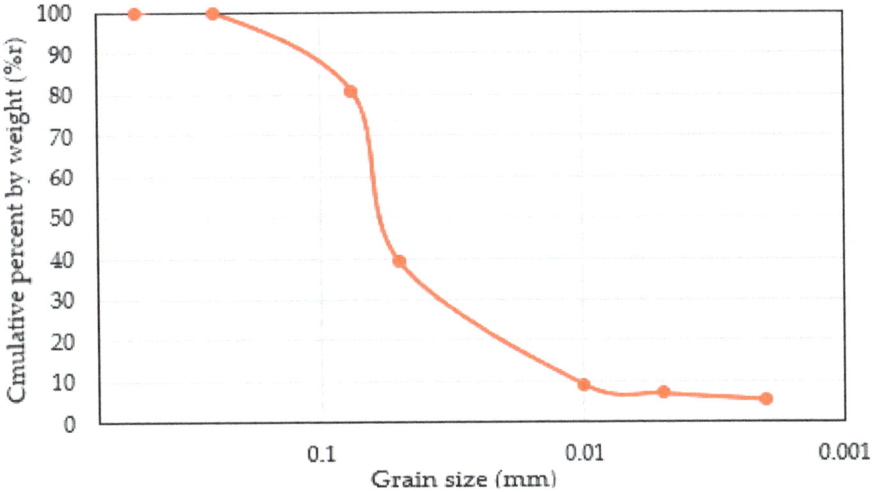

Figure 1. Grading curve of sediment size.

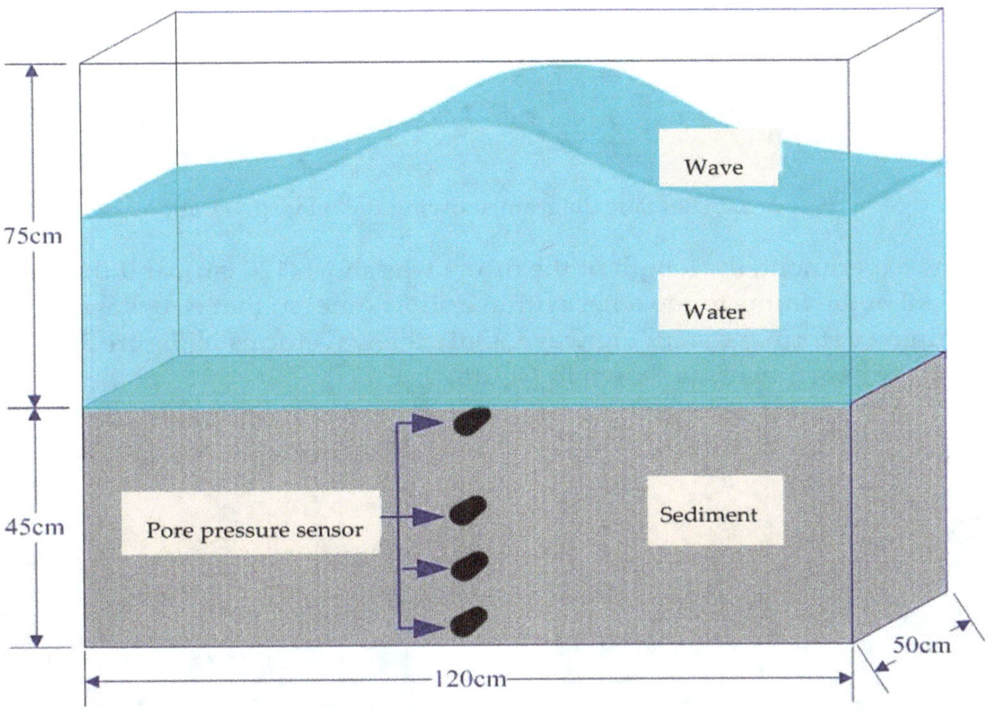

Figure 2. Schematic diagram of wave flume.

2.2. *Wave Generation*

A hollow cylinder with a diameter of 10 cm and height of 40 cm was used in the experiment to artificially create waves in the flume. The cylinder was placed horizontally on one end of the flume and then it was moved up and down the water surface to create the first wave as shown in Figure 3a; when the first wave propagated to the other end of the flume, a reflected wave was formed and then propagated in the opposite direction. The cylinder was moved up and down the water surface to produce the second wave as shown in Figure 3b; when two waves collided in the middle of the flume, a superposition occurred (Figure 3c); then, the two waves continued to advance in their respective directions (Figure 3d), and a reflected wave was formed simultaneously at each side wall of the flume.

When the first wave was located at the hollow cylinder, the cylinder was moved up and down at the frequency of the wave to increase the amplitude of the wave or to supplement the energy lost during the propagation (Figure 3e). The second wave collided with the first wave in the middle of the flume after being reflected by the side wall of the flume and then they separated (Figure 3f); we supplemented the energy lost during the propagation in the same way. Continuous waves with different frequencies were formed in the flume in this way.

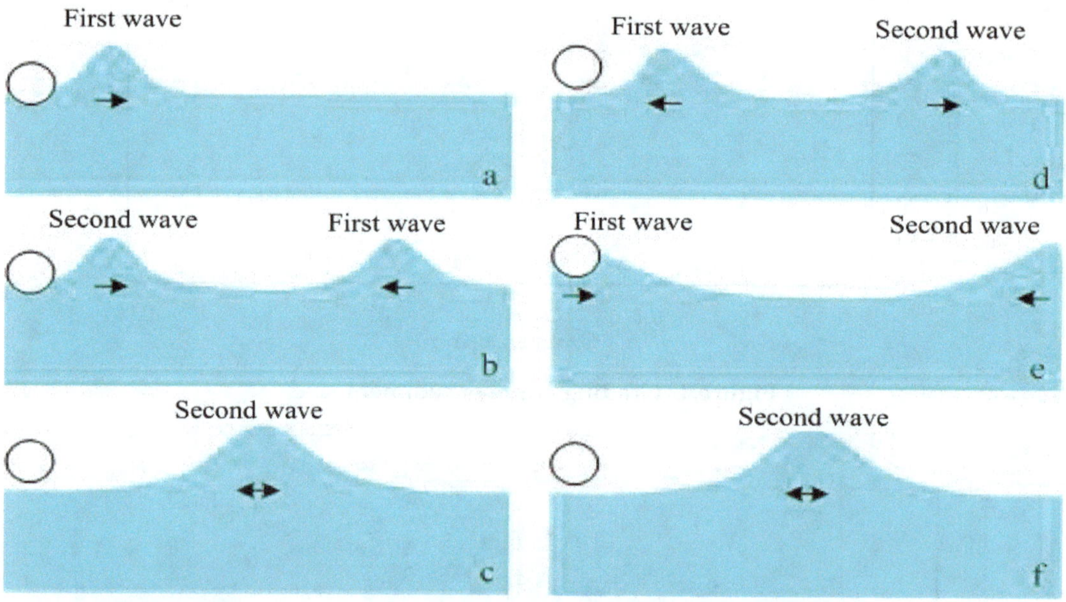

Figure 3. Schematic diagram showing the wave generation.

In the flume experiment, the length of the flume was short (120 cm), and the wave propagated from one side wall of the flume to the other within a short time, so that two waves were superposed together in the flume to form the standing wave. Only the waveforms in Figure 3e,f could be seen in the flume during the experiment, as shown in Figure 4.

(**a**) wave crest

(**b**) wave trough

Figure 4. Experimental photographs showing the wave generation.

3. Results and Discussion

3.1. Slide Surface and Ripples

No significant movement of the soil mass occurred within 5 min upon the wave application action, while the water body began to become turbid, and the 15-cm-thick sediment started to exhibit a weak waveform oscillation motion with a frequency and phase consistent with those of the water wave when the wave acted for approximately 3 min. The soil mass showed distinct arc-shaped sliding surface in 20 min. The sliding surfaces located in the middle of the flume and near the wall of the flume were higher, and those at 30 cm and 90 cm approximately from the left wall of the flume were lower. The sediment on the sliding surface reciprocated the wave, and the lower part of the sliding surface was almost stationary. As the duration of the wave action increased, the sliding surface constantly expanded deeper, and the thickness of the moving soil mass increased continuously, forming a distinct "W" shaped sliding surface. When the wave application was stopped, the concentration of suspended sediments in the water gradually decreased. In addition, fine-grained sedimentary formation occurred on the surface layer (Figure 5a).

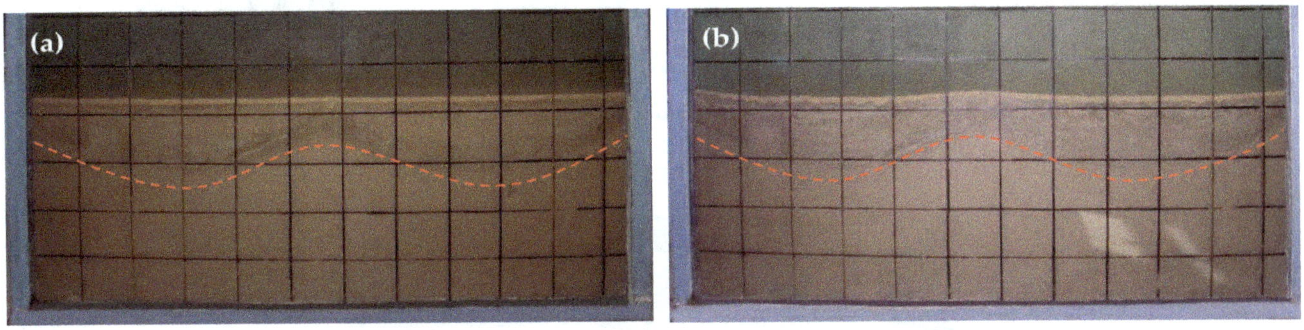

Figure 5. Experimental photographs showing the slide surface. (**a**) First day; (**b**) Seventh day. The dotted lines are used to show the slide surface.

Similar phenomenon still occurred in the course of the wave application in subsequent days, while the maximum depth of the sliding surface gradually decreased, and the sliding surface gradually moved upwards; the surface of the soil mass started to undergo local erosion and sedimentation, and ripples appeared. Similar experimental phenomena were also observed in other experiments [22,23], but "V"-shaped sliding surfaces occurred, rather than "W"-shaped. The maximum expansion depth of the sliding surface was close to the interface of water and the soil when the last wave application was made, and the soil mass did not further oscillate. Finally, a continuous array of ripples with the wavelength of 5–10 cm formed on the surface of the soil mass (Figure 6). Additionally, sedimentation occurred in the middle position of the flume and the position near the wall of the flume, and the erosion occurred at sites approximately 30 cm and 90 cm away from the left wall. The surface shape of the soil mass tended to be consistent with the sliding surface, and W-shaped forms appeared.

Figure 6. Experimental photographs showing the ripples. (**a**) Top view; (**b**) Side view.

3.2. Gas Distribution

A sample portion of the soil approximately 10 cm in the upper layer was taken after the end of the experiment. It was found that there was a large number of gas holes in the vertical section of the soil sample (Figure 7). The gas holes in the soil were distributed in a horizontal or inclined and zonal manner rather than being uniform (Figure 8), and even a gas channel existed. Within a depth ranging from 3 cm to 4 cm and from 8 cm to 10 cm, the number of pores was greater than 100, while the holes in other depth ranges were less than 30. The area where the pores were densely distributed was consistent with the sliding surface of the soil mass.

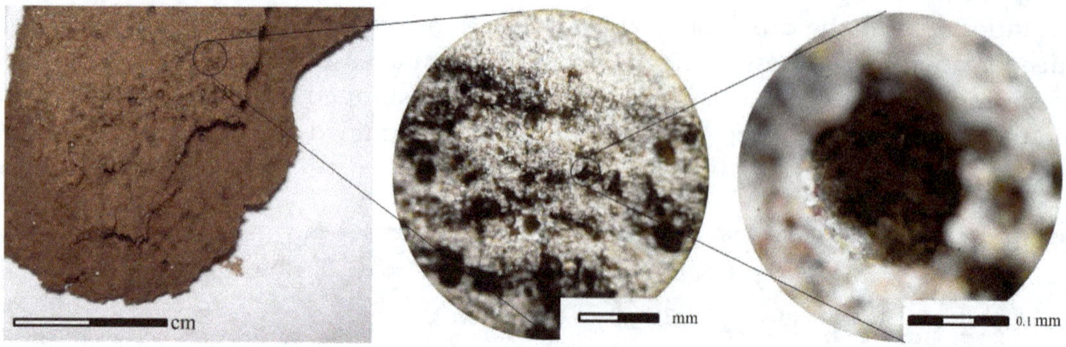

Figure 7. Photographs showing the gas holes in different scales.

Figure 8. Photographs showing the ribbon of gas holes.

The gas holes in the soil sample were distributed in an oblique and zonal manner at a distance of 100–110 cm from the left wall of the flume, at an angle of approximately 30°–40° with the horizontal plane. The closer they were to the surface of the soil, the smaller the inclination. The distribution of gas holes (Figure 9) was substantially consistent with that of sliding surfaces. Previous studies confirmed fine-particle migration due to external actions, such as waves [22,24,25]. However, this experiment offered the first evidence of possible gas migration due to wave action. The presence of gas may serve as a primer for submarine slope failure [26].

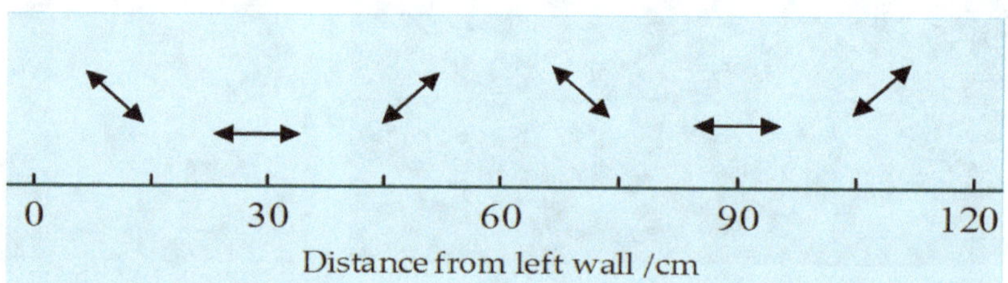

Figure 9. Schematic diagram showing the direction of gas distribution.

3.3. Pore Pressure

The pore pressure is an important indicator of the properties of soil masses [27,28]. A total of seven wave application processes each with a duration of approximately 1 hour were performed. It was found from Figure 10 that the pore pressure in the seabed soil body significantly accumulated and could reach the maximum cumulative pore pressure within a short time (approximately 5 min) and remained stable. The fine-grained seabed had low permeability, and it was difficult to quickly dissipate the excess pore water pressure caused by the wave.

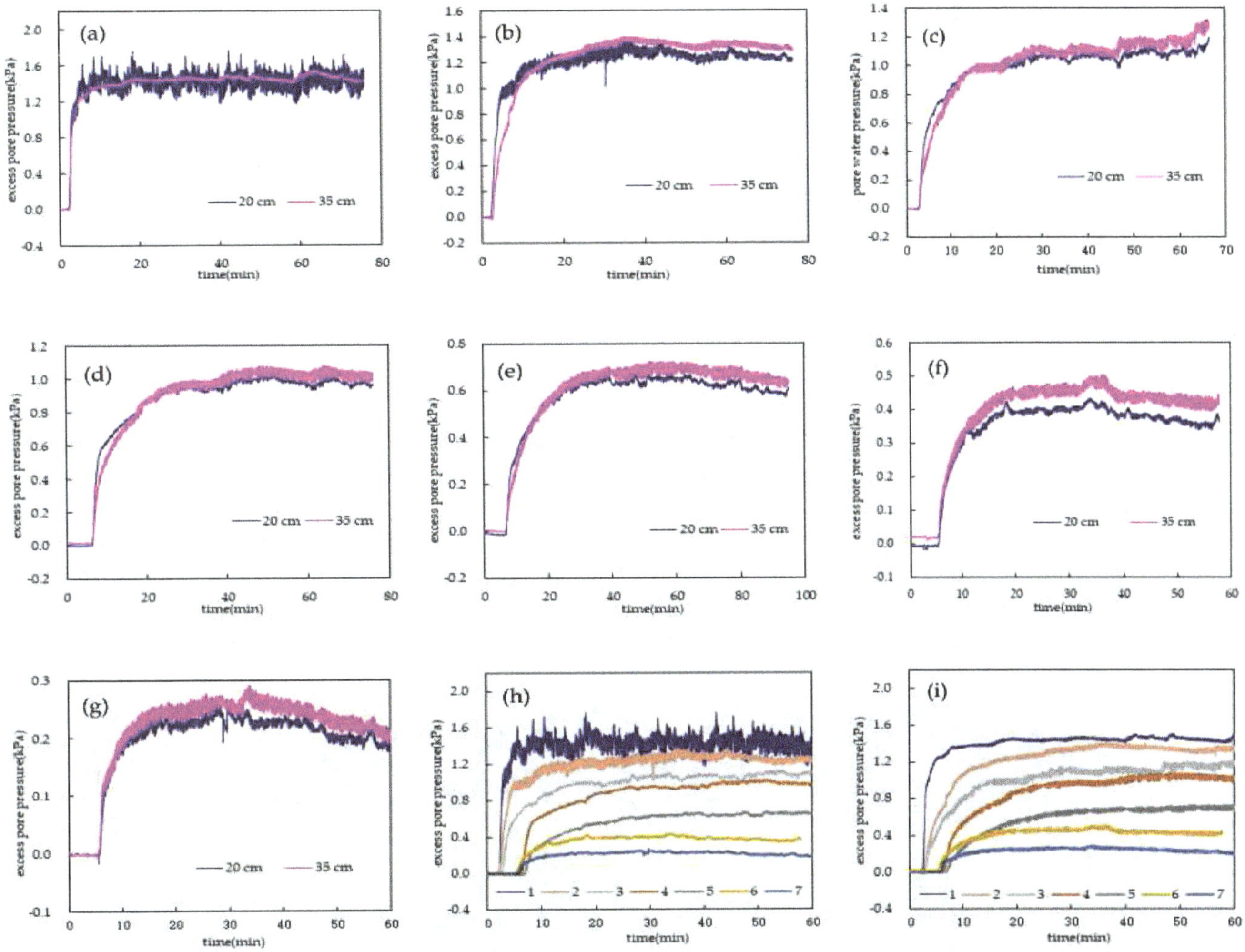

Figure 10. Excess pore pressure during the wave action. The curves thorough from (**a**–**g**) represent the excess pore pressure curves for the first to seventh wave application process; (**h**) excess pore pressure of the seabed for different wave application processes at a depth of 20 cm; (**i**) excess pore pressure of the seabed at a depth of 35 cm from different wave application processes.

Figure 10 shows the excess pore pressure curve at a depth of 20 cm and 35 cm in the seabed during seven wave application processes. The figures from 10a to 10g show that the excess pore pressure at a depth of 35 cm was slightly greater than that at a depth of 20 cm, indicating that the greater the depth was, the harder it was for the excess pore pressure to dissipate. The excess pore pressure data from the seven wave applications at the same depth (Figure 10h,i) indicated that the maximum excess pore pressure formed from the process of the last wave application was lower than that formed from previous wave applications. For example, the maximum pore pressure at a depth of 20 cm of the seabed hit approximately 1.4 kPa in the first wave application, approximately 1.0 kPa in the third wave application, and approximately 0.2 kPa in the seventh wave application. The maximum excess pore pressure inside the soil mass showed a decrease trend with the increase in the number of wave actions, indicating that the whole soil mass tended to be stable. It was consistent with the experimental phenomenon that the sliding surface gradually moved up and became shallow with wave application process, and that the thickness of the oscillation layer reduced. Each wave application process might lead to an increase in the excess pore pressure of the seabed sediment. However, the liquefaction resistance of the sediment improved upon each wave action process, resulting in a decrease in the excess pore pressure caused by the next process of wave action. This might be due to the fact that the wave action can promote the rearrangement of sediment particles. The experiment results show that the seabed is the most unstable in the initial stage of rapid deposition of sediment, and it is most likely to be destroyed under wave action. Due to the limitations of laboratory experiments [29,30], more field work is required to confirm these experimental results.

3.4. Sediment Strength

The overlying water mass was discharged after the seventh wave action, and the entire soil sample was subject to a penetration strength experiment with a micro penetration instrument. The diameter of the mini penetrometer was 3 cm, the horizontal separation distance between test points was 15 cm, and the vertical depth interval was 5 cm. As shown in Figure 11, the penetration strength of the soil mass was substantially greater than 50 kN in the depth range from 0 cm to 20 cm, and the penetration strength of the soil mass ranging from 8 cm to 13 cm was greater, with a mean value greater than 80 kN, indicating there was a high-strength formation at this depth. The soil mass had a relatively large variation in strength gradient, with the strength of the soil mass decreasing from 60 kN to 30 kN in this 7-cm-thick range. The penetration strengths of the soil mass with the depth ranging from 20 cm to 40 cm were less than 30 kN, and most of the values were approximately 10 kN, being less than the penetration strength in the hard formation of the surface strength. By comparing the strength distribution law for sliding surface and soil penetration, it was found that the depth of the high-strength formation was slightly less than the depth (15 cm) of maximum sliding surface, that is, the high-strength formation was located above the sliding surface. This indicates that exiting surface sediment undergoes multiple wave actions, and its strength is improved in the absence of new deposition conditions.

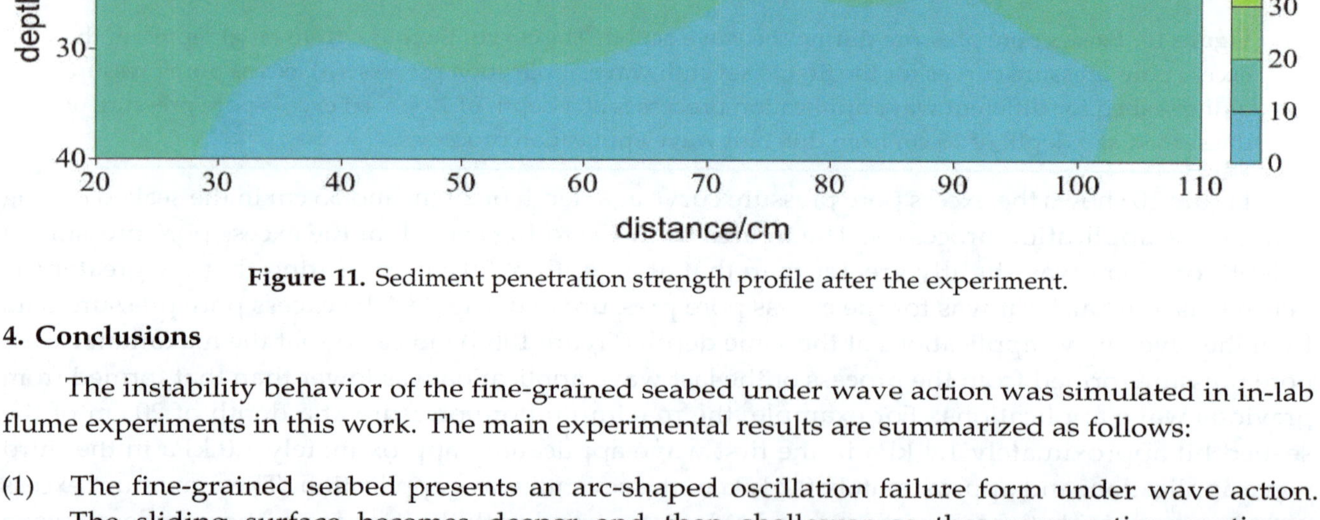

Figure 11. Sediment penetration strength profile after the experiment.

4. Conclusions

The instability behavior of the fine-grained seabed under wave action was simulated in in-lab flume experiments in this work. The main experimental results are summarized as follows:

(1) The fine-grained seabed presents an arc-shaped oscillation failure form under wave action. The sliding surface becomes deeper and then shallower as the wave action continues. The distribution of the gas pores is substantially the same as that of the sliding surfaces.

(2) We found the first evidence for possible gas migration due to wave action. The presence of gas may serve as a primer for submarine slope failures, which suggests a new mechanism for seafloor instability in the Yellow River Delta.

(3) Each wave process may cause the perforation pressure of the sediment to increase, and the liquefaction resistance capacity of the sediment increases during each wave process, leading to a decrease in the excess pore pressure caused by the next wave process. The seabed is the most unstable and most susceptible to damage under wave action in the initial stage of rapid deposition of the sediment.
(4) Without new depositional sediments, the existing surface sediments can form high-strength formation under wave actions.

Author Contributions: Conceptualization, H.L.; Data curation, X.W. and C.Z.; Formal analysis, X.W. and C.Z.; Funding acquisition, H.L.; Writing—original draft, X.W. and C.Z.; Writing—review & editing, X.W., C.Z., and H.L. The final manuscript has been approved by all the authors.

Acknowledgments: Authors acknowledge the experimental and technical support provided by Minsheng Zhang.

References

1. Wang, H.; Yang, Z.; Li, G.; Jiang, W. Wave climate modeling on the abandoned Huanghe (Yellow River) delta lobe and related deltaic erosion. *J. Coast. Res.* **2006**, *22*, 906–918. [CrossRef]
2. Saito, Y.; Yang, Z.S.; Hori, K. The Huanghe (Yellow River) and Changjiang (Yangtze River) deltas: A review on their characteristics, evolution and sediment discharge during the Holocene. *Geomorphology* **2001**, *41*, 219–231. [CrossRef]
3. Yang, Z.; Ji, Y.; Bi, N.; Lei, K.; Wang, H. Sediment transport off the Huanghe (Yellow River) delta and in the adjacent Bohai Sea in winter and seasonal comparison. *Estuar. Coast. Shelf Sci.* **2011**, *93*, 173–181. [CrossRef]
4. Bi, N.; Yang, Z.; Wang, H.; Fan, D.; Sun, X.; Lei, K. Seasonal variation of suspended-sediment transport through the southern Bohai Strait. *Estuar. Coast. Shelf Sci.* **2011**, *93*, 239–247. [CrossRef]
5. Johnson, M.E.; Gudveig Baarli, B. Geomorphology and coastal erosion of a quartzite island: Hongdo in the Yellow Sea off the SW Korean Peninsula. *J. Geol.* **2013**, *121*, 503–516. [CrossRef]
6. Prior, D.B.; Suhayda, J.N.; Lu, N.Z.; Bornhold, B.D. Storm wave reactivation of a submarine landslide. *Nature* **1989**, *341*, 47–50. [CrossRef]
7. Guo, Z.; Zhou, W.; Zhu, C.; Yuan, F.; Rui, S. Numerical Simulations of Wave-Induced Soil Erosion in Silty Sand Seabeds. *J. Mar. Sci. Eng.* **2019**, *7*, 52. [CrossRef]
8. Zhu, C.; Liu, X.; Shan, H.; Zhang, H.; Shen, Z.; Zhang, B.; Jia, Y. Properties of suspended sediment concentrations in the Yellow River delta based on observation. *Mar. Georesour. Geotech.* **2018**, *36*, 139–149. [CrossRef]
9. Xu, G.; Sun, Y.; Wang, X.; Hu, G.; Song, Y. Wave-induced shallow slides and their features on the subaqueous Yellow River delta. *Can. Geotech. J.* **2009**, *46*, 1406–1417. [CrossRef]
10. Jia, Y.; Zhu, C.; Liu, L.; Wang, D. Marine Geohazards: Review and Future Perspective. *Acta Geol. Sin. Engl.* **2016**, *90*, 1455–1470. [CrossRef]
11. Sun, Y.; Dong, L.; Song, Y. Analysis of characteristics and formation of disturbed soil on subaqueous delta of Yellow River. *Rock Soil Mech.* **2008**, *29*, 1494–1499.
12. Zhu, C.; Cheng, S.; Li, Q.; Shan, H.; Shen, Z.; Liu, X.; Jia, Y. Giant Submarine Landslide in the South China Sea: Evidence, Causes and Implications. *J. Mar. Sci. Eng.* **2019**, *7*, 152. [CrossRef]
13. Chang, F. Study on Mechanism of Wave-Induced Submarine Landslide at the Yellow River Estuary, China. Ph.D. Thesis, Ocean University of China, Qingdao, China, 30 June 2009.
14. Zheng, C. Shengli Chengdao Oilfield Submarine Power Cable Fault Type and Causes Analysis. *Saf. Health Environ.* **2016**, *16*, 6–10.
15. Du, F. Research on the Geological Causes of Sheng Li Well Workover Platform III Overturning Accident. Ph.D. Thesis, Ocean University of China, Qingdao, China, 30 June 2013.
16. Jia, Y.; Zhang, L.; Zheng, J.; Liu, X.; Jeng, D.; Shan, H. Effects of wave-induced seabed liquefaction on sediment re-suspension in the Yellow River Delta. *Ocean Eng.* **2014**, *89*, 146–156. [CrossRef]
17. Keller, G.H.; Prior, D.B. Sediment dynamics of the Huanghe (Yellow River) delta and neighboring gulf of Bohai, People's Republic of China: Project overview. *Geo Mar. Lett.* **1986**, *6*, 63–66. [CrossRef]
18. Sumer, B.M.; Hatipoglu, F.; Fredsoe, J.; Sumer, S.K. The sequence of sediment behaviour during wave-induced liquefaction. *Sedimentology* **2006**, *53*, 611–629. [CrossRef]

19. Tong, L.; Zhang, J.; Sun, K.; Guo, Y.; Zheng, J.; Jeng, D. Experimental study on soil response and wave attenuation in a silt bed. *Ocean Eng.* **2018**, *160*, 105–118. [CrossRef]
20. Wang, Z.; Jia, Y.; Liu, X.; Wang, D.; Shan, H.; Guo, L.; Wei, W. In situ observation of storm-wave-induced seabed deformation with a submarine landslide monitoring system. *Bull. Eng. Geol. Environ.* **2018**, *77*, 1091–1102. [CrossRef]
21. Wang, H.; Liu, H.; Zhang, M. Pore pressure response of seabed in standing waves and its mechanism. *Coast. Eng.* **2014**, *91*, 213–219. [CrossRef]
22. Liu, X.; Jia, Y.; Zheng, J.; Wen, M.; Shan, H. An experimental investigation of wave-induced sediment responses in a natural silty seabed: New insights into seabed stratification. *Sedimentology* **2017**, *64*, 508–529. [CrossRef]
23. Xu, G.; Liu, Z.; Sun, Y.; Wan, X.; Lin, L.; Ren, Y. Experimental characterization of storm liquefaction deposits sequences. *Mar. Geol.* **2016**, *382*, 191–199. [CrossRef]
24. Zhang, S.; Jia, Y.; Wen, M.; Wang, Z.; Zhang, Y.; Zhu, C.; Li, B.; Liu, X. Vertical migration of fine-grained sediments from interior to surface of seabed driven by seepage flows—'sub-bottom sediment pump action'. *J. Ocean Univ. China* **2017**, *16*, 15–24. [CrossRef]
25. Yang, Q.; Cheng, K.; Wang, Y.; Ahmad, M. Improvement of semi-resolved CFD-DEM model for seepage-induced fine-particle migration: Eliminate limitation on mesh refinement. *Comput. Geotech.* **2019**, *110*, 1–18. [CrossRef]
26. Li, A.; Davies, R.J.; Yang, J. Gas trapped below hydrate as a primer for submarine slope failures. *Mar. Geol.* **2016**, *380*, 264–271. [CrossRef]
27. Yang, Q.; Ren, Y.; Niu, J.; Cheng, K.; Hu, Y.; Wang, Y. Characteristics of soft marine clay under cyclic loading: A review. *Bull. Eng. Geol. Environ.* **2018**, *77*, 1027–1046. [CrossRef]
28. Zhang, Y.; Jeng, D.S.; Zhao, H.Y.; Zhang, J.S. An integrated model for pore pressure accumulations in marine sediment under combined wave and current loading. *Geomech. Eng.* **2016**, *10*, 387–403. [CrossRef]
29. Barry, M.A.; Boudreau, B.P.; Johnson, B.D. Gas domes in soft cohesive sediments. *Geology* **2012**, *40*, 379–382. [CrossRef]
30. Williams, S.C.P. Skimming the surface of underwater landslides. *Proc. Natl. Acad. Sci. USA* **2016**, *113*, 1675–1678. [CrossRef]

5

Comparative Study on Seismic Response of Pile Group Foundation in Coral Sand and Fujian Sand

Qi Wu [1], Xuanming Ding [1,2,*], Yanling Zhang [1] and Zhixiong Chen [1,2]

1 School of Civil Engineering, Chongqing University, Chongqing 400045, China; wuqi_cqu@163.com (Q.W.); ylz_cqu@163.com (Y.Z.); chenzhixiong@cuq.edu.cn (Z.C.)

2 National Joint Engineering Research Center of Geohazards Prevention in the Reservoir Areas, Chongqing 400045, China

* Correspondence: dxmhhu@163.com

Abstract: The physical and mechanical properties of coral sand are quite different from those of common terrestrial sands due to the special marine biogenesis. Shaking table tests of three-story structures with nine-pile foundation in coral sand and Fujian sand were carried out in order to study the dynamic response characteristics of pile-soil-structure system in coral sand under earthquake. The influence of shaking intensity on the dynamic response of the system was taken into consideration. The results indicated that the peak value of the excess pore pressure ratio of coral sand was smaller than that of Fujian sand under two kinds of shaking intensities; moreover, the development speed of excess pore pressure ratio of coral sand was smaller than that of Fujian sand. The liquefaction of coral sand was more difficult than Fujian sand under the same relative density and similar grain-size distribution. The horizontal displacement, settlement, column bending moment, and pile bending moment of coral sand were smaller than those of Fujian sand, respectively. The magnification effect of column bending moment of buildings in coral sand was less than that in Fujian sand with increasing shaking intensity. This study can provide some supports for the seismic design of coral reef projects.

Keywords: coral sand; Fujian sand; shaking table test; dynamic response; pile group

---

1. Introduction

Coral sand is a special geotechnical medium that is rich in calcium carbonate and deposited by marine protozoan skeletons. Coral sand is widely distributed in the coasts and reefs of Australia, the Gulf of Mexico, and the South China Sea [1]. When compared with the common terrigenous sands, coral sand has special physical properties, such as multi-voids, irregular particle shape, and fragility. These properties lead to the difference of compressibility, shear strength, and permeability between the coral sands and the common terrigenous sands [2–5]. With the expansion of construction scale on coral reefs recent years, the seismic safety of coral reefs has attracted considerable attention [6–9]. Historically, a magnitude 8.1 earthquake struck Guam in the Pacific Ocean on August 8, 1993, which caused severe liquefaction of coral sandy sites in dredger fills and lake sediments [10]. Large-scale liquefaction of coral sand also occurred in the Hawaii earthquake in 2006 and Haiti earthquake in 2010, which caused irreparable losses to local infrastructure and people's lives and property [11–13].

Xiao et al. analyzed the dynamic strength, cyclic deformation, and pore pressure of saturated coral sand and microbial reinforced coral sand by the dynamic triaxial test in order to study the dynamic response characteristics of pile-soil-structure system in coral sand under earthquake [1].

Salem et al. pointed out that the calcareous sand has higher dynamic strength than siliceous sands and suggested the dynamic resistance ratio-confining pressure-relative density relationship of calcareous sand from North Coast Dabaa [14]. Xu et al. studied the transmission law of explosive stress wave in saturated coral sand and quartz sand, and concluded that saturated coral sand has a stronger absorption and attenuation effect on explosive stress wave than quartz sand [15]. Sandoval et al. found the differences of dynamic response between Puerto Rico coral sand and Ottawa siliceous sand through the result of dynamic triaxial tests [16]. Other scholars have also carried out some research on the dynamic characteristics of coral sands [17,18]. However, the former researches mainly focused on the strength, deformation, and pore pressure characteristics of coral sands in small-scale models, and it does not involve the dynamic response while considering the interaction between structures and soil in large-scale coral sands sites.

The shaking table test is an important method for studying the dynamic response of liquefiable sites and the structures on them under earthquakes. Tang et al. performed the failure mode of pile foundation of bridge under earthquakes while considering the pile-soil interaction [19]. Dashti et al. carried out a series of liquefaction model tests of low-rise buildings through the centrifuge shaking table test; the mechanism of building settlement that was caused by liquefaction was discussed [20]. Chen et al. studied the dynamic characteristics and damage law of subway station [21]. Other scholars have also undertaken a lot of research on the dynamic response of various structures and foundations under earthquake using shaking table test and found that the depth and compactness of soil, the existence of structures, and the input parameters of shaking excitation have obvious effects on site liquefaction and dynamic response of structures and soil [22–30]. However, the above studies are aimed at the terrigenous sands, and there are few studies on coral sand at present.

Important military buildings for large equipment and heavy machinery on coral sand sites often use the low-rise concrete structures that are supported by pile group foundations. A series of shaking table tests of buildings with nine-pile foundation in coral sand were carried out in order to study the seismic response of coral sand and the structures on it. Shaking table tests on quartz sand sites in the same situation were also performed as comparative tests. The differences and similarities of dynamic characteristics of coral sand and Fujian sand were compared and analyzed based on the results of pore water pressure, acceleration, displacement, and dynamic bending moment.

2. Shaking Table Test

The test was performed using ANCO shaking table, which is sourced from ANCO company, Los Angeles, America. The shaking table can simultaneously carry out horizontal and vertical shaking, with a dimension of 1200 mm × 1200 mm in plane, as shown in Figure 1. The maximum proof model mass, horizontal displacement, and base excitation is 1000 kg, 100 mm, and 2.0 g, respectively. A soil container with size of 950 mm in length, 850 mm in width, and 550 mm in depth was used. The soil container changes from a laminar shear type to a rigid type when the controls at the four corners of the soil container are tightened (Figure 1). Two DHDAS acquisition instruments with 96 channels were used to collect sensor signal data simultaneously in this test. The DHDAS acquisition instruments is sourced from Donghua Testing Technology Co., Ltd., Jingjiang, China.

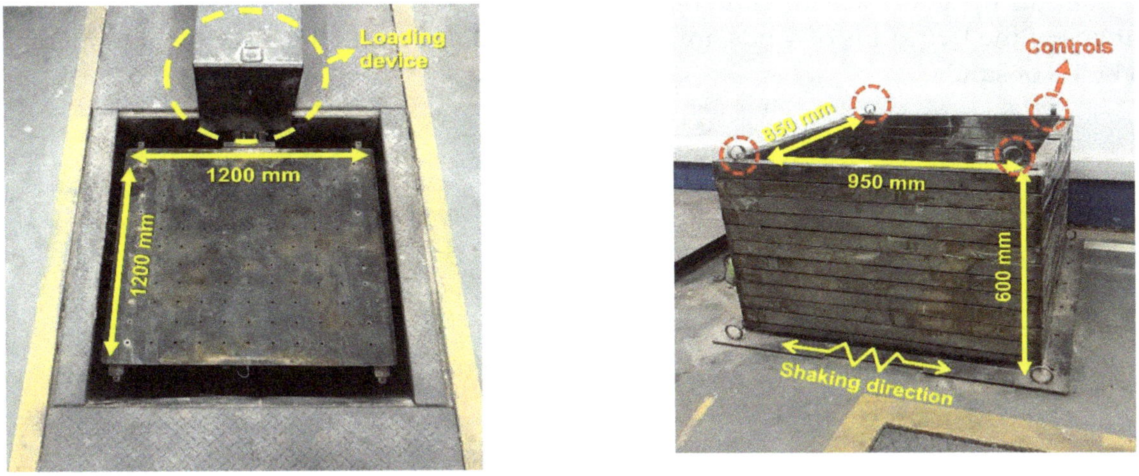

(**a**) Shaking table (**b**) Soil container

Figure 1. Shaking table facility.

2.1. Similitude Ratio

The geometric similarity ratio is set as 1:40 based on the maximum load of the shaking table and the size of the soil container. According to the Buckingham theory [31,32] and scaling laws [33], length l, elastic modulus E, and equivalent density ρ were chosen as the basic physical quantities. The similarity ratio of elastic modulus S_E is obtained by comparing the elastic modulus of the model building material and the prototype building material. The similarity ratio of equivalent density S_ρ is calculated by Equation (1), when the preset acceleration similarity ratio S_a is 1. The other physical quantities, such as time, acceleration, and displacement, were derived based on the similarity relations, as shown in Table 1.

$$S_a = \frac{S_E}{S_\rho \cdot S_l} \quad (1)$$

Table 1. Similitude laws of shaking table test.

Parameters	Similitude Relation	Similitude Ratio
Length l	S_l	1:40
Equivalent density ρ	S_ρ	6:1
Elastic modulus E	S_E	3:20
Acceleration a	$S_a=S_E S_\rho{}^{-1} S_l{}^{-1}$	1
Duration t	$S_t=S_E{}^{-0.5} S_\rho{}^{0.5} S_l$	0.158
Frequency ω	$S_\omega=S_t{}^{-1}$	6.325
Stress σ	$S_\sigma=S_E$	3:20
Linear displacement r	$S_r=S_l$	1:40

2.2. Preparation of the Model

The soil container was separated into two halves along the orientation of shaking, half of which was for preparing the coral sand foundation and the other half was for preparing Fujian sand foundation. The rigid soil container was used in the test to prevent the two sands from interacting with each other. Relatively compressible foam cushions with thicknesses of 100 mm were attached to the inner walls of the soil container perpendicular to the shaking direction to reduce the energy that was reflected by the container. The foam was made of polystyrene. The density, water absorption, and compressive strength of the foam were 30 kg·m^{-3}, 1%, and 150 kPa. A foam board was installed at the middle of the soil container to prevent the mix of the two kinds of sands. Coral sand that was used in the test was taken from a reef in the South China Sea, and the quartz sand used was Fujian standard sand.

Figure 2 presents the grain-size distribution of the two kinds of sands, which shows that the grain-size distributions of the two kinds of sands are similar. Table 2 illustrates the basic physical parameters of the two kinds of sands.

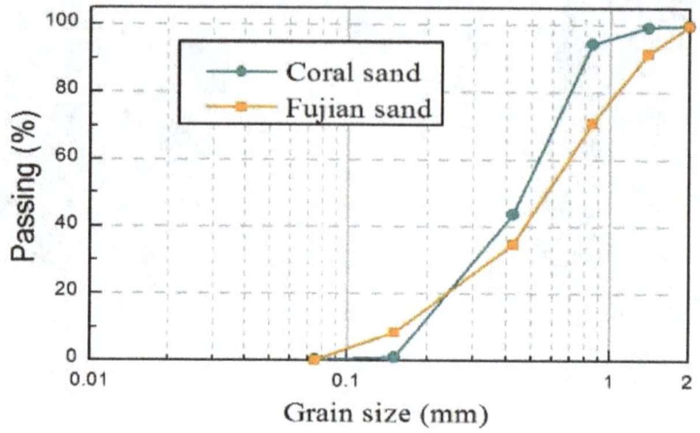

Figure 2. Grain-size distribution curves of coral sand and Fujian sand.

Table 2. Physical properties of sand.

The Category of Sand	Specific Gravity	Maximum Dry Density (g·cm^{-3})	Minimum Dry Density (g·cm^{-3})	Coefficient of Uniformity, C_u	Mean Grain Size, (D_{50}: mm)
Coral sand	2.80	1.48	1.15	2.67	0.48
Fujian sand	2.63	1.64	1.35	4.50	0.60

The model foundation consisted of two parts along the vertical direction, above and below the water level. The soil layer below the water level was prepared while using the water sedimentation method. The water surface is about 10 cm above the sand surface throughout the process. The main factor affecting the relative density of the model foundation that was prepared by the water sedimentation method is the fall distance [34]. The two kinds of sands fall to the water surface at the same height during sample preparation, and other influencing factors, such as the speed and flow of the sand ejection head, should be consistent, in order to make the relative density of the coral sand and quartz sand sites approximately the same. During the preparation of the model foundation, a calibrated aluminum box was used for sampling analysis in time to ensure that the uniformity and relative density of the two sand model foundations were approximately similar. The soil layer above the water level with a thickness of 30 mm was prepared to keep consistent with the actual engineering situation. The relative density of the whole model foundation is 0.67.

The prototype of the structures were three-story concrete frame buildings with a nine-pile foundation. The buildings are used to store large equipment and heavy machinery on coral sand sites. The organic glass was selected to prepare the model buildings and piles, because the geometric dimensions of the model building were too small after shrinking according to the similar law and there were practical operation difficulties in concrete pouring. The three-story model building with side length of 180 mm, net height of 100 mm for the bottom floor, and 90 mm for the other floors is made of organic glass, as shown in Figure 3. The organic glass plates with the thickness of 5 mm were used as slabs, under which rectangular organic glass bars with geometry of 5 mm × 5 mm × 160 mm were installed to simulate the beam. The cross-sectional dimension of the model column was 10 mm × 10 mm, and the outer edge of which was leveled with the outer edge of the model beam. The geometry of the model raft was 220 mm in length and 15 mm in thickness. The diameter and length of the model pile were 20 mm and 400 mm, respectively. The organic glass can be changed into liquid state by dropping acetone on it. Each component of the model building was dissolved in order to connect by the special adhesive of organic glass. In the connection process, the verticality among the components was ensured by the triangular rule and the integrity of model was guaranteed by the connection after the organic glass dissolved.

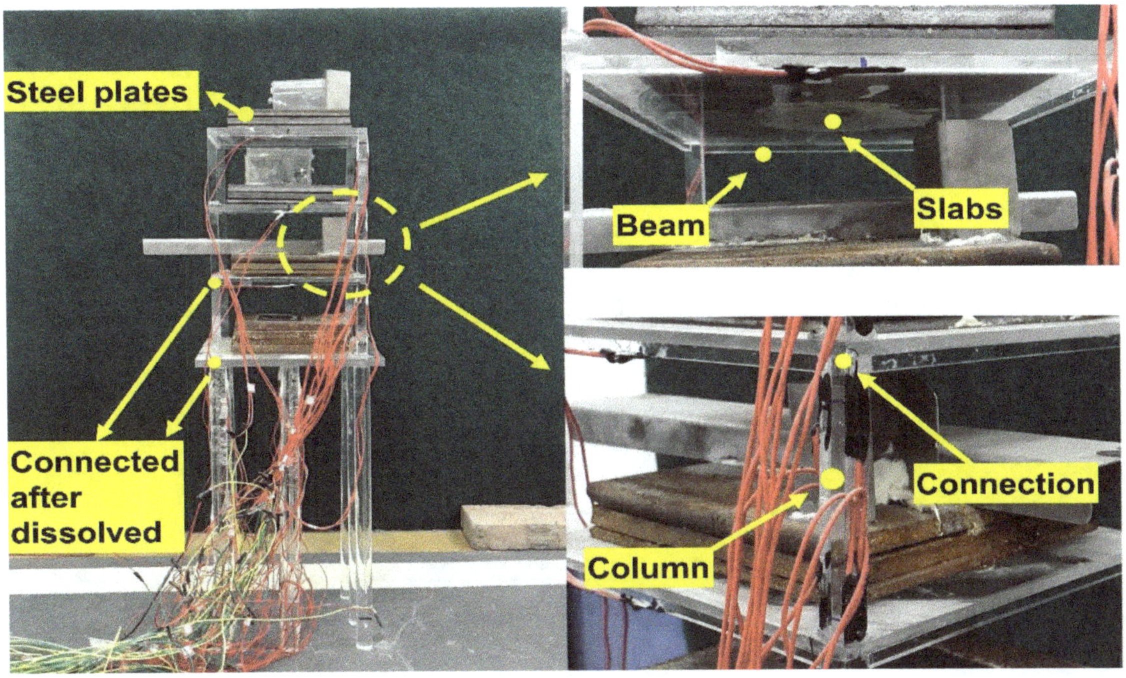

Figure 3. Model structure details.

While considering the gravity effect on the prototype structure, steel plates weighing 3.5 kg with geometry of 150 mm in length, 150 mm in width, and 20 mm in height were glued on each floor of the model structure, and steel plates weighing 6.2 kg with geometry of 150 mm in length, 150 mm in width, and 35.5 mm in height was glued on the model raft. A total additional mass of 16.7 kg or 81% of the enough artificial mass was placed on the model structure. The density of the model building is increased by adding enough artificial mass to meet the similarity rate. The gravity effect of pile foundation was ignored in this test.

2.3. *Instrumentation and Experimental Program*

The coral sand and Fujian sand sites adopt the same sensor arrangement, as shown in Figure 4. Laser displacement sensors with heights of 150 mm and 330 mm from the ground of model foundation were installed on the shaking table using the rigid brackets, respectively, and the rigid targets point were installed on the floors of the structure. The horizontal displacement sensor was 280 mm from the vertical center line of the structure. The model foundation stood for 24 h before the test. The capillarity action is considered and the water level is consistent with Figure 4b during 24 h. The experimental program was arranged as a comparative study of the dynamic response of pile-soil-structure system in coral sand and Fujian sand sites, while considering the influence of shaking intensity. Table 3 summarizes the specific experimental program. Figure 5 shows the time history curves of sinusoidal wave excitation. The sinusoidal wave has a simpler law than the seismic wave, and it is easy to analyze the dynamic response of the model foundation and structure, many scholars have used sinusoidal wave as excitation, especially in the liquefaction condition [35,36]. The white noise with an amplitude of 0.02 g and a duration of 20 s was input before and after each sinusoidal wave excitation input.

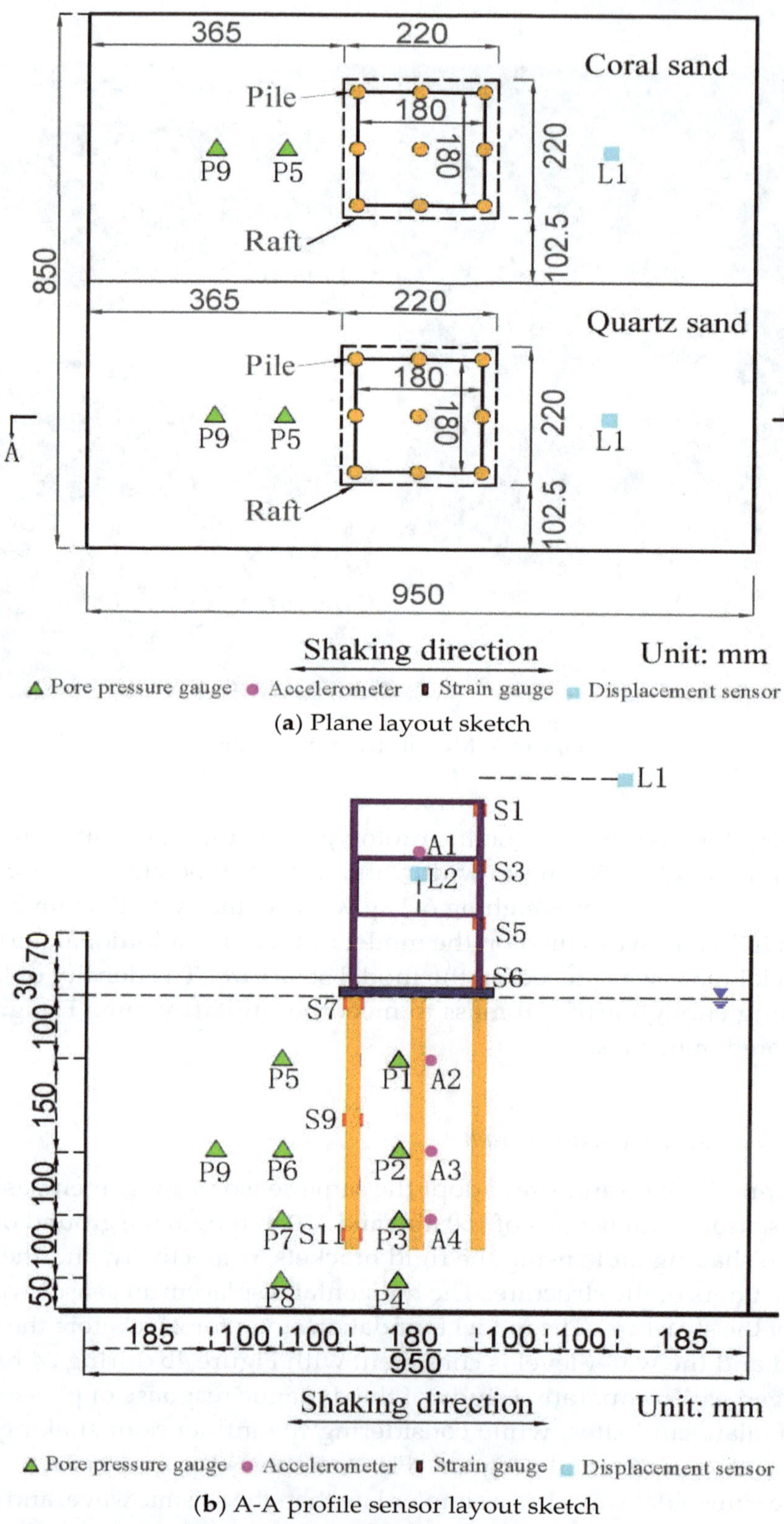

Figure 4. General configuration of model tests.

Table 3. Summary of shake table tests conducted.

Case	Relative Density	Input Motions	Peak Acceleration (g)	Duration (s)	Frequency (Hz)
1	0.67	Sine wave	0.1	10	5
2	0.67	Sine wave	0.2	10	5

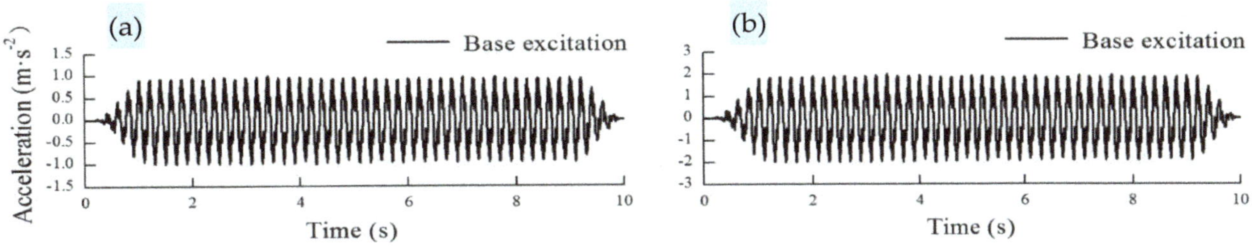

Figure 5. Time history curves of base excitation: (**a**) 0.1 g; and, (**b**) 0.2 g.

3. Macroscopic Phenomena of Soil and Structure

Figure 6a shows the surfaces of coral sand and Fujian sand sites. When the 0.1 g sinusoidal wave excitation was input, the building on the coral sand site began to shake slightly, and no water was discharged from the model soil. The phenomenon of Fujian sand site was similar to that of the coral sand site. Figure 6b presents the site condition after test. When the 0.2 g sinusoidal wave excitation was input, the shaking degree of buildings in the two kinds of sand sites increased and reached the maximum at about 4.8 s, and then the shaking degree suddenly decreased. With the input of shaking excitation, the shaking degree gradually increased again. The buildings subsided and inclined, and the soil on both sides of the building rose. For coral sand, the surface of model soil was gradually getting wet, and little water accumulated after test, as shown in Figure 6c. The water of Fujian sand site increased from the surrounding of the soil container and accumulated a little on the surface of the site (Figure 6d).

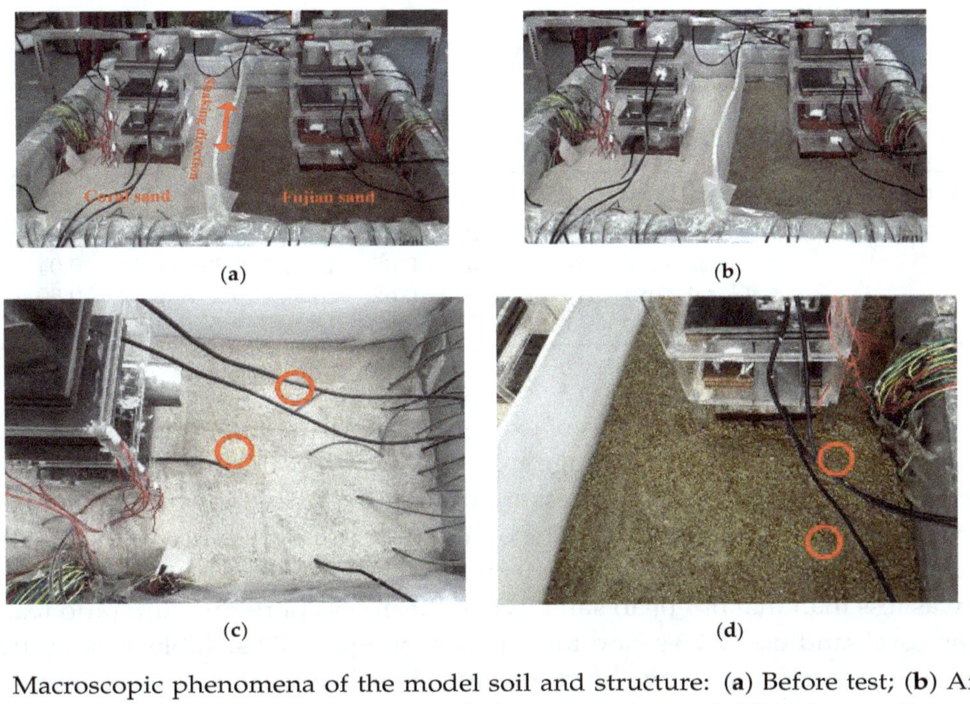

Figure 6. Macroscopic phenomena of the model soil and structure: (**a**) Before test; (**b**) After 0.1 g shaking excitation; (**c**) Coral sand site after 0.2 g shaking excitation; and, (**d**) Fujian sand site after 0.2 g shaking excitation.

4. Result and Discussion

4.1. Pore Water Pressure Response

The excess pore pressure ratio (r_u) was defined here as the ratio of the difference of pore water pressure in a specified stage and initial pore water pressure over the vertical effective stress to detect the occurrence of soil liquefaction.

Under 0.1 g shaking intensity, Figure 7 shows the time history curves of excess pore pressure ratio directly under the buildings (P1, P2, P3, and P4) of coral sand and Fujian sand. The signal of pore water pressure gauge was lost at P4 position in Fujian sand site. During the period of shaking (10 s), the excess pore pressure ratio of two kinds of sand sites gradually increased and the growth rate of coral sand was significantly less than that of Fujian sand. After 2 s of shaking, the excess pore pressure ratio of coral sand at P1 position was approximately 0.02, which of Fujian sand was about 0.04, and the excess pore pressure ratio of coral sand was less than that of Fujian sand. During the whole shaking period, Table 4 shows the peak values of excess pore pressure ratio. With the decrease of depth, the peak values of the two kinds of sand sites gradually increased. The peak values of excess pore pressure ratio of coral sand were less than that of Fujian sand. From top to bottom (P1–P3), the peak values of coral sand were about 0.86, 0.67, and 0.80 times of that of Fujian sand.

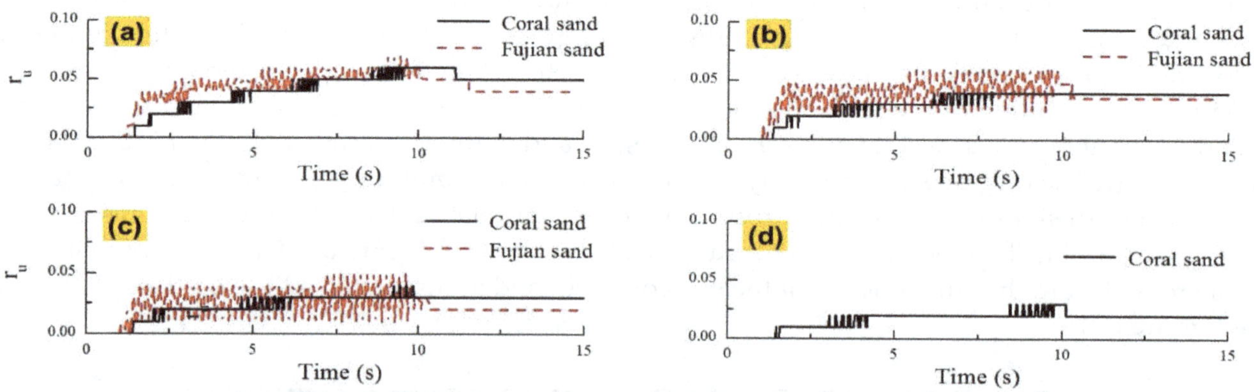

Figure 7. Excess pore pressure ratio time history curves under 0.1 g shaking intensity: (**a**) P1 position; (**b**) P2 position; (**c**) P3 position; and, (**d**) P4 position.

Table 4. Comparison of peak excess pore pressure ratios.

Shaking Intensity	Sand Type	P1	P2	P3	P4	P5	P6	P7	P8
0.1 g	Coral sand	0.06	0.04	0.04	0.03	0.08	0.04	0.04	0.03
	Fujian sand	0.07	0.06	0.05	Lost	0.09	0.06	0.04	0.04
0.2 g	Coral sand	0.94	0.68	0.62	0.5	1.1	0.72	0.59	0.54
	Fujian sand	1.2	1.09	0.98	Lost	1.5	1.24	1.12	0.81

Figure 8 shows the time history curves of excess pore pressure ratio under 0.2 g shaking intensity, and the signal of pore water pressure gauge was lost at P4 position in the Fujian sand site. The development patterns of the excess pore pressure ratio of the two kinds of sand sites were the same with time during the shaking period (10 s). The excess pore pressure ratio reached peak value after a sharp increase of about 4 s, and then began to decrease. The growth rate of excess pore pressure ratio of coral sand was less than that of Fujian sand, when the excess pore pressure ratio reached 0.5 at the P1 position, the coral sand uses 2.73 s, and the Fujian sand uses 2.22 s. Table 4 shows the peak values of the excess pore pressure ratio of two kinds of sand sites. The peak values of excess pore pressure ratio of coral sand were less than that of Fujian sand. From top to bottom (P1–P3), the peak values of excess pore pressure ratio of coral sand were 0.78, 0.62, and 0.63 times of that of Fujian sand. With the increase of depth, the peak values of the excess pore pressure ratio of two kinds of sand sites gradually decreased. The liquefaction degree of coral sand site is less than that of the Fujian sand site.

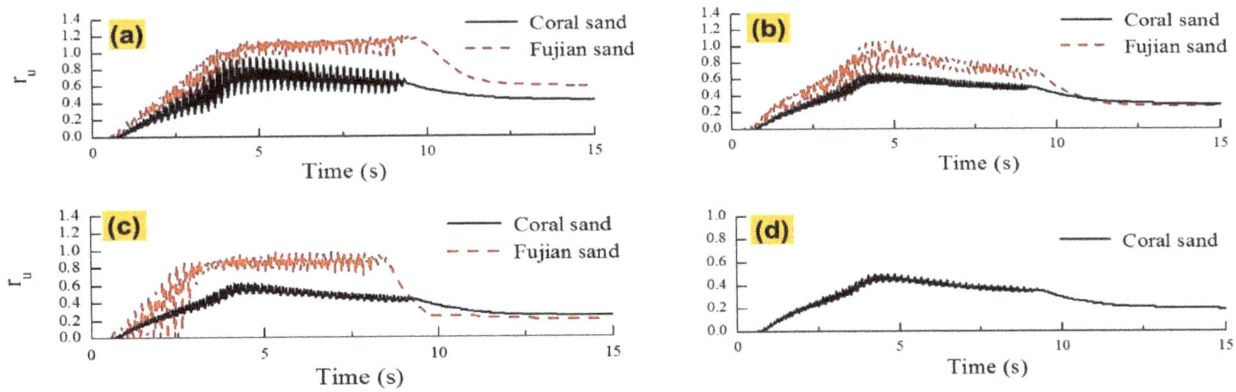

Figure 8. Excess pore pressure ratio time history curves under 0.2 g shaking intensity: (**a**) P1 position; (**b**) P2 position; (**c**) P3 position; and, (**d**) P4 position.

4.2. Acceleration Response

Figure 9 shows the acceleration time history curves of the coral sand and Fujian sand site. Under 0.1 g shaking intensity, the shape of acceleration time history curves of two kinds of sand sites was similar to that of the input sinusoidal wave excitation, which indicated that the soil was basically in an elastic state, and there was a near-linear amplification effect on the input sinusoidal wave excitation. The acceleration amplification factors got larger when the depth decreased, as shown in Figure 10. The acceleration amplification factors of coral sand were less than that of Fujian sand, which were approximately 0.54–0.90 times of that of Fujian sand. The difference of peak values of the acceleration between two kinds of sand sites was the greatest at the A1 position and the smallest at the A3 position.

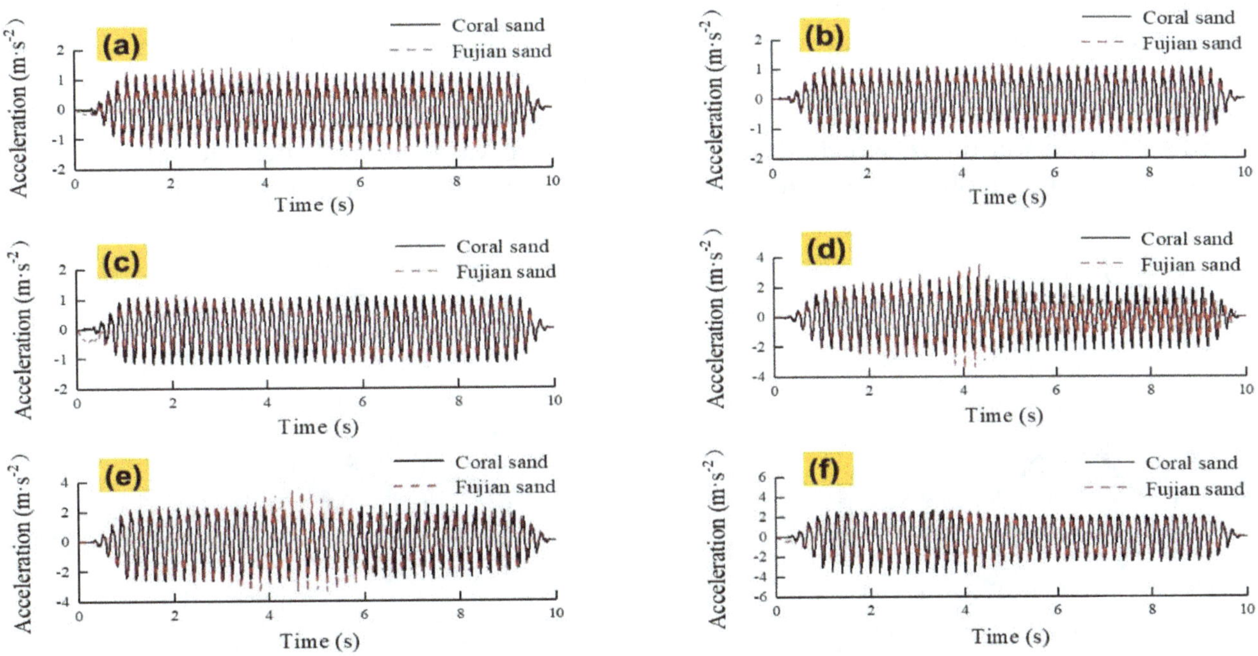

Figure 9. Acceleration time history curves: (**a**) A1 position under 0.1 g intensity; (**b**) A2 position under 0.1 g intensity; (**c**) A3 position under 0.1 g intensity; (**d**) A1 position under 0.2 g intensity; (**e**) A2 position under 0.2 g intensity; and, (**f**) A3 position under 0.2 g intensity.

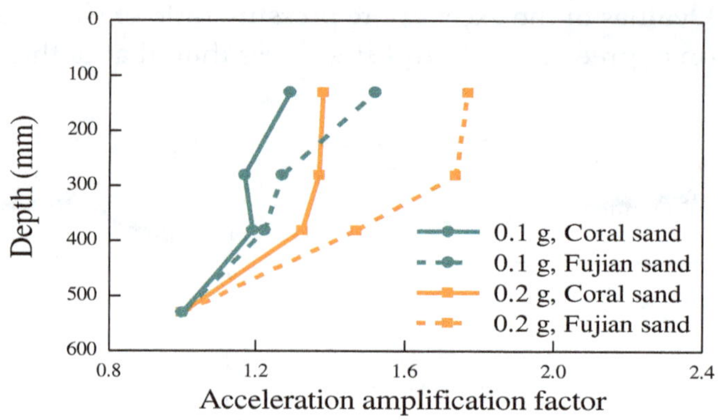

Figure 10. Acceleration amplification factors at different depth.

Under 0.2 g shaking intensity, the acceleration of Fujian sand gradually increased with time, reaching peak values at about 4.5 s, and then suddenly decreased. The acceleration change of coral sand with time was not that obvious, like Fujian sand. The shape of time history curves of coral sand was basically similar to that of input sinusoidal wave excitation. The difference of shape of acceleration time history curves between coral sand and Fujian sand was due to the liquefaction of the Fujian sand site. The shear strength of Fujian sand site sharply decreased, and the soil had an obvious attenuation effect on input acceleration excitation with the onset of liquefaction. The coral sand site was in a high pore pressure state at this time, but there was no obvious liquefaction phenomenon, the attenuation effect of soil on input acceleration was less obvious than that of Fujian sand. Figure 10 shows acceleration amplification factors at different depths of two kinds of sand sites. Acceleration amplification factors increased with the decrease of depth, which is consist with the acceleration response of general liquefaction sites. The acceleration amplification factors of coral sand were less than that of Fujian sand, which were approximately 0.79–0.90 times of that of Fujian sand.

Figure 11 shows the spectrum analysis with a damping ratio of 0.05 for the white noise at the A1 position. The dominant frequencies of the coral sand and Fujian sand sites were 10 Hz before the sinusoidal wave excitation, and the spectrum distribution of two kinds of sand sites was similar, which indicated that the initial state of two kinds of sand sites was close. After 0.2 g sinusoidal wave excitation, the high frequency component attenuation and low frequency component amplification occurred in both coral sand and Fujian sand sites although the dominant frequency of the two kinds of sand sites were still 10 Hz, which illustrated that the two kinds of sand sites had softened and the stiffness of model foundation was reduced when compared with the initial state.

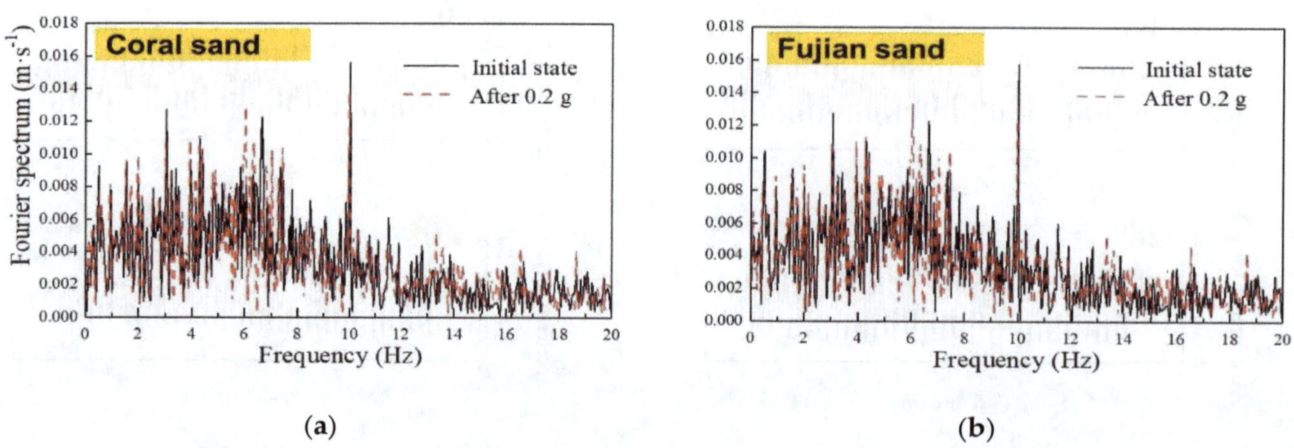

Figure 11. Fourier analysis of white noise: (**a**) coral sand; and, (**b**) Fujian sand.

4.3. Displacement Response

Figure 12 shows the horizontal displacement time history curves of buildings in coral sand and Fujian sand sites. Under 0.1 g shaking intensity, the variation law of horizontal displacement oscillation amplitude of buildings in the two kinds of sand sites was basically similar with time. Both of the building horizontal displacements of coral sand and Fujian sand experienced a rapid increase in a short time (about 1.5 s) and remained stable. The variation law of building oscillation amplitude with time was consistent with the input sinusoidal wave excitation. At this time, there was no liquefaction in the two kinds of sand sites, no obvious reduction of soil stiffness and shear strength, and the horizontal displacement oscillation amplitude of the building changed with the input sinusoidal wave excitation. When the shaking ended, the horizontal displacement of building in coral sand site was 0.02 mm, and that in the Fujian sand site was 0.22 mm. The horizontal displacement in coral sand site was less than that in Fujian sand site, which was approximately 0.09 times of that in the Fujian sand site.

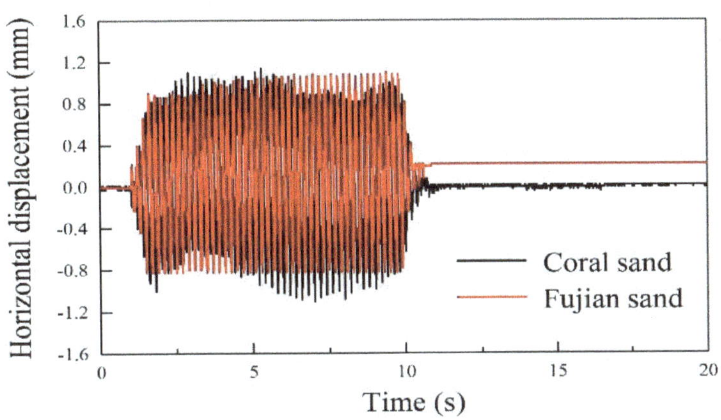

(**a**) 0.1 g shaking intensity

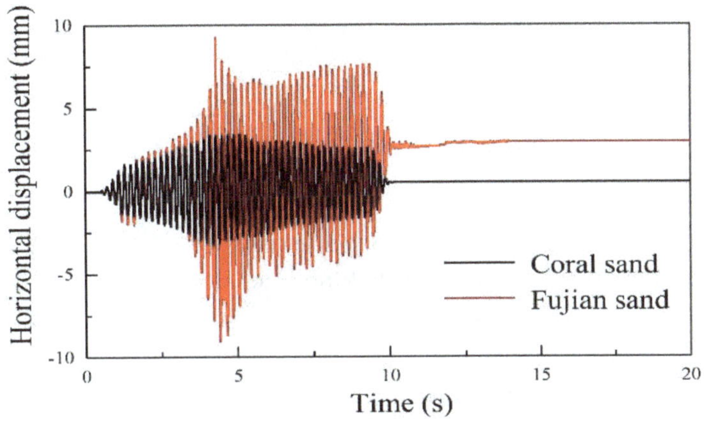

(**b**) 0.2 g shaking intensity

Figure 12. Horizontal displacement time history curves of building.

The horizontal displacement of buildings in the two kinds of sand sites changed with time similarly under 0.2 g shaking intensity, while the horizontal displacement oscillation amplitude in the coral sand site was obviously smaller than that in the Fujian sand site. The horizontal displacement oscillation amplitude in the two kinds of sand sites decreased abruptly at around 4.8 s, at this moment, the excess pore pressure ratio reached its peak value (Figure 8). The coral sand site was in high pore pressure state, and the research of Chen et al. [37] showed that the soil under that condition exhibited shear thinning non-Newtonian fluid characteristics, even if the soil did not liquefy, so the attenuation of horizontal displacement oscillation amplitude of buildings was related to a certain reduction of soil to input shaking excitation and horizontal force of pile that was caused by stiffness degradation of the coral sand site. The attenuation of building oscillation amplitude in the Fujian sand site was due to the sharp decrease of shear strength of soil that was caused by liquefaction. When shaking ended, the horizontal displacement of building in coral sand site was 0.53 mm, and that in Fujian sand site was 2.96 mm. The horizontal displacement of building in coral sand site was less than that in Fujian sand site, which was approximately 0.18 times of that in the Fujian sand site.

Figure 13 shows the horizontal displacement time history curves of buildings in coral sand and Fujian sand sites. Under 0.1 g shaking intensity, the building settlement in two kinds of sand sites increased slowly with time. The building settlement in coral sand site was 0.07 mm, which in the Fujian sand site was 0.43 mm. The building settlement in coral sand site was less than that in the Fujian sand site, which was about 0.16 times of that in Fujian sand site. Under 0.2 g shaking intensity, the building settlement in coral sand site increased linearly with time. The building settlement development trend in the Fujian sand site was similar to that in coral sand site within 4.8 s from the beginning of shaking, while the building settlement rate in Fujian sand site suddenly increased at 4.8 s, when compared with the results of the excess pore pressure ratio of Fujian sand, as illustrated in Figure 8, it could be seen that the excess pore pressure ratio of Fujian sand reached 1 at this time and the soil was in the initial liquefaction state. The effective stress between the soil particles of Fujian sand was close to 0, which led to the soil having almost no shear strength and the bearing capacity of the foundation decreasing, consequently, the building subsided sharply. After shaking, the building settlement in coral sand site was 1.29 mm, which in the Fujian sand site was 7.98 mm. The building settlement in coral sand site was less than that in Fujian sand site, which was approximately 0.16 times of that in the Fujian sand site.

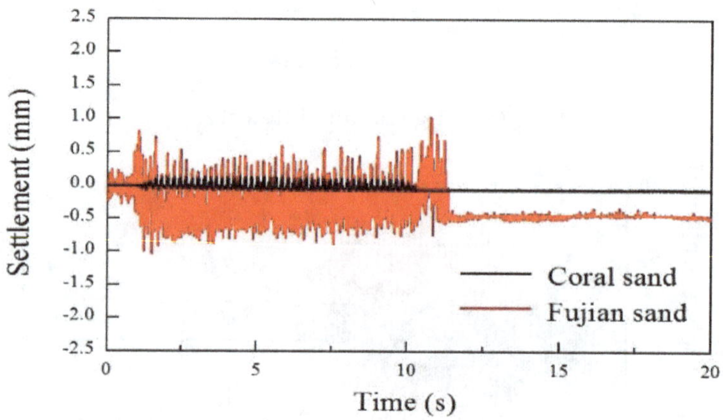

(**a**) 0.1 g shaking intensity

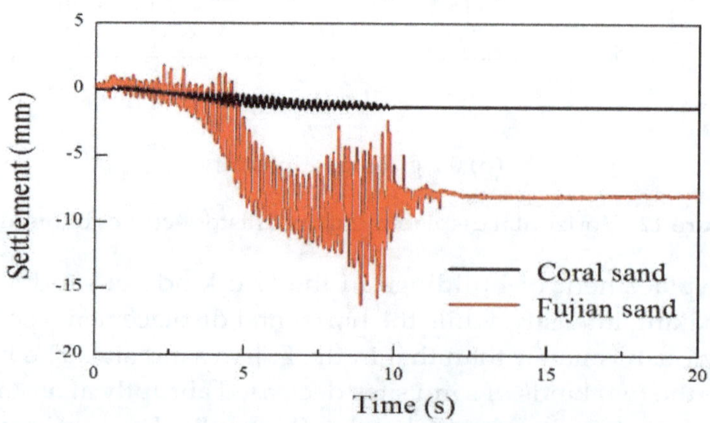

(**b**) 0.2 g shaking intensity

Figure 13. Settlement time history curves of building.

4.4. Dynamic Bending Moment Response

The bending moment of the column and pile foundation is obtained by the following equation:

$$M = \frac{EI(\varepsilon_t - \varepsilon_c)}{h} \quad (2)$$

where M is the bending moment, ε_t and ε_c are tensile and compressive strain, respectively, h is the length of the section side for square cross-section columns, and h is the diameter of the pile for circular cross-section piles.

Figure 14 shows the peak values of the dynamic column bending moments in the coral sand and Fujian sand sites. The dynamic column moments in two kinds of sand sites were the largest at the bottom of column, followed by the second story column, which was consistent with the general law of dynamic moment response of building columns under earthquake. Under 0.1 g shaking intensity, the peak column moments in the coral sand site were smaller than that in the Fujian sand site. From top to bottom (S1–S4), the peak column moments in coral sand site were approximately 0.98, 0.88, 0.82, and 0.98 times of that in Fujian sand site. The peak column moments in the coral sand site were also smaller than that in Fujian sand site under 0.2 g shaking intensity. When compared with 0.1 g shaking intensity, the dynamic column moments in the coral sand site increased by 3.11–3.61 times, and by 4.41-5.93 times in Fujian sand site under 0.2 g shaking intensity. The dynamic moment amplification effect of building columns in coral sand site was smaller than that in Fujian sand site when the shaking intensity increased.

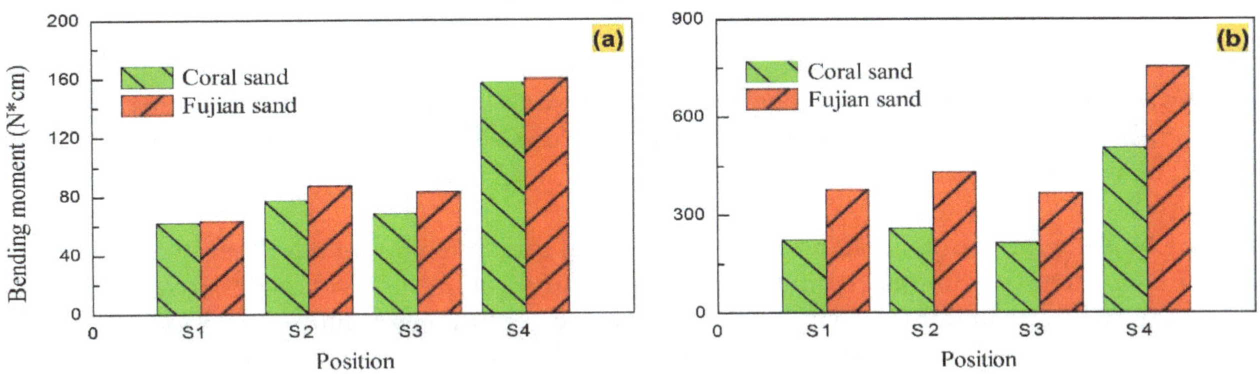

Figure 14. Peak column bending moments at different height: (**a**) 0.1 g shaking intensity; and, (**b**) 0.2 g shaking intensity.

Figure 15 shows the peak values of the dynamic pile moments in coral sand and Fujian sand sites. The dynamic moments of corner pile, edge pile, and center pile similarly varied with buried depth. The peak values of moments were the largest at the top and the smallest at the bottom of pile. When 0.1 g shaking excitation was input, the peak moments of corner piles in coral sand site were less than that in the Fujian sand site, which were approximately 0.69–0.94 times of that in the Fujian sand site. The peak moments of edge piles and center piles in coral sand site were also less than that in Fujian sand site, respectively. When 0.2 g shaking excitation was input, the moments of pile groups in coral sand site were less than that in Fujian sand site, which were about 0.62–0.93 times of that in Fujian sand site.

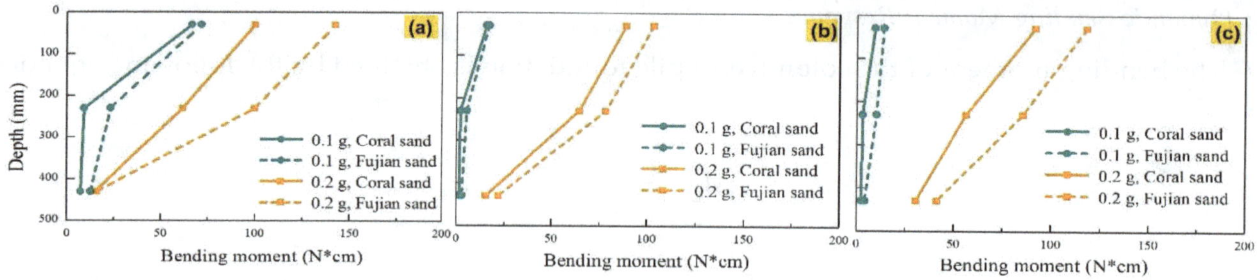

Figure 15. Peak pile bending moments at different depth: (**a**) corner pile; (**b**) edge pile; and (**c**) center pile.

5. Summary and Conclusions

Shaking table tests of three-story buildings with nine-pile foundation in the coral sand and Fujian sand sites were carried out in this research. The similarities and differences of dynamic responses of coral sand and Fujian sand sites were studied through testing and analyzing the physical quantities, such as pore water pressure, acceleration, displacement, and dynamic bending moment. The following conclusions are drawn:

(1) The peak values of excess pore pressure ratio of coral sand and Fujian sand were far less than 1 under 0.1 g shaking intensity, there was no liquefaction in two kinds of sand sites. The peak values of excess pore pressure ratio of coral sand were basically less than that of Fujian sand, which were approximately 0.67–1.00 times of that of Fujian sand.

(2) The development rate of excess pore pressure ratio of coral sand was smaller than that of Fujian sand and the peak values of excess pore pressure ratio of coral sand were less than that of Fujian sand, which were about 0.53–0.78 times of that of Fujian sand. The coral sand sites were more difficult to liquefy than Fujian sand sites under the same relative density and similar grain-size distributions.

(3) The acceleration amplification coefficient of coral sand and Fujian sand sites increased with the decrease of depth, and the acceleration amplification factors of coral sand was smaller than that of Fujian sand.

(4) The building horizontal displacement in the coral sand site was smaller than that in the Fujian sand site. The building horizontal displacement in coral sand site was 0.09 times of that in the Fujian sand site under 0.1 g shaking intensity and 0.18 times of that in Fujian sand site under 0.2 g shaking intensity. The building settlement in coral sand site was also smaller than that in the Fujian sand site.

(5) The dynamic bending moments of building columns in coral sand site were smaller than that in the Fujian sand site, and the magnification effect of increasing shaking intensity on the building column moment in coral sand site was smaller than that in the Fujian sand site. The peak values of dynamic bending moments of pile groups in the coral sand site were smaller than that in the Fujian sand site. The buildings in coral sand will withstand earthquakes better than buildings in silica sand.

Author Contributions: Conceptualization, Q.W. and X.D.; Data curation, Q.W.; Formal analysis, Y.Z.; Funding acquisition, X.D.; Investigation, X.D.; Methodology, Q.W.; Project administration, X.D.; Resources, X.D.; Supervision, X.D.; Validation, Q.W., Y.Z. and Z.C.; Visualization, Y.Z.; Writing—original draft, Q.W.; Writing—review & editing, Z.C. All authors have read and agreed to the published version of the manuscript.

Acknowledgments: The authors appreciate the assistant of our group members in the experiments.

References

1. Xiao, P.; Liu, H.; Xiao, Y.; Stuedlein, A.W.; Evans, T.M. Liquefaction resistance of bio-cemented calcareous sand. *Soil Dyn. Earthq. Eng.* **2018**, *107*, 9–19. [CrossRef]
2. Hyodd, M.; Hyde, A.; Aramaki, N. Liquefaction of crushable soils. *Geotechnique* **1998**, *48*, 527–543. [CrossRef]
3. Xiao, Y.; Liu, H.; Chen, Q.; Ma, Q.; Xiang, Y.; Zheng, Y. Particle breakage and deformation of carbonate sands with wide range of densities during compression loading process. *Acta Geotech.* **2017**, *12*, 1177–1184. [CrossRef]
4. Ren, Y.; Wang, Y.; Yang, Q. Effects of particle size distribution and shape on permeability of calcareous sand. *Rock Soil Mech.* **2018**, *39*, 491–497.
5. Ye, W.; Yin, W.; Qing, Y. Experiment on drag force coefficient of calcareous sand in liquid considering the effect of particle shape. *Rock Soil Mech.* **2018**, *39*, 3201–3212.
6. Schwiderski, E.W. On charting global ocean tides. *Rev. Geophys.* **1980**, *18*, 243–268. [CrossRef]
7. Singh, S.C.; Carton, H.; Tapponnier, P.; Hananto, N.D.; Chauhan, A.P.; Hartoyo, D.; Bayly, M.; Moeljopranoto, S.; Bunting, T.; Christie, P. Seismic evidence for broken oceanic crust in the 2004 Sumatra earthquake epicentral region. *Nat. Geosci.* **2008**, *1*, 777. [CrossRef]
8. Wang, X.; Jiao, Y.; Wang, R.; Hu, M.; Meng, Q.; Tan, F. Engineering characteristics of the calcareous sand in Nansha Islands, South China Sea. *Eng. Geol.* **2011**, *120*, 40–47. [CrossRef]
9. Choi, S.; Park, S.; Wu, S.; Chu, J. Methods for calcium carbonate content measurement of biocemented soils. *J. Mater. Civ. Eng.* **2017**, *29*, 6017015. [CrossRef]
10. Potirakis, S.M.; Hayakawa, M.; Schekotov, A. Fractal analysis of the ground-recorded ULF magnetic fields prior to the 11 March 2011 Tohoku earthquake (M W = 9): Discriminating possible earthquake precursors from space-sourced disturbances. *Nat. Hazards* **2017**, *85*, 59–86. [CrossRef]
11. Medley, E.W. Geological Engineering Reconnaissance of Damage caused by the October 15, 2006 Hawaii Earthquakes. *Int. J. Geoeng. Case Hist.* **2007**, *1*, 89–135.
12. Green, R.A.; Olson, S.M.; Cox, B.R.; Rix, G.J.; Rathje, E.; Bachhuber, J.; French, J.; Lasley, S.; Martin, N. Geotechnical aspects of failures at Port-au-Prince seaport during the 12 January 2010 Haiti earthquake. *Earthq. Spectra* **2011**, *27*, S43–S65. [CrossRef]
13. Williams, T.A.; Shepherd, D.A. Building resilience or providing sustenance: Different paths of emergent ventures in the aftermath of the Haiti earthquake. *Acad. Manag. J.* **2016**, *59*, 2069–2102. [CrossRef]
14. Salem, M.; Elmamlouk, H.; Agaiby, S. Static and cyclic behavior of North Coast calcareous sand in Egypt. *Soil Dyn. Earthq. Eng.* **2013**, *55*, 83–91. [CrossRef]
15. Xu, X.; Wang, R.; Hu, M.; Meng, Q. Experimental study of dynamic characteristics of saturated calcareous soil explosion compaction. *Rock Soil Mech.* **2012**, *15*, 402–406.
16. Sandoval, E.A.; Pando, M.A. Experimental assessment of the liquefaction resistance of calcareous biogenous sands. *Earth Sci. Res. J.* **2012**, *16*, 55–63.
17. Brandes, H. Simple shear behavior of calcareous and quartz sands. *Geotech. Geol. Eng.* **2011**, *29*, 113–126. [CrossRef]
18. Sharma, S.S.; Fahey, M. Evaluation of cyclic shear strength of two cemented calcareous soils. *J. Geotech. Geoenviron. Eng.* **2003**, *129*, 608–618. [CrossRef]
19. Tang, L.; Zhang, X.; Ling, X.; Su, L.; Liu, C. Response of a pile group behind quay wall to liquefaction-induced lateral spreading: A shake-table investigation. *Earthq. Eng. Eng. Vib.* **2014**, *13*, 741–749. [CrossRef]
20. Dashti, S.; Bray, J.D.; Pestana, J.M.; Riemer, M.; Wilson, D. Centrifuge testing to evaluate and mitigate liquefaction-induced building settlement mechanisms. *J. Geotech. Geoenviron. Eng.* **2009**, *136*, 918–929. [CrossRef]
21. Chen, G.; Chen, S.; Qi, C.; Du, X.; Wang, Z.; Chen, W. Shaking table tests on a three-arch type subway station structure in a liquefiable soil. *Bull. Earthq. Eng.* **2015**, *13*, 1675–1701. [CrossRef]
22. Jiang, L.; Chen, J.; Li, J. Seismic response of underground utility tunnels: Shaking table testing and FEM analysis. *Earthq. Eng. Eng. Vib.* **2010**, *9*, 555–567. [CrossRef]
23. Rasouli, R.; Towhata, I.; Hayashida, T. Mitigation of seismic settlement of light surface structures by installation of sheet-pile walls around the foundation. *Soil Dyn. Earthq. Eng.* **2015**, *72*, 108–118. [CrossRef]
24. Luan, L.; Zheng, C.; Kouretzis, G.; Ding, X. Dynamic analysis of pile groups subjected to horizontal loads considering coupled pile-to-pile interaction. *Comput. Geotech.* **2020**, *117*, 103276. [CrossRef]

25. Luan, L.; Ding, X.; Zheng, C.; Kouretzis, G.P.; Wu, Q. Dynamic response of pile groups subjected to horizontal loads. *Can. Geotech. J.* **2019**. [CrossRef]
26. Lv, Y.; Liu, J.; Zuo, D. Moisture effects on the undrained dynamic behavior of calcareous sand at high strain rates. *Geotech. Test. J.* **2018**, *42*, 725–746. [CrossRef]
27. Lv, Y.; Wang, Y.; Zuo, D. Effects of particle size on dynamic constitutive relation and energy absorption of calcareous sand. *Powder Technol.* **2019**, *356*, 21–30. [CrossRef]
28. Ni, P.; Song, L.; Mei, G.; Zhao, Y. Predicting excavation-induced settlement for embedded footing: Case study. *Int. J. Geomech.* **2018**, *18*, 5018001. [CrossRef]
29. Cui, C.Y.; Meng, K.; Wu, Y.J.; Chapman, D.; Liang, Z.M. Dynamic response of pipe pile embedded in layered visco-elastic media with radial inhomogeneity under vertical excitation. *Geomech. Eng.* **2018**, *16*, 609–618. [CrossRef]
30. Wu, W.; Liu, H.; Yang, X.; Jiang, G.; El Naggar, M.H.; Mei, G.; Liang, R. New method to calculate apparent phase velocity of open-ended pipe pile. *Can. Geotech. J.* **2020**, *57*, 127–138. [CrossRef]
31. Iai, S. Similitude for shaking table tests on soil-structure-fluid model in 1g gravitational field. *Soils Found.* **1989**, *29*, 105–118. [CrossRef]
32. Chen, S.; Tang, B.; Zhao, K.; Li, X.; Zhuang, H. Seismic response of irregular underground structures under adverse soil conditions using shaking table tests. *Tunn. Undergr. Space Technol.* **2020**, *95*, 103145. [CrossRef]
33. Zhou, Z.; Lei, J.; Shi, S.; Liu, T. Seismic response of aeolian sand high embankment slopes in shaking table tests. *Appl. Sci.* **2019**, *9*, 1677. [CrossRef]
34. Kheradi, H.; Morikawa, Y.; Ye, G.; Zhang, F. Liquefaction-Induced Buckling Failure of Group-Pile Foundation and Countermeasure by Partial Ground Improvement. *Int. J. Geomech.* **2019**, *19*, 4019020. [CrossRef]
35. Ebeido, A.; Elgamal, A.; Tokimatsu, K.; Abe, A. Pile and Pile-Group Response to Liquefaction-Induced Lateral Spreading in Four Large-Scale Shake-Table Experiments. *J. Geotech. Geoenviron. Eng.* **2019**, *145*, 4019080. [CrossRef]
36. Teparaksa, J.; Koseki, J. Effect of past history on liquefaction resistance of level ground in shaking table test. *Géotech. Lett.* **2018**, *8*, 256–261. [CrossRef]
37. Chen, Y.; Liu, H.; Shao, G.; Zhao, N. Laboratory tests on flow characteristics of liquefied and post-liquefied sand. *Chin. J. Geotech. Eng.* **2009**, *31*, 1408–1412.

6

Improved Double-Layer Soil Consolidation Theory and its Application in Marine Soft Soil Engineering

Deqiang Chen [1], Junhui Luo [2,*], Xianlin Liu [2], Decai Mi [2] and Longwang Xu [2]

1 College of Civil Engineering and Architecture, Guangxi University, Nanning 530004, China; dqchen94@163.com
2 Guangxi Communication Design Group Co., LTD, Nanning 530029, China; Chen_GXU17@163.com (X.L.); Chen_GXU17@126.com (D.M.); longwangxu_gx@163.com (L.X.)
* Correspondence: jhluo85@hotmail.com

Abstract: Marine soft soil foundation is a double-layer foundation structure with a crust layer and soft substratum. Moreover, it is common that there are various forms of drainage. Accordingly, based on Terzaghi's consolidation theory and the continuous drainage boundary conditions theory of controllable drainage conditions, an improved double-layer soil consolidation theory considering continuous drainage boundary conditions was proposed. To improve the computational efficiency and accuracy, the Laplace transform and the Stehfest algorithm was used to deduce the numerical solution of the improved double-layer soil consolidation theory considering continuous drainage boundary conditions and to compile a computer program. Subsequently, the theory was validated and analyzed by the degenerated model of the perfectly permeable boundary conditions and the semi-permeable boundary conditions, respectively, which showed that this theory has higher accuracy. Simultaneously, the analysis of double-layer consolidation settlement under continuous drainage boundary conditions for marine soft soil foundation of Guangxi Binhai Highway was carried on. The result showed that the consolidation settlement calculated by the improved double-layer consolidation theory presented is basically consistent with the field measurement results, and that the correlation coefficient between them is higher. Accordingly, the research results can provide useful basic information for marine soft foundation engineering.

Keywords: marine soft soil; double-layer foundation; consolidation theory; drainage boundary

---

1. Introduction

Marine soft soil is widely distributed throughout the world, and the performance of the soil varies as a result of differences in geological origin, occurrence law, and composition. Therefore, the engineering characteristics that emerged from these factors reflect substantial spatial and temporal variability and regionality. For example, the Canadian coastline exceeds 200,000 km [1], and therefore marine soft soil represented by Leda soft soil is widely distributed in Canada, which has the significant characteristics of marine soft soil [2]. Norway's Drammen marine soft soil also has characteristics of marine soft soil, but Sangrey showed that the engineering properties of Drammen soft soil were different from those of Leda soft soil [2–4]. Other soft soils, such as London soft soil [5], Mexican soft soil [6], Japanese soft soil [7], Busan soft soil [8], Shenzhen soft soil, Shanghai soft soil and so on, all exhibit different marine soft soil characteristics [9–16]. The engineering characteristics of marine soft soil make the problem of consolidation and settlement of marine soft soil foundation complicated, but the one-dimensional consolidation theory by Terzaghi is not suitable for the analysis of complex marine soft soil foundation consolidation problems. The main reason for this is that Terzaghi's one-dimensional consolidation theory assumes that the drainage boundary is perfectly permeable or perfectly impervious, and the pore water pressure equations were as follows [17]:

$$u(0,t) = 0 \tag{1}$$

and:

$$u_t(h,t) = 0 \tag{2}$$

In reality, the sand layer of the top surface of the foundation treatment and the lower layer of the bottom surface are neither perfectly permeable nor perfectly impervious, and they are often somewhere in between. Previous studies showed that permeability of drainage boundary will have a major impact on the final calculation results [18,19]. In view of the unreasonable drainage boundary conditions of consolidation theory, Gray [20] took the lead in conducting research. Based on one-dimensional consolidation theory by Terzaghi, Gray [20] proposed a semi-permeable boundary theory, and the pore water pressure equation was as follows:

$$u_t(h,t) + hu(h,t) = 0 \tag{3}$$

Later, Schiffiman and Stein [21] conducted a detailed study on the one-dimensional consolidation problem with a semi-permeable boundary. Huang [22] discussed and promoted the semi-permeable boundary theory. While the one-dimensional consolidation theory of semi-permeable boundary is more practical than one-dimensional consolidation theory by Terzaghi, it is relatively simple, and it assumes that the soil layer is homogeneous elastomer without considering multi-layer soil. According to this, Xie [23] proposed a one-dimensional consolidation theory for a double-layer foundation with a semi-permeable boundary and solved and discussed it. The results showed that the semi-permeable boundary conditions and the stratification of soil can bring the foundation consolidation analysis closer to the actual situation [23]. Based on the research of Xie [23], Hu and Xie [24] studied the one-dimensional consolidation problem of the semi-permeable boundary under the gradual load of a multi-layer elastic foundation and systematically analyzed the layered soil. Hu and Xie [24] studied the effect of semi-permeable boundary conditions, soil properties, and loading rate of external load on pore pressure distribution and total average consolidation. However, the above scholars considered the foundation soil to be a homogeneous elastic soil layer and a static load [23,24]. For this reason, Fang et al. [25], Lin et al. [26], Wang et al. [27], Wang et al. [28], Cai et al. [29], Li et al. [30], Wang and Xia [31], and Zheng et al. [32] carried out consolidation analysis of viscoelastic soil layers under semi-permeable conditions and dynamic load. According to the one-dimensional consolidation equation of the semi-permeable boundary proposed by Gray [20], scholars carried out consolidation analysis under different loading modes and non-Darcy's law conditions [33,34]. However, the one-dimensional consolidation equation based on the semi-permeable boundary proposed by Gray [20] is difficult to solve and not easy to generalize. Combined with the above problems, Mei et al. [35] proposed and solved a one-dimensional consolidation equation for continuous drainage boundary in order to solve the contradiction between the boundary conditions and the initial conditions of the one-dimensional consolidation equation by Terzaghi. The pore water pressure equations by Mei et al. [35] are as follows:

$$u(0,t) = q(t)e^{-bt} \text{ (top drainage)}, \tag{4}$$

and

$$u(2h,t) = q(t)e^{-ct} \text{ (bottom drainage)}. \tag{5}$$

The continuous drainage boundary is a time-dependent interface boundary that varies between perfectly pervious and impervious conditions. The boundary pore pressure changes exponentially with time. By adjusting the parameters b and c related to the undisturbed soil, the drainage properties of the top and bottom drainage surfaces of the soil layer can be controlled. For example, b can tend to infinity, and thus the top surface of the soil layer is undrained. Subsequently, scholars carried out research on continuous drainage boundary theory. Zong et al. [36] and Zheng et al. [37] established a generalized Terzaghi's consolidation theory under double-sided asymmetric continuous drainage boundary conditions. Cai et al. [38] developed a subroutine within ABAQUS program to verify the solution of the one-dimensional consolidation equation of the continuous drainage boundary. Wang et al. [39] showed that the continuous boundary has good applicability and extends to the semi-analytical solution of one-dimensional consolidation of unsaturated soil. Feng et al. [40] established the one-dimensional consolidation equation for the continuous drainage boundary and studied the contribution of a soil's self-weight stress. Sun et al. [41] established a general analytical solution for the one-dimensional consolidation of soil for the continuous drainage boundary under a ramp load. Zhang et al. [42] analyzed the excess pore water pressure and the average degree of consolidation under the continuous drainage boundary conditions and discussed the effect of the drainage capacity of the top surface, the smear effect, and the well resistance on consolidation. However, there are few reports on the one-dimensional consolidation theory with the continuous drainage boundary of double-layer soil and its application in marine soft soil engineering.

Domestic and foreign scholar have done a lot of research on the algorithm for the mathematical modeling of consolidated seepage. In 1949, Van Everdingen and Hurst [43] proposed the Laplace transform method. In 1968, Dubner and Abate [44] proposed the Laplace numerical inversion method. Subsequently, Durbin [45] improved the Dubner and Abate algorithm. In 1970, Stehfest [46,47] proposed the Stehfest algorithm because the algorithm is easy to program, has few parameters, has fast calculation, does not involve complex numbers, and has high stability. It has thus been widely used in engineering. It is worth mentioning that the Stehfest algorithm looks like an empirical formula, but it actually has a complicated mathematical background and theoretical derivation. Then, Crump [48] proposed the Crump algorithm, which is based on the Fourier series. Duffy [49] improved the Crump algorithm, avoiding the trigonometric function term. However, because the Crump algorithm does not put forward a method to determine the appropriate attenuation index and truncation term number, it is difficult to apply widely. Therefore, according to the advantages and disadvantages of the above algorithm, an appropriate method should be adopted to carry out the calculation according to the characteristics of the consolidation theoretical model.

Based on an analysis of the above literature, aiming at the characteristics of the marine soft soil double-layer foundation structure and complex drainage conditions, a numerical solution of the improved double-layer foundation consolidation theory considering continuous drainage boundary conditions is presented. In Section 2, the basic equations of the improved consolidation theory is introduced and deduced in detail, and the Laplace transform and Stehfest algorithm are applied in the derivation process, and the improved model is compiled into a program by this paper. In Section 3, the improved model is validated and analyzed by three examples in the literature. This section includes degradation analysis of perfectly permeable boundary conditions and semi-permeable boundary conditions by the improved model. Finally, in Section 4, the improved model is applied to an actual marine soft soil foundation project in Guangxi for application analysis. The settlement and consolidation degree of soft soil foundation are analyzed and compared with the measured data for verification. The conclusions can provide scientific guidance for consolidation analysis of marine soft soil foundation. They also have certain theoretical value and practical significance.

2. Improved Double-Layer Soil Consolidation Theory Considering Continuous Drainage Boundary Conditions

In addition to the load, the same basic assumptions as in the one-dimensional consolidation theory by Terzaghi were made. Equations (1)–(5) are the basic assumptions of one-dimensional consolidation theory by Terzaghi:

(1) The soil layer is homogeneous and fully saturated.
(2) Soil particles and water are incompressible.

(3) Water seepage and compression of the soil layer occur only in one direction (vertical).
(4) The seepage of water obeys Darcy's law.
(5) In the osmotic consolidation, the permeability coefficient and the compression coefficient of the soil are constants.
(6) The external load is applied at two levels of average speed.
(7) Additional stress of soil does not decrease with depth under the large area load of highway roadbed.

Based on the above assumptions, the double-layer soil consolidation equations considering continuous drainage boundary conditions were proposed. Figure 1a presents a simplified diagram of the theoretical calculation of an improved double-layered soil foundation consolidation considering continuous drainage boundary conditions. We take the ground table as the coordinate origin. In the figure, $q(t)$ is an arbitrary loading function. h_1, E_{s1}, and k_1 are the parameters of topsoil. h_2, E_{s2}, and k_2 are the parameters of subsoil. The load simplification is applied in two stages. q_1 and q_2 are the primary and secondary load increments, respectively, as shown in Figure 1b.

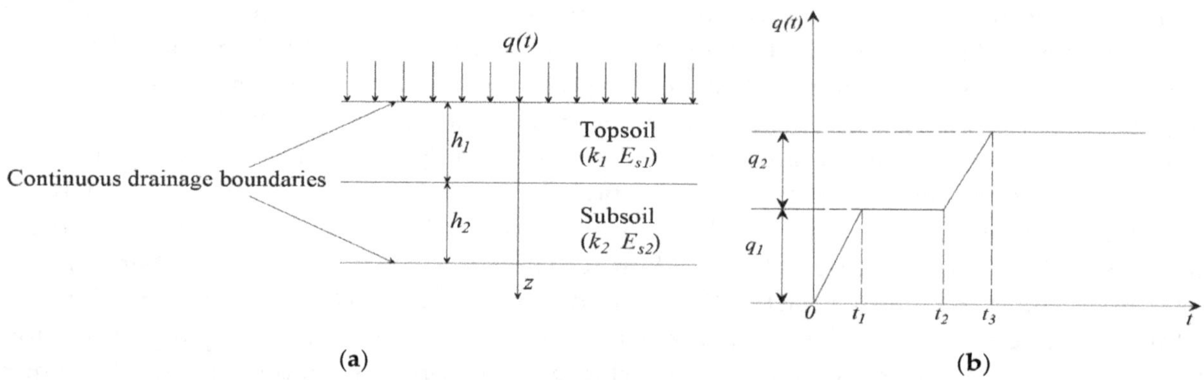

Figure 1. Schematic diagram: (**a**) Double-layer soil foundation calculation model, and (**b**) the curve of loading.

Therefore, one-dimensional consolidation differential equations for a double-layer soil foundation can be obtained:

$$c_{v1}\frac{\partial^2 u_1(z,t)}{\partial z^2} = \frac{\partial u_1(z,t)}{\partial t} - \frac{\partial q(t)}{\partial t}(0 \le z \le h_1) \tag{6}$$

$$c_{v2}\frac{\partial^2 u_2(z,t)}{\partial z^2} = \frac{\partial u_2(z,t)}{\partial t} - \frac{\partial q(t)}{\partial t}(h_1 \le z \le h_1 + h_2) \tag{7}$$

where u_1 and u_2 are the first and second layers of excess pore water pressure, respectively; and c_{v1} and c_{v2} are the first and second layers of consolidation coefficients, respectively. The latter are calculated by:

$$c_{v1} = \frac{k_1 E_{s1}}{\gamma_w},\ c_{v2} = \frac{k_2 E_{s2}}{\gamma_w}.$$

The pore pressures in Equations (6) and (7) can be expressed by the effective stress and converted into the following equations:

$$\frac{\partial \sigma_1'(z,t)}{\partial t} = c_{v1}\frac{\partial^2 \sigma_1'(z,t)}{\partial z^2}(0 \le z \le h_1) \tag{8}$$

$$\frac{\partial \sigma_2'(z,t)}{\partial t} = c_{v2}\frac{\partial^2 \sigma_2'(z,t)}{\partial z^2}(h_1 \le z \le h_1 + h_2) \tag{9}$$

The initial condition is:

$$\sigma_{1,2}'(z,0) = 0 \tag{10}$$

The boundary conditions are:

$$\sigma_1'(0,t) = q(t) - q(t)\exp(-bt) \tag{11}$$

$$\sigma_2'(h_1 + h_2, t) = q(t) - q(t)\exp(-ct) \tag{12}$$

These boundary conditions are continuous drainage boundary conditions [35,36], and b and c are parameters related to soil drainage properties. They are interface parameters that reflect the drainage properties of the top and bottom drainage surfaces of the soil layer [41]. As shown in Table 1, the permeability of the drainage boundary can be controlled by adjusting parameters b and c.

Table 1. Description of parameters b (top surface) and c (bottom surface).

Parameter \ Value	0	(0, +∞)	+∞
b	impermeable	semi-permeable	perfectly permeable
c	impermeable	semi-permeable	perfectly permeable

The continuous condition between layers is:

$$\sigma_1'\big|_{z=h_1} = \sigma_2'\big|_{z=h_1} \tag{13}$$

The continuous flow condition is:

$$\frac{k_1}{\gamma_w}\frac{\partial \sigma_1'}{\partial z}\bigg|_{z=h_1} = \frac{k_2}{\gamma_w}\frac{\partial \sigma_2'}{\partial z}\bigg|_{z=h_1} \tag{14}$$

Equations (8)–(14) constitute the improved double-layer soil consolidation theory equation considering continuous drainage boundary conditions. A Laplace transform is performed on Equations (8)–(14):

$$s\overline{\sigma_1'}(z,s) - \overline{\sigma_1'}(z,0) = c_{v1}\frac{\partial^2 \overline{\sigma_1'}(z,s)}{\partial z^2}(0 \le z \le h_1) \tag{15}$$

$$s\overline{\sigma_2'}(z,s) - \overline{\sigma_2'}(z,0) = c_{v2}\frac{\partial^2 \overline{\sigma_2'}(z,s)}{\partial z^2}(h_1 \le z \le h_1 + h_2) \tag{16}$$

$$\overline{\sigma_{1,2}'}(z,0) = 0 \tag{17}$$

$$\overline{\sigma_1'}(0,s) = \overline{q}(s) - \overline{q}(s+b) \tag{18}$$

$$\overline{\sigma_2'}(h_1+h_2,s) = \overline{q}(s) - \overline{q}(s+c) \tag{19}$$

$$\overline{\sigma_1'}\big|_{z=h_1} = \overline{\sigma_2'}\big|_{z=h_1} \tag{20}$$

$$\frac{k_1}{\gamma_w}\frac{\partial \overline{\sigma_1'}}{\partial z}\bigg|_{z=h_1} = \frac{k_2}{\gamma_w}\frac{\partial \overline{\sigma_2'}}{\partial z}\bigg|_{z=h_1} \tag{21}$$

The general solutions of Equations (8) and (9) obtained by Equations (15)–(17) are as follows in the Laplace transform domain:

$$\overline{\sigma_1'}(z,s) = A_{11}\exp(r_1 z) + A_{12}\exp(-r_1 z)(0 \le z \le h_1) \tag{22}$$

$$\overline{\sigma_2'}(z,s) = A_{21}\exp(r_2 z) + A_{22}\exp(-r_2 z)(h_1 \le z \le h_1 + h_2) \tag{23}$$

Here:

$r_1^2 = \frac{s}{c_{v1}}, r_2^2 = \frac{s}{c_{v2}}$.

Bands Equations (18)–(21) substituted into Equations (22) and (23) yields the following:

$$\begin{cases} A_{11} + A_{12} = \overline{q}(s) - \overline{q}(s+b) \\ A_{21}e^{r_2(h_1+h_2)} + A_{22}e^{-r_2(h_1+h_2)} = \overline{q}(s) - \overline{q}(s+c) \\ A_{11}e^{r_1h_1} + A_{12}e^{-r_1h_1} = A_{21}e^{r_2h_1} + A_{22}e^{-r_2h_1} \\ \frac{k_1}{\gamma_w}\left(A_{11}r_1e^{r_1h_1} + A_{12}(-r_1)e^{-r_1h_1}\right) = \frac{k_2}{\gamma_w}\left(A_{21}r_2e^{r_2h_1} + A_{22}(-r_2)e^{-r_2h_1}\right) \end{cases} \tag{24}$$

where:

$$\overline{q}(s) - \overline{q}(s+b) = \int_o^{+\infty} q(t)\left(1 - e^{-bt}\right)e^{-st}dt,$$
$$\overline{q}(s) - \overline{q}(s+c) = \int_o^{+\infty} q(t)\left(1 - e^{-ct}\right)e^{-st}dt.$$

$q(t)$ can be expressed as follows (Figure 1b):

$$q(t) = \begin{cases} \frac{t}{t_1}q_1 & (0 \le t \le t_1) \\ q_1 & (t_1 \le t \le t_2) \\ q_1 + \frac{t-t_2}{t_3-t_2}q_2 & (t_2 \le t \le t_3) \\ q_1 + q_2 & (t_3 \le t) \end{cases} \tag{25}$$

According to Equations (24) and (25), $\overline{q}(s) - \overline{q}(s+b)$, $\overline{q}(s) - \overline{q}(s+c)$, A_{11}, A_{12}, A_{21}, A_{22} can be obtained, see the Appendix A for details.

By putting A_{11}, A_{12}, A_{21}, and A_{22} into Equations (22) and (23) and then performing Laplace inverse transformation, the solutions of consolidation Equations (6) and (7) can be obtained. However, for the complex Laplace solution, it is difficult to carry out the Laplace inverse transform. Then it needs to be solved by the numerical solution of Laplace inverse transform. According to the prior literature, the Stehfest algorithm has better stability and has the advantage of fewer computational parameters [50]. In this case, the Stehfest algorithm is used to write the corresponding program for numerical inversion of Equations (22) and (23). The Stehfest inversion equation is as follows:

$$f(T) = \frac{\ln 2}{T}\sum_{i=1}^{N} V_i\overline{f}\left(\frac{\ln 2}{T}i\right) \tag{26}$$

where $\overline{f}(s)$ is the Laplace function of $f(t)$, $\overline{f}(s) = L[f(t)] = \int_0^{\infty} f(t)e^{-st}dt$, and $V_i = (-1)^{N/2+i}\sum_{k=[\frac{i+1}{2}]}^{Min(i,N/2)} \frac{k^{N/2}(2k)!}{(N/2-k)!k!(k-1)!(i-k)!(2k-i)!}$, where N must be a positive even number. Stehfest [46,47] recommended taking N as between 4 and 32. Through repeated verification and reference to the relevant literature [50], we found that 8 was the best choice for N.

The Stehfest algorithm can be used to invert the numerical solution of the improved double-layer soil consolidation equation considering the continuous drainage boundary conditions. Then the total consolidation settlement of the double-layer soil can be calculated as follows:

$$S_t = \int_0^{h_1} \frac{\sigma_1'(z,t)}{E_{s1}}dz + \int_{h_1}^{h_1+h_2} \frac{\sigma_2'(z,t)}{E_{s2}}dz \tag{27}$$

The average consolidation degree of the double-layer soil is:

$$U = \frac{S_t}{S_\infty} = \frac{S_t}{\left(\int_0^{h_1} \frac{p(t)}{E_{s1}}dz + \int_{h_1}^{h_1+h_2} \frac{p(t)}{E_{s2}}dz\right)} \tag{28}$$

3. Improved Double-Layer Consolidation Theory Model Verification Analysis

3.1. Degradation Analysis of Perfectly Permeable Boundary Conditions by the Improved Model

Xie made remarkable contributions to the theoretical research of the double-layer consolidation model. Therefore, in order to verify the improved model, it is used the case in Xie's paper [51] to conduct the degradation analysis of perfectly permeable boundary conditions by the improved model. When b tends to ∞ and c tends to 0, the improved equations can be degraded to Xie's double-layer soil consolidation equation (single-sided permeability). In order to verify the conclusion, the corresponding program is compiled by this paper for calculation using the data for the examples of Xie's paper [51]. The parameters of example in paper [51] are shown in Table 2.

Table 2. Study data of perfectly permeable boundary conditions.

Layer	Layer Thickness h (m)	Permeability Coefficient k (10^{-8}m/s)	Compression Modulus E_s (MPa)
Topsoil	1	1.014	8
Subsoil	9	2.028	4

Question 1: When the load is applied instantaneously, how long will it take for the average consolidation degree of the foundation to reach 60%?

Answer to question 1: According to the data in Table 2, analysis of the double-layer soil consolidation degree is carried out according to the corresponding program compiled by this paper. The results by the proposed model are compared with those of Xie's model. Figure 2 is the solution graph of Question 1. When the load is applied instantaneously, it takes 55 days to reach the average consolidation degree of 60% according to the proposed method, which is basically consistent with the calculation results of Xie's model. Further analysis shows that the relation curves between time and consolidation degree (the t–U curve) obtained by the proposed method is slightly different from those of Xie's model. This difference shows that the solution by the proposed model is slightly larger in the early stage and is slightly smaller in the later stage. The analysis shows that this reason is based on the Stehfest algorithm. While the algorithm requires fewer parameters and provides higher accuracy, it also leads to some errors in the inversion data. According to Figure 2, the consolidation curve obtained by the proposed method is basically consistent with Xie's method.

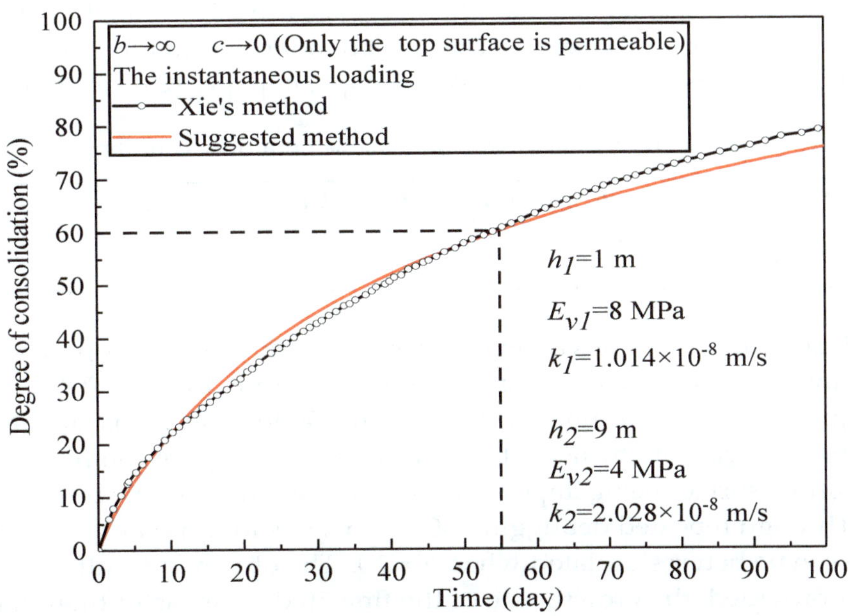

Figure 2. Consolidation curve for the instantaneous loading.

Question 2: When the single-stage constant-speed load is 70 days, what is the average consolidation degree of the foundation at 140 days?

Answer to question 2: Figure 3 is the solution graph of Question 2 and shows that when the single-stage constant-speed load is 70 days, the average consolidation degree of the foundation after 140 days is about 80%, which is consistent with Xie's solution. Similarly, the analysis of the whole curve shows that the solution by the proposed model is slightly larger in the early stage and is slightly smaller in the later stage. The specific reasons for this have already been explained as those above mention. Through the answers to Questions 1 and 2, the double-layer soil continuous drainage boundary consolidation theory based on the Stehfest algorithm is found to have higher accuracy.

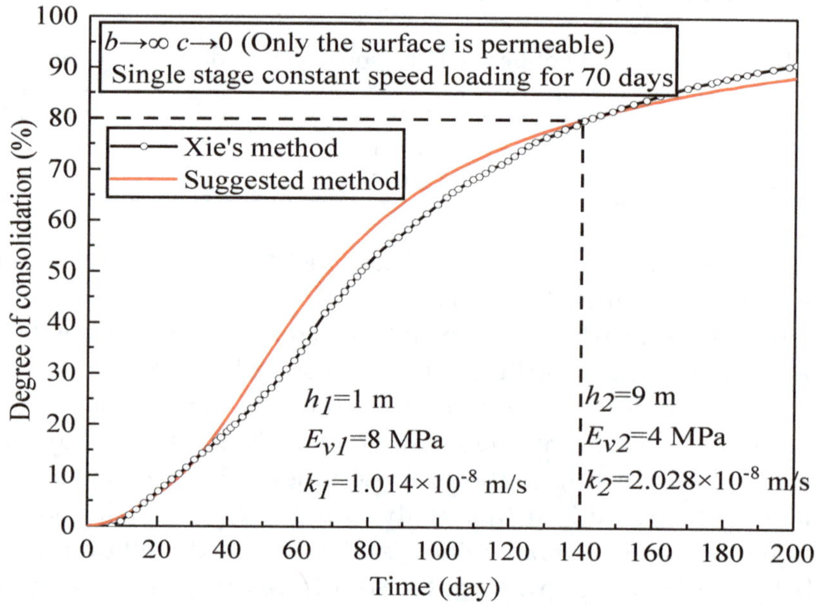

Figure 3. Consolidation curve for single-stage constant-speed loading.

3.2. Degradation Analysis of Semi-Permeable Boundary Conditions by the Improved Model

Gray [20] pioneered the study of the semi-permeable boundary of the consolidation theory model, but the stratum studied was homogeneous. On this basis, Xie [23] studied the theoretical model of double-layer soil consolidation of the semi-permeable boundary. Therefore, we again use the example in Xie's study [23] for analysis. The parameters of example in paper [23] are shown in Table 3.

Table 3. Study data of semi-permeable boundary conditions.

Layer	Layer Thickness h (m)	Permeability Coefficient k (10^{-9} m/s)	Compression Modulus E_s (MPa)
Topsoil	3	1	8
Subsoil	3	5	1.6

Question 3: Under the conditions of an impervious bottom and semi-permeable top, how long will it take for the average consolidation degree of the foundation to reach 70%?

Answer to question 3: According to the data in Table 3, analysis of the double-layer soil consolidation degree is carried out by using the corresponding program compiled by this paper. $c \to 0$ can simulate the bottom surface being impervious, but the semi-permeable boundary of top surface is a fuzzy concept. Through repeated debugging of b value, it is found that semi-permeable boundary of the top surface can be better simulated when $b = 10$. The obtained results are shown in Figure 4. Using the method provided, the growth rate of the first 50 days is faster than that in Xie's method, and then the consolidation degree gradually became consistent with Xie's, indicating that the method provided is suitable for double-layer soil and the prediction has high accuracy. The time required for the proposed method and Xie's method to calculate the total average consolidation degree to 70% is 1115 and 1185 days, respectively, and the error is about 6%. The main reasons for the errors are as follows: (1) The proposed method is to control the permeability of the drainage boundary by adjusting parameters b and c. However, the parameter values do not easily fully correspond to those in Xie's method. (2) The Stehfest algorithm requires fewer parameters and has higher accuracy, but it also has some errors in the inversion process.

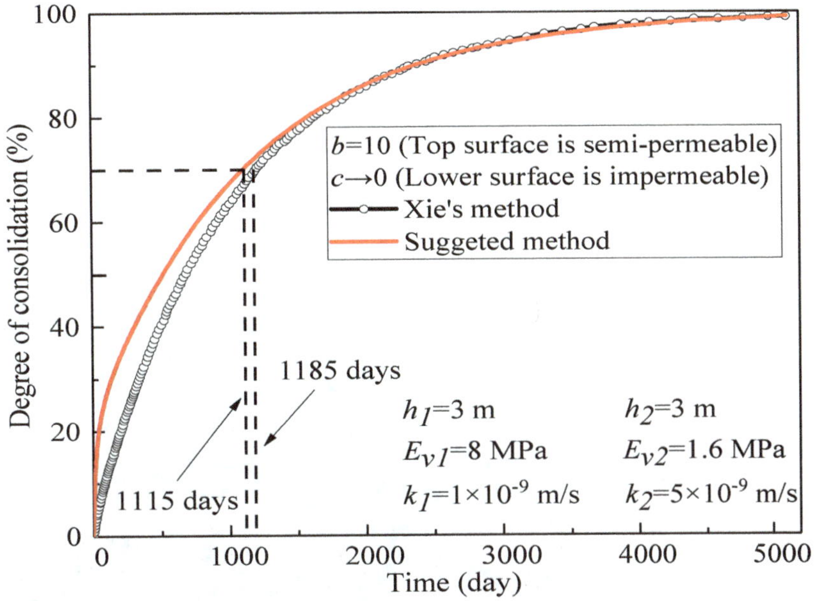

Figure 4. Consolidation curve for the semi-permeable boundary.

4. Engineering Case Analysis

4.1. Project Overview

The Guangxi Binhai Highway is located along the coastline of Beibu Gulf (Figure 5). According to the on-site investigation, the completed survey and design section, the soft land base section exceeds 200 km, accounting for 70% of the total length of the route.

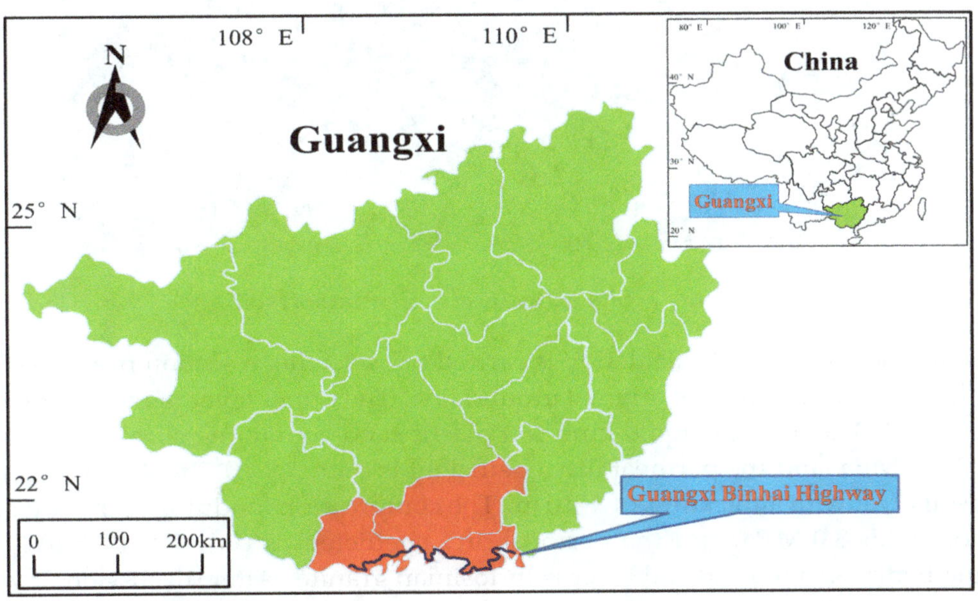

Figure 5. Location of the Guangxi Binhai Highway within China.

The Guangxi Binhai Highway starts from Dongxing City, passes through Fangchenggang, Qinzhou, and Beihai, and ends in Shankou. The main line is 314.2 km in length. The project relies on Xiniujiao town to Dafengjiang section of the Guangxi Binhai Highway, with a total length of 10.9 km. The road grade of the project is Grade I, and the width of the roadbed is 24.5 m (Figure 6). Geological research shows that the project section was originally a tidal zone between the high tide and low tide of the sea. After the construction of a flood control seawall, it was gradually reclaimed as paddy fields, shrimp ponds, or dry land. The surface is mostly distributed with typical coastal sedimentary soft soil or soft soil to saturate. Silt clay, silt, and fine sand are the main components, and the coarse sand and gravel sand are partially sandwiched between thin layers or lens bodies. Figure 7 is a picture of the excavation site of marine soft soil.

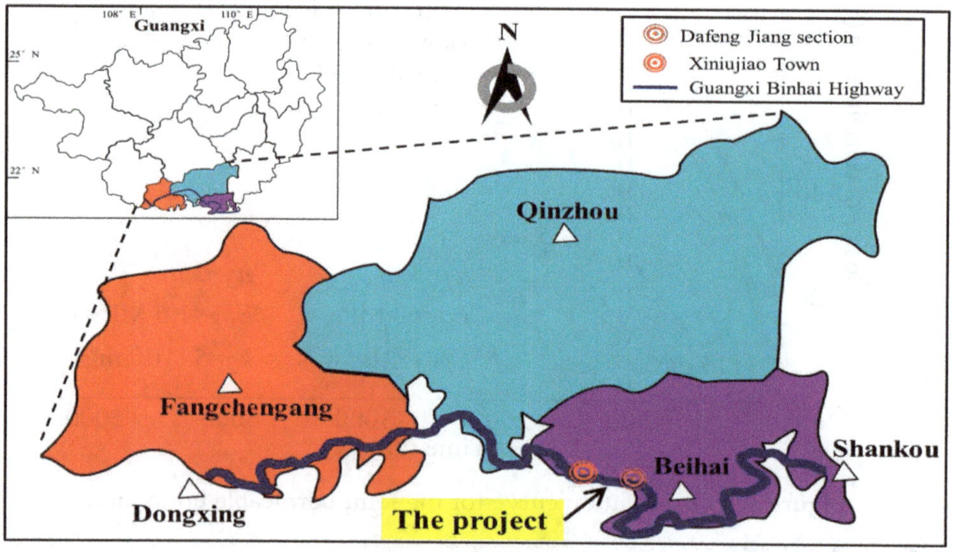

Figure 6. Relying on the engineering location map.

Figure 7. Site excavation of marine soft soil.

The consolidation settlement calculation is carried out selecting a section from K8+000 to K9+000. The water table level is 1 m below the original ground line. The upper layer of the section is a crust layer with a thickness of 2–3 m, the high liquid limit soft plastic sand-containing clay has a high compression modulus of 15–20 MPa, and the permeability coefficient is only 3×10^{-6} to 4×10^{-6} cm/s. The lower layer is a soft soil layer with a thickness of 5–10 m. The compression modulus of the soft soil containing sand is low, only 4.5–8.0 MPa, and the permeability coefficient is only 2×10^{-6} to 3.5×10^{-6} cm/s. The underlying bedrock in the soft soil layer is Indosinian granite. After discussion, it was decided to replace the crust layer with the middle-decomposed granite ($E_s = 1200$ MPa, $k = 5 \times 10^{-2}$ cm/s). Figure 8 is a schematic diagram of the soft soil foundation treated by the replacement method. According to the data provided in the geotechnical engineering investigation report of the Soft Soil Foundation Treatment project of Guangxi Binhai Highway area (provided by Guangxi Communication Design Group Co., Ltd., Guangxi, China), in combination with in-situ test data and the laboratory soil test data sampled, the data obtained are shown in Table 4. The preloading was then carried out and monitored continuously for 205 days. The upper layer's drainage boundary is neither perfectly permeable nor perfectly impervious (actually, it is a semi-permeable boundary). Due to the strong dispersion of rock and soil, the consolidation settlement of this kind of soft soil foundation can be effectively predicted by adjusting the boundary parameters b and c in the proposed method.

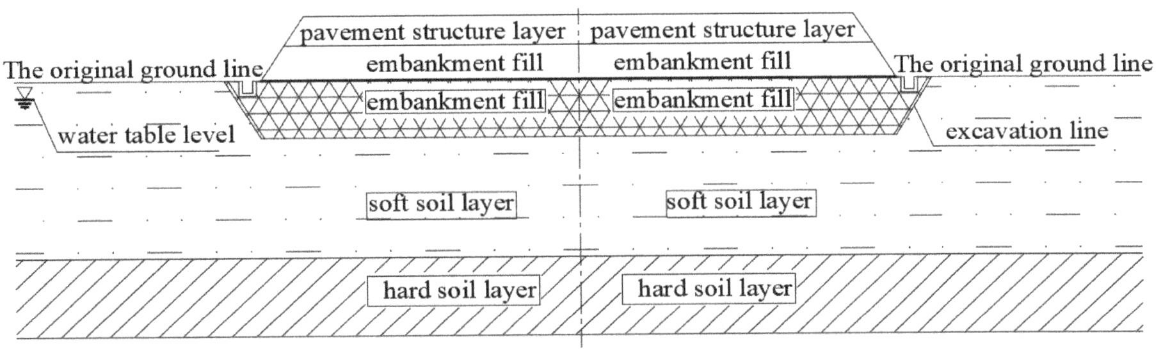

Figure 8. Schematic diagram of disposal of soft soil foundation by the replacement method.

Table 4. Soil parameters of subgrade settlement monitoring section.

Layer	Layer Thickness (m)	Bulk Unit Weight (kN/m³)	Water Content (%)	Liquid Limit (%)	Cohesion (kPa)	Friction Angle (°)	Compression Modulus (MPa)	Permeability Coefficient (10^{-8}m/s)	SPT
Crust layer	3	18.8	20.6	35.7	25.2	18.7	15	3.8	7
Soft soil layer	7	16.7	45.2	36.7	4.5	8.6	7.0	3.0	6
Filling in granite	-	26	-	-	150	45	1200	5×10^4	>50

4.2. Settlement and Consolidation Analysis

By making b equal to 100 and c tend to 0, it is better to simulate the consolidation settlement of the monitored section. The proposed method simplifies the actual load of multi-stage loading mode to the secondary load. The calculated settlement and measured settlement are plotted in Figure 9. The calculated settlement curve obtained by using the replacing soil is basically consistent with the development trend of the measured settlement, which indicates that the proposed method is suitable for analyzing soft soil foundations with a crust layer.

In the process of highway construction, an engineer is typically more concerned with the change in consolidation degree than the settlement of the foundation. The comparison between the calculated and the measured values of consolidation degree is shown in Figure 10. The calculated consolidation degree is found to be basically consistent with the measured consolidation degree. At 205 days, the calculated consolidation degree and the measured consolidation degree are approximately 55%.

The comparison of the calculated and measured results shows that the proposed method is reliable, and the calculated results are basically consistent with the measured results. Therefore, this proposed method is further used to calculate the settlement and consolidation degree of foundation before the replacement, compared with the foundation after the replacement, and study the improvement of foundation performance before and after the replacement. Here, for the convenience of analysis, the conditions of the first layer of soil are different before and after the replacement, and the other conditions are the same. The proposed method is used to calculate the difference in consolidation settlement before and after the replacement (Figure 11). The settlement of the foundation after the replacement is found to be greatly reduced, and the differential settlement is 21.5 mm after 205 days. Comparison of the change in consolidation degree of the foundation before and after the replacement (Figure 12) reveals that the consolidation degree of the foundation after the replacement is relatively high. As same as the settlement of the foundation, the difference in consolidation degree before and after the replacement becomes greater with increasing time. At 205 days, the consolidation degree differed by about 12%. As shown in Figures 11 and 12, the performance of the foundation is improved after the replacement.

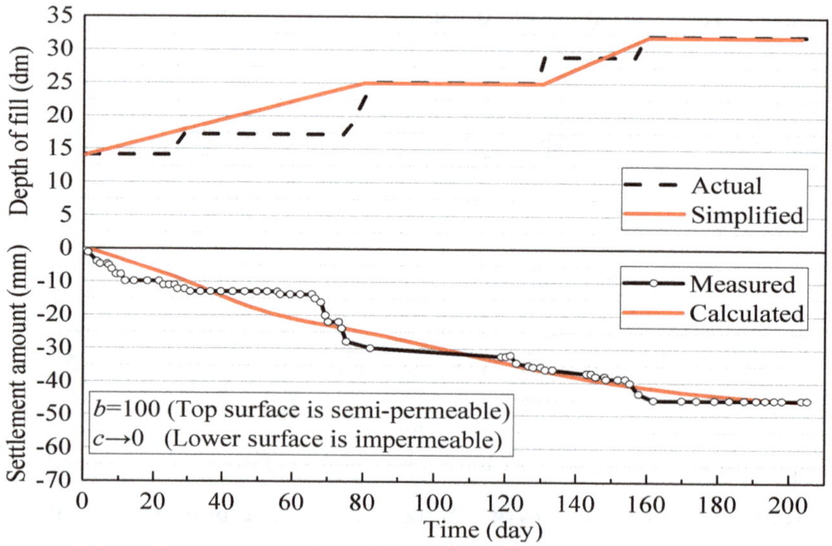

Figure 9. Comparison of measured and calculated values of settlement.

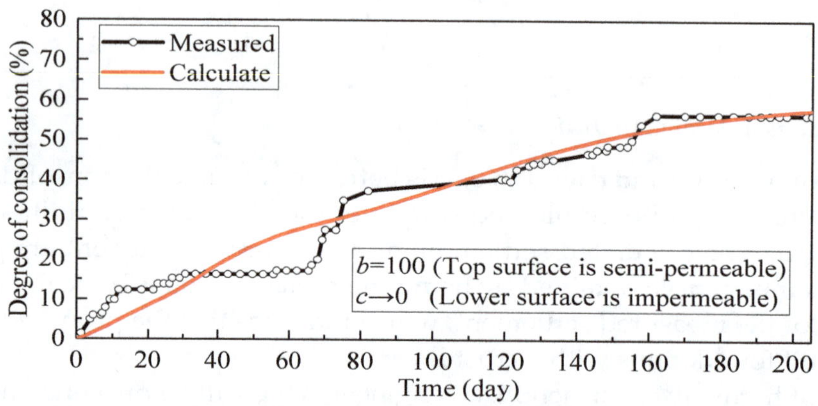

Figure 10. Comparison of the calculated and measured consolidation degrees.

As shown in Table 5, the measured final settlement after the replacement is 80 mm. The calculated settlement after the replacement is 78.7 mm, the error is about 1.6%. The settlement and consolidation degree are calculated for 205 days, the errors are about 0.7% and 2.3%, respectively. Comparison of the consolidation settlement before and after the replacement by calculated revealed that the final settlement after the replacement reduced by 68.9 mm, and the reduction rate is 87.5%. The consolidation settlement after the replacement of 205 days is reduced by 21.5 mm, and the reduction rate over 205 days is 47.3%. After the replacement, the consolidation degree increased by 12.4%, and the improvement rate increased by 21.5%, which further showed that the performance of the foundation after the replacement is greatly improved.

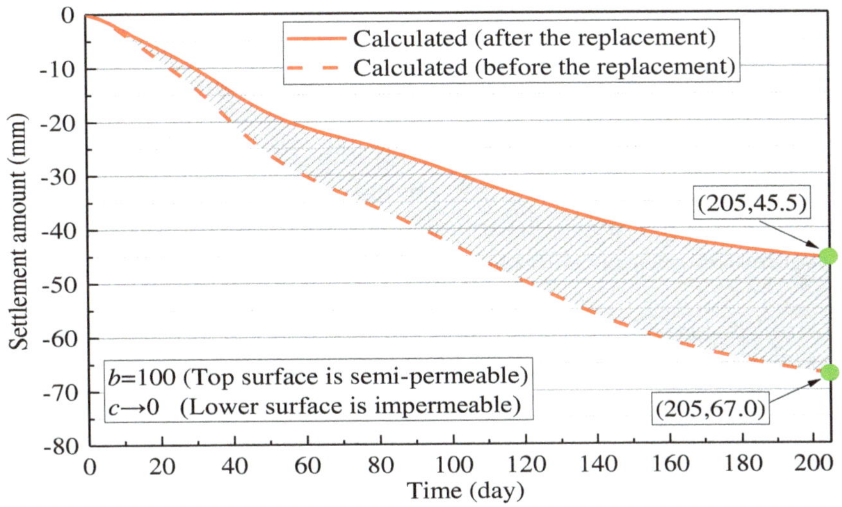

Figure 11. Comparison of the calculated settlement curve before and after the replacement.

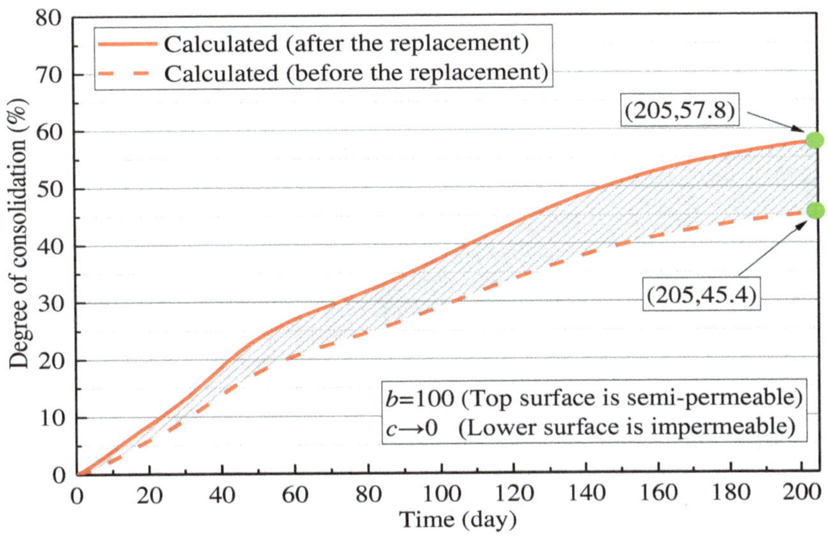

Figure 12. Comparison of the calculated consolidation curve before and after the replacement.

Table 5. Calculation parameters of the replacement foundation of marine soft soil in Guangxi.

Data Source	Final Settlement (mm)	Settlement (mm) After 205 Days	Consolidation Degree (%) After 205 Days
Measured data	80.0	45.2	56.5
Calculated data (after the replacement)	78.7	45.5	57.8
Calculated data (before the replacement)	147.6	67.0	45.4

5. Conclusions

Relying on marine soft soil foundation of complex drainage conditions, the continuous drainage boundary conditions are introduced into the double-layer soil consolidation theory model. Combined with the Laplace transform and Stehfest algorithm, the equations of the improved double-layer soil consolidation theory are deduced and solved. The degradation model of the theory is validated and analyzed by the perfectly permeable boundary conditions and the semi-permeable boundary conditions, respectively. Then, the consolidation analysis of coastal soft soil in Guangxi is carried out. We made some useful conclusions:

(1) Considering the complex drainage boundary conditions, combining the continuous drainage boundary conditions theory, the improved double-layer soil consolidation theory is derived by using the Laplace transform and Stehfest algorithm with high computational efficiency.

(2) Based on the improved double-layer soil consolidation theory, the perfectly permeable boundary case and the semi-permeable boundary case are used to verify this theory. The calculation results are basically consistent with Xie's solution, and it is concluded that the improved double-layer soil consolidation model proposed in the present paper had higher accuracy.

(3) The improved double-layer soil consolidation theory is used in an actual engineering case of marine soft soil in Guangxi. Compared with the measured data, the error of calculated data is 0.7–2.3%. The calculated results are basically consistent with the measured results, indicating that this theory is suitable for the analysis of consolidation and settlement of marine soft soil foundation with complex drainage conditions.

(4) It is difficult to quantitatively control the drainage parameters of the improved double-layer soil consolidation theory, which requires a large number of practical cases to determine the parameters. Only the double-layer foundation has been studied here, and soft foundation with more layers needs further study.

Author Contributions: D.C. and J.L. conducted the calculation analysis and wrote the paper. D.M., X.L. and L.X. motivated this study with good and workable ideas.

Acknowledgments: This work was supported by the Guangxi traffic science and technology project ("Research on standardization of soft soil foundation treatment technology in Guangxi highway" Grant No. 20140203). We thank LetPub (www.letpub.com) for its linguistic assistance during the preparation of this manuscript.

Nomenclature

Symbol	Description	Units
A_{11}, A_{12}	The undetermined constant in Equation (17)	[-]
A_{21}, A_{22}	The undetermined constant in Equation (18)	[-]
b	Related to the drainage property of topsoil	[-]
c	Related to the drainage property of subsoil	[-]
c_{v1}, c_{v2}	Consolidation coefficients of the first/second layer	[m^2/s]
E_{s1}, E_{s2}	Compression modulus of the first/second layer	[Pa]
$\overline{f}(s)$	Laplace function of $f(t)$, $\overline{f}(s) = L[f(t)] = \int_0^\infty f(t)e^{-st}dt$	-
$f(t)$	Time domain function of $\overline{f}(s)$	-
$f(T)$	An approximation of the inverse Laplace transform of $\overline{f}(s)$	-
h_1, h_2	Thickness of the first/second layer	[m]
i	The process variable of accumulation equation	-
k	The process variable of accumulation equation	-
k_1, k_2	Permeability coefficients of the first/second layer	[m/s]
N	Positive even number	[-]
$q(t)$	Arbitrary loading function	[Pa]
q_1, q_2	Load increments at the first/second levels	[Pa]
S_t	Total consolidation settlement at time t of foundation	[m]
S_∞	Final total consolidation settlement of foundation	[m]
r_1, r_2	Intermediate variable, no real meaning	[/m]
s	Complex variable involved in the Laplace transform	[/s]
t	Time	[s]
t_1	Loading time of first class load	[s]
t_2	Starting time of the second stage load	[s]

Symbol	Description	Units
t_3	Completion time of the second stage load	[s]
$u(h,t)$	Excess pore water pressure	[Pa]
U	Average degree of consolidation of foundation	[-]
$u_1(h,t)$, $u_2(h,t)$	Excess pore water pressure of the first/second layer	[Pa]
$u_t(h,t)$	Derivative of $u(h,t)$ with respect to t	-
V_i	Dependent on N only	[-]
z	Foundation depth, from the ground up	[m]
γ_w	Water unit weight, take 10^4 N/m^3	[N/m^3]
$\sigma_1'(h,t)$, $\sigma_2'(h,t)$	Effective stress of the first/second layer	[Pa]
$\sigma_{1,2}'(h,t)$	Effective stress at any point in the foundation at any time	[Pa]

Appendix A

$$\begin{aligned}\bar{q}(s)-\bar{q}(s+b) &= \int_o^{+\infty} q(t)\left(1-e^{-bt}\right)e^{-st}dt = \int_0^{t_1}\tfrac{t}{t_1}q_1\left(1-e^{-bt}\right)e^{-st}dt + \int_{t_1}^{t_2} q_1\left(1-e^{-bt}\right)e^{-st}dt \\ &\quad + \int_{t_2}^{t_3}\left(q_1+\tfrac{t-t_2}{t_3-t_2}q_2\right)\left(1-e^{-bt}\right)e^{-st}dt + \int_{t_3}^{\infty}(q_1+q_2)\left(1-e^{-bt}\right)e^{-st}dt \\ &= \tfrac{q_1}{t_1}\left(\tfrac{1-e^{-st_1}}{s^2}+\tfrac{e^{-(s+b)t_1}-1}{(s+b)^2}\right)+\tfrac{q_2}{(t_3-t_2)}\left(\tfrac{e^{-st_2}-e^{-st_3}}{s^2}+\tfrac{e^{-(s+b)t_3}-e^{-(s+b)t_2}}{(s+b)^2}\right)\end{aligned}$$

$$\begin{aligned}\bar{q}(s)-\bar{q}(s+c) &= \int_o^{+\infty} q(t)\left(1-e^{-ct}\right)e^{-st}dt = \int_0^{t_1}\tfrac{t}{t_1}q_1\left(1-e^{-ct}\right)e^{-st}dt + \int_{t_1}^{t_2} q_1\left(1-e^{-ct}\right)e^{-st}dt \\ &\quad + \int_{t_2}^{t_3}\left(q_1+\tfrac{t-t_2}{t_3-t_2}q_2\right)\left(1-e^{-ct}\right)e^{-st}dt + \int_{t_3}^{\infty}(q_1+q_2)\left(1-e^{-ct}\right)e^{-st}dt \\ &= \tfrac{q_1}{t_1}\left(\tfrac{1-e^{-st_1}}{s^2}+\tfrac{e^{-(s+c)t_1}-1}{(s+c)^2}\right)+\tfrac{q_2}{(t_3-t_2)}\left(\tfrac{e^{-st_2}-e^{-st_3}}{s^2}+\tfrac{e^{-(s+c)t_3}-e^{-(s+c)t_2}}{(s+c)^2}\right)\end{aligned}$$

$$A_{11} = -\frac{k_1P_1r_1\left(e^{-h_1r_1-h_2r_2}-e^{h_2r_2-h_1r_1}\right)+k_2P_1r_2\left(e^{h_2r_2-h_1r_1}+e^{-h_1r_1-h_2r_2}\right)-2k_2P_2r_2}{k_1r_1\left(e^{h_1r_1+h_2r_2}+e^{h_2r_2-h_1r_1}-e^{h_1r_1-h_2r_2}-e^{-h_1r_1-h_2r_2}\right)+k_2r_2\left(e^{h_1r_1+h_2r_2}+e^{h_1r_1-h_2r_2}-e^{h_2r_2-h_1r_1}-e^{-h_1r_1-h_2r_2}\right)}$$

$$A_{12} = \frac{k_1P_1r_1\left(e^{h_1r_1+h_2r_2}-e^{h_1r_1-h_2r_2}\right)+k_2P_1r_2\left(e^{h_1r_1+h_2r_2}+e^{h_1r_1-h_2r_2}\right)-2k_2P_2r_2}{k_1r_1\left(e^{h_1r_1+h_2r_2}+e^{h_2r_2-h_1r_1}-e^{h_1r_1-h_2r_2}-e^{-h_1r_1-h_2r_2}\right)+k_2r_2\left(e^{h_1r_1+h_2r_2}+e^{h_1r_1-h_2r_2}-e^{h_2r_2-h_1r_1}-e^{-h_1r_1-h_2r_2}\right)}$$

$$A_{21} = \frac{k_1P_2r_1\left(e^{-h_1(r_1+r_2)}+e^{h_1(r_1-r_2)}\right)+k_2P_2r_2\left(e^{h_1(r_1-r_2)}-e^{-h_1(r_1+r_2)}\right)-2k_1P_1r_1e^{-(h_1+h_2)r_2}}{k_1r_1\left(e^{h_1r_1+h_2r_2}+e^{h_2r_2-h_1r_1}-e^{h_1r_1-h_2r_2}-e^{-h_1r_1-h_2r_2}\right)+k_2r_2\left(e^{h_1r_1+h_2r_2}+e^{h_1r_1-h_2r_2}-e^{h_2r_2-h_1r_1}-e^{-h_1r_1-h_2r_2}\right)}$$

$$A_{22} = \frac{k_1P_2r_1\left(e^{h_1(r_1+r_2)}+e^{h_1(r_2-r_1)}\right)+k_2P_2r_2\left(e^{h_1(r_2-r_1)}-e^{-h_1(r_1+r_2)}\right)-2k_1P_1r_1e^{(h_1+h_2)r_2}}{k_1r_1\left(e^{h_1r_1+h_2r_2}+e^{h_2r_2-h_1r_1}-e^{h_1r_1-h_2r_2}-e^{-h_1r_1-h_2r_2}\right)+k_2r_2\left(e^{h_1r_1+h_2r_2}+e^{h_1r_1-h_2r_2}-e^{h_2r_2-h_1r_1}-e^{-h_1r_1-h_2r_2}\right)}$$

where $P_1 = \bar{q}(s)-\bar{q}(s+b)$, $P_2 = \bar{q}(s)-\bar{q}(s+c)$.

References

1. Boak, E.H.; Turner, I.L. Shoreline Definition and Detection: A Review. *J. Coast. Res.* **2005**, *21*, 688–703. [CrossRef]
2. Penner, E. A Study of Sensitivity in Leda Clay. *Can. J. Earth Sci.* **1965**, *2*, 425–441. [CrossRef]
3. Gillott, J.E. *Clay, Engineering Geology*; Springer: New York, NY, USA, 1984.
4. Sangrey, D.A. Naturally Cemented Sensitive Soils. *Géotechnique* **1972**, *22*, 139–152. [CrossRef]
5. Gasparre, A.; Nishimura, S.; Jardine, R.J.; Coop, M.R.; Minh, N.A. The Stiffness of Natural London Clay. *Géotechnique* **2007**, *57*, 33–47. [CrossRef]
6. Dìaz-Rodrìguez, J.A.; Leroueil, S.; Alemàn, J.D. Yielding of Mexico City Clay and Other Natural Clays. *J. Geotech. Eng.* **1992**, *118*, 981–995. [CrossRef]
7. Aoki, S.; Oinuma, K. The Distribution of Clay Minerals in Recent Sediments of the Okhotsk Sea. *Deep Sea Res. Oceanogr. Abstr.* **1974**, *21*, 299–310. [CrossRef]
8. Kim, Y.T.; Nguyen, B.P.; Yun, D.H. Effect of Artesian Pressure on Consolidation Behavior of Drainage-Installed Marine Clay Deposit. *ASCE's J. Mater. Civ. Eng.* **2018**, *30*, 04018156. [CrossRef]
9. Yin, L.H.; Wang, X.M.; Zhang, L.J. Probabilistical Distribution Statistical Analysis of Tianjin Soft Soil Indices. *Rock Soil Mech.* **2010**, *31*, 462–469. (In Chinese)
10. Li, X.G.; Xu, R.Q.; Wang, X.C.; Rong, X.N. Assessment of Engineering Properties for Marine and Lacustrine Soft Soil in Hangzhou. *J. Zhejiang Univ. (Eng. Sci.)* **2013**. (In Chinese)

11. Liang, S.J.; Zhu, H.D. Statistical Analysis on Soil Indices of Lagoon Facies Soft Soil Layer in Hangzhou. *Subgrade Eng.* **2012**. (In Chinese)
12. Zheng, Y.Y.; Zhu, J.F.; Liu, G.B.; Jia, B. Probability and Correlation between Physical and Mechanical Parameters of Soft Clays in Ningbo Rail Transit Engineering. *China Sciencepaper* **2013**, *8*, 367–373. (In Chinese)
13. Chen, X.P.; Huang, G.Y.; Liang, Z.S. Study on Soft Soil Properties of the Pearl River Delta. *Chin. J. Rock Mech. Eng.* **2003**, *22*, 137–141. (In Chinese)
14. Lei, D.; Jiang, Y.J.; Chen, D.C.; Zhang, X.Y. Mechanical Properties and Correlation Analysis of Soft Soil in Guangzhou. *Railway Eng.* **2011**, *10*, 75–78. (In Chinese)
15. Zhu, W.D. Engineering Properties and Comparison Analysis of Wen Zhou Soft Clay and Tai Zhou Soft Clays. *Zhejiang Univ.* **2003**. (In Chinese)
16. Luo, Y.D.; Yang, G.H.; Zhang, Y.C.; Liu, P. Engineering Characteristics and Its Statistical Analysis of Soft Marine Clay on Shenzhen West Coast. *J. Civ. Eng. Manag.* **2012**, *29*, 79–86. (In Chinese)
17. Terzaghi, K. *Theoretical Soil Mechanics*; John Wiley and Sons: New York, NY, USA, 1943.
18. Xie, K.H.; Xie, X.Y.; Gao, X. Theory of One Dimensional Consolidation of Two Layered with Partially Drained Boundaries. *Comput. Geotech.* **1999**, *24*, 265–278. [CrossRef]
19. Zhou, W.H.; Zhao, L.S. One-Dimensional Consolidation of Unsaturated Soil Subjected to Time-Dependent Loading with Various Initial and Boundary Conditions. *Int. J. Geomech.* **2014**, *14*, 291–301. [CrossRef]
20. Gray, H. Simultaneous Consolidation of Contiguous Layers of Unlike Compressible Soils. *Am. Soc. Civ. Eng.* **1945**, *110*, 1327–1356.
21. Schiffman, R.L.; Stein, J.R. One-Dimensional Consolidation of Layered Systems. *JSM FD ASCE* **1970**, *96*, 1499–1504.
22. Huang, W.X. Application of Consolidation Theory to Earth Dams Built by Sluicing-Siltation Method. *J. Hydraul. Eng.* **1982**, *9*, 13–23. (In Chinese)
23. Xie, K.H. One Dimensional Consolidation Analysis of Layered Soils with Impeded Boundaries. *J. Zhejiang Univ. (Eng. Sci.)* **1996**. (In Chinese)
24. Hu, L.H.; Xie, K.H. On One Dimensional Consolidation Behavior of Layered Soil with Partial Drainage Boundaries. *Bull. Sci. Technol.* **2005**. (In Chinese)
25. Feng, Y.; Xu, C.J.; Cai, Y.Q. One-Dimensional Consolidation Analysis of Saturated Gibson Soils with Semi-Pervious Boundaries. *J. Zhejiang Univ. (Eng. Sci.)* **2003**, *37*, 646–651. (In Chinese)
26. Lin, Z.Y.; Pan, W.X.; Xu, C.J. One-Dimensional Consolidation Analysis of Saturated Layered Gibson Soils with Semi-Pervious Boundaries. *J. Zhejiang Univ. (Eng. Sci.)* **2003**, *37*, 570–575. (In Chinese)
27. Wang, K.H.; Xie, K.H.; Zeng, G.X. A Study on 1D Consolidation of Soils Exhibiting Rheological Characteristics with Impeded Boundaries. *Chin. J. Geotech. Eng.* **1998**. (In Chinese)
28. Wang, L.; Li, L.Z.; Xu, Y.F.; Xia, X.H.; Sun, D.A. Analysis of One-Dimensional Consolidation of Fractional Viscoelastic Saturated Soils with Semi-Permeable Boundary. *Rock Soil Mech.* **2018**, *39*, 234–240. (In Chinese)
29. Cai, Y.Q.; Liang, X.; Zheng, Z.F.; Pan, X.D. One-Dimensional Consolidation of Viscoelastic Soils Layer with Semi-Permeable Boundaries under Cyclic Loadings. *China Civ. Eng. J.* **2003**, *36*, 86–90. (In Chinese)
30. Li, X.B.; Xie, K.H.; Wang, K.H.; Zhuang, Y.C. Analytical Solution of 1-D Visco-Elastic Consolidation of Soils with Impeded Boundaries under Cyclic Loadings. *Eng. Mech.* **2004**, *21*, 103–108. (In Chinese)
31. Wang, W.G.; Xia, S.W. On Analytical Solutions to Viscoelastic One-Dimensional Consolidation of Impeded Boundaries under Cyclic Loading. *Shanxi Archit.* **2011**, *37*, 67–69. (In Chinese)
32. Zheng, Z.F.; Cai, Y.Q.; Xu, C.J.; Zhan, H. One-Dimensional Consolidation of Layered and Visco-Elastic Ground under Arbitrary Loading with Impeded Boundaries. *J. Zhejiang Univ. (Eng. Sci.)* **2005**, *39*, 1234–1237. (In Chinese)
33. Cheng, Z.H.; Chen, Y.M.; Ling, D.S.; Lv, F.R. Analytical Solutions for Consolidation Problems with Impeded Boundaries. *Acta Mech. Solida Sin.* **2004**, *25*, 93–97. (In Chinese)
34. Liu, J.C.; Lei, G.G.; Wang, Y.X. One-Dimensional Consolidation of Soft Ground Considering non-Darcy Flows. *Chin. J. Geotech. Eng.* **2011**, *33*, 1117–1122. (In Chinese)
35. Mei, G.X.; Xia, J.; Mei, L. Terzaghi's One-Dimensional Consolidation Equation and its Solution based on Asymmetric Continuous Drainage Boundary. *Chin. J. Geotech. Eng.* **2011**, *1*, 28–31. (In Chinese)
36. Zong, M.F.; Wu, W.B.; Mei, G.X.; Liang, R.Z. An Analytical Solution for One-Dimensional Nonlinear Consolidation of Soils with Continuous Drainage Boundary. *Chin. J. Rock Mech. Eng.* **2018**, *37*, 2829–2838. (In Chinese)

37. Zheng, Y.; Mei, G.X.; Mei, L. Generalized Continuous Drainage Boundary Applied in One-Dimensional Consolidation Theory. *J. Nanjing Univ. Technol. (Nat. Sci. Ed.)* **2010**, *32*, 54–58. (In Chinese)
38. Cai, F.; He, L.J.; Zhou, X.P.; Mei, G.X. Finite Element Analysis of One-Dimensional Consolidation Problem with Continuous Drainage Boundaries in Layered Ground. *J. Cent. South Univ. (Sci. Technol.)* **2013**, *44*, 315–323. (In Chinese)
39. Wang, L.; Sun, D.A.; Qin, A.F. Semi-Analytical Solution to One-Dimensional Consolidation for Unsaturated Soils with Exponentially Time-Growing Drainage Boundary Conditions. *Int. J. Geomech.* **2018**, *18*, 04017144. [CrossRef]
40. Feng, J.X.; Ni, P.P.; Mei, G.X. One-Dimensional Self-Weight Consolidation with Continuous Drainage Boundary Conditions: Solution and Application to Clay-Drain Reclamation. *Int. J. Numer. Anal. Methods Geomech.* **2019**, 1–19. [CrossRef]
41. Sun, M.; Zong, M.F.; Ma, S.J.; Wu, W.B.; Liang, R.Z. Analytical Solution for One-Dimensional Consolidation of Soil with Exponentially Time-Growing Drainage Boundary under a Ramp Load. *Math. Probl. Eng.* **2018**. [CrossRef]
42. Zhang, Y.; Wu, W.B.; Mei, G.X.; Duan, L.C. Three-Dimensional Consolidation Theory of Vertical Drain Based on Continuous Drainage Boundary. *J. Civ. Eng. Manag.* **2019**, *25*, 145–155. [CrossRef]
43. Van Everdingen, A.F.; Hurst, W. The Application of the Laplace Transformation to Flow Problems in Reservoirs. *J. Pet. Technol.* **1949**, *186*, 305–324. [CrossRef]
44. Dubner, H.; Abate, J. Numerical Inversion of Laplace Transforms by Relating Them to the Finite Fourier Cosine Transform. *J. ACM* **1968**, *15*, 115–123. [CrossRef]
45. Durbin, F. Numerical Inversion of Laplace Transforms: An Efficient Improvement to Dubner and Abate's Method. *Comput. J.* **1974**, *17*, 371–376. [CrossRef]
46. Stehfest, H. Algorithm 368: Numerical Inversion of Laplace Transforms [D5]. *Commun. ACM* **1970**, *13*, 47–49. [CrossRef]
47. Stehfest, H. Remark on Algorithm 368: Numerical Inversion of Laplace Transforms. *Commun. ACM* **1970**, *13*, 624. [CrossRef]
48. Crump, K.S. Numerical Inversion of Laplace Transforms Using a Fourier Series Approximation. *J. ACM* **1976**, *23*, 89–96. [CrossRef]
49. Duffy, D.G. On the Numerical Inversion of Laplace Transforms: Comparison of Three New Methods on Characteristic Problems from Applications. *ACM Trans. Math. Softw.* **1993**, *19*, 333–359. [CrossRef]
50. Wang, X.H.; Miao, L.C.; Gao, J.K.; Zhao, M. Solution of One-Dimensional Consolidation for Double-Layered Ground by Laplace Transform. *Rock Soil Mech.* **2005**, *26*, 833–836. (In Chinese)
51. Xie, K.H. One Dimensional Consolidation Theory of Double-Layered Ground. *Chin. J. Geotech. Eng.* **1994**, *5*, 24–35. (In Chinese)

7

The Hydrodynamic Dispersion Characteristics of Coral Sands

Xiang Cui [1,2], Changqi Zhu [1,*], Mingjian Hu [1], Xinzhi Wang [1] and Haifeng Liu [1]

1 State Key Laboratory of Geomechanics and Geotechnical Engineering, Institute of Rock and Soil Mechanics, Chinese Academy of Sciences, Wuhan 430071, China
2 University of Chinese Academy of Sciences, Beijing 100049, China
* Correspondence: cqzhu@whrsm.ac.cn

Abstract: Dispersion characteristics are important factors affecting groundwater solute transport in porous media. In marine environments, solute dispersion leads to the formation of freshwater aquifers under islands. In this study, a series of model tests were designed to explore the relationship between the dispersion characteristics of solute in calcareous sands and the particle size, degree of compactness, and gradation of porous media, with a discussion of the types of dispersion mechanisms in coral sands. It was found that the particle size of coral sands was an important parameter affecting the dispersion coefficient, with the dispersion coefficient increasing with particle size. Gradation was also an important factor affecting the dispersion coefficient of coral sands, with the dispersion coefficient increasing with increasing d_{10}. The dispersion coefficient of coral sands decreased approximately linearly with increasing compactness. The rate of decrease was −0.7244 for single-grained coral sands of particle size 0.25–0.5 mm. When the solute concentrations and particle sizes increased, the limiting concentration gradients at equilibrium decreased. In this study, based on the relative weights of molecular diffusion versus mechanical dispersion under different flow velocity conditions, the dispersion mechanisms were classified into five types, and for each type, a corresponding flow velocity limit was derived.

Keywords: coral sands; porous media; model test; dispersion; mechanical dispersion; molecular diffusion

---

1. Introduction

Coral sands are a kind of biogenic soil in marine environments, originating from the fracturing and sedimentation of coral skeletons by wind and waves [1]. The primary features of coral sands include a high content of calcium carbonate—often more than 90%—and a relatively short sediment transport distance, which results in particles frequently retaining the inner pore structure of coral bones, and are characterized by irregular shapes, high angularity, and fragility [2–4]. The sediments formed by coral sands have a high void ratio and display hydrogeological characteristics different from those of terrigenous sediments. Islands in the South China Sea are all composed of coral sediments, with these sediments being the only carriers in the formation of fresh groundwater aquifers in the South China Sea islands and reefs. Given that the dispersion coefficient of coral sands is a key factor affecting the conservation of fresh groundwater, uncovering the dispersion pattern of groundwater solute in coral sands will provide the basic parameters and theoretical basis for the numerical simulation of the formation and evolution of fresh groundwater aquifers in the South China Sea islands and reefs, as well as the conservation and utilization of these aquifers.

Solute dispersion in hydrated porous media has been extensively studied. The earliest work was performed by Taylor, who used a capillary tube model to investigate this topic and proposed a method for calculating the longitudinal dispersion coefficient of porous media [5]. Taylor considered only convection, and therefore derived the molecular diffusion coefficient by measuring the longitudinal dispersion coefficient [5]. Klotzd et al. explored the relationship between the longitudinal dispersion coefficient and average pore flow velocity, the fluid viscosity coefficient, and the characteristic parameters of soil media by conducting a large number of field and laboratory experiments [6]. Gupta proposed a solute transport mechanism for unsaturated porous media, suggesting that two types of pores exist in unsaturated soils, namely, "backbones" and "dead ends," with solute mainly moving by convection in the former and by diffusion in the latter [7]. De Arcangelis et al. studied dispersion calculation methods for network models of porous media, deduced the precise law of tracer motion under the combined action of molecular diffusion and convection, and introduced an effective probability propagation algorithm, which permitted an exact calculation of the distribution of the first-passage-time of the tracer as it flowed through the medium [8]. Sahimi studied hydrodynamic dispersion in two types of heterogeneous porous media, one in which a fraction of the pores did not allow material transport to take place, and the other in which the permeabilities of various regions of the pore space were fractally distributed [9]. Lowe calculated the dispersion coefficient of tracer particles in the fluid of a porous medium randomly filled with spheres, finding that at high Peclet numbers, the tracer motion was mainly determined by convection, and the dispersion process was abnormal with a divergent dispersion coefficient [10]. Zhang proposed a calculation method for hydrodynamic dispersion parameters of adsorptive solutes, deriving formulas for calculating hydrodynamic dispersion coefficients of saturated and unsaturated soils [11]. LI improved the flexible-wall permeameter to make it suitable for finding the dispersion parameters of low-permeability soils, namely, determining the dispersion coefficients through numerical inversion of the breakthrough curve [12]. Shao conducted a one-dimensional dispersion test with silt loam, calculated the hydrodynamic dispersion coefficient of unsaturated silty loam using the soil water and salt dynamics measured in a vertical soil column solute transport experiment, and established the relationship between hydrodynamic dispersion coefficients and pore flow velocities for this type of soil [13]. Jensen adopted the nonlinear least-squares optimization code CXTFIT developed by Parker and van Genuchten to perform curve fitting, thereby obtaining parameters for different forms of the convection–dispersion equation (CDE) [14].

The underground freshwater of islands are the basis for the normal operation of their ecosystems. The generation of underground fresh water is closely related to island size and stratum characteristics. Restricted by the scale of research objects and experimental conditions, the research in this aspect is mostly conducted by means of numerical simulation. The permeability coefficient, dispersion coefficient, and specific yield are three important parameters that must be assigned to strata in numerical simulations. All three parameters can be obtained using field tests. However, the data obtained represent only the island studied, and are not universally representative. In addition, the artificial islands in the South China Sea are not open to the public, making it difficult to conduct field experiments there. In our study, with coral sands as the research subject, which have different granular morphologies and sediment characteristics than terrigenous sediments, laboratory model tests were employed to uncover the dispersion mechanisms of groundwater solute in coral sands, as well as the main factors influencing these mechanisms. According to the results of our study, the empirical value of dispersion coefficients can be provided for calcareous soils with common gradation and compactness. In this way, a more stratigraphic collocation design can be considered in the numerical simulation, so as to find the optimal stratigraphic design scheme that can promote the formation of underground fresh water.

In addition, the study of hydrodynamic dispersion presented in this paper can also be applied to many other aspects. In the simulation and prediction of groundwater pollution, the dispersion coefficient is an important parameter needed for simulation, which provides a quantitative basis for groundwater resource management and groundwater pollution reconstruction. Regarding the intrusion of seawater into coastal aquifers, the study of hydrodynamic dispersion is helpful in studying the migration of the transition zone of brackish water. In terms of water and salt transport in the vade-zone, the study of hydrodynamic dispersion is helpful in solving the problem of the effect of fertilizers and pesticides on underground water quality in islands. Finally, in terms of sewage treatment, hydrodynamic dispersion is helpful in solving the problem of the impact of sewage discharge from living and production on the underground freshwater quality in islands.

2. Test Scheme

The coral sands used for testing were taken from a natural reef in the South China Sea, whose gradation characteristics are shown in Figure 1. The coefficient of curvature C_c was 2.12, which is within the range of 1–3. The coefficient of uniformity C_u was 45.45, which is much greater than 5, indicating that the reef consisted of a type of coral sand with good gradation. The coral sand was screened to obtain six kinds of coral sand with a single particle size, as shown in Figure 2. More specifically, the test scheme was composed of three steps: (1) using a self-designed one-dimensional dispersion test device, solute dispersion tests were conducted on a total of 14 groups of coral sands of a single particle size and dry density under various conditions to explore how the dispersion coefficient varies as a function of particle size and dry density; (2) using a custom-designed pore tortuosity test device, where pore tortuosity tests were conducted on a total of five groups of coral sands of a single particle size under various conditions to explore how the pore tortuosity varies as a function of particle size; and (3) using a custom-designed molecular diffusion and mechanical dispersion test device, where for a total of six test groups, molecular diffusion tests were conducted with different concentrations of injected solutions under the condition of the pore fluid having a flow velocity. In addition, molecular diffusion and mechanical dispersion tests were conducted with pore flow velocity increasing in a stepwise manner in order to explore how the weight of molecular diffusion versus mechanical diffusion varies with a stepwise increase in pore flow velocity. The test scheme is shown in Table 1.

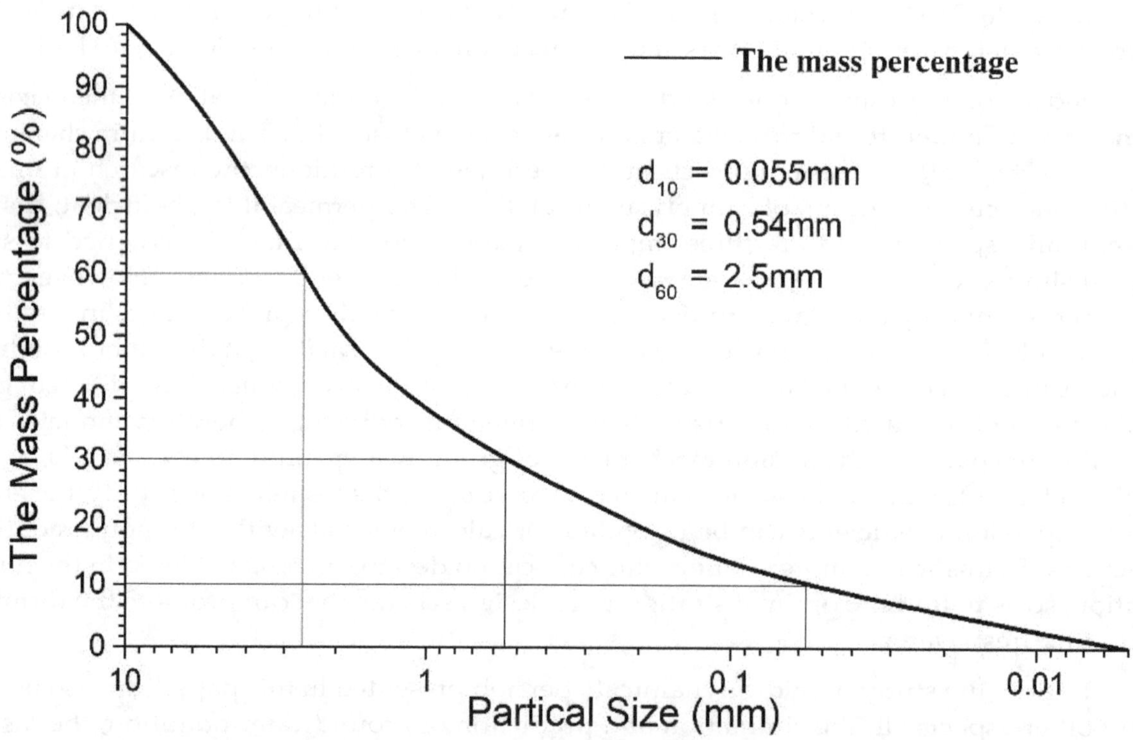

Figure 1. Gradation curve of coral sands used in the tests.

Figure 2. Calcareous sand of various grain sizes.

Table 1. Test scheme.

Test 1: One-Dimensional Dispersion Test			
Group	**Gradation**	**Dry Density (g/cm^3)**	**C (NaCl) (g/L)**
1	0–0.1 mm (100%)	1.3	20
2	0.1–0.25 mm (100%)	1.3	20
3	0.25–0.5 mm (100%)	1.3	20
4	0.5–1 mm (100%)	1.3	20
5	1–2 mm (100%)	1.3	20
6	2–5 mm (100%)	1.3	20
7	0.25–0.5 mm (100%)	1.2	20
8	0.25–0.5 mm (100%)	1.3	20
9	0.25–0.5 mm (100%)	1.4	20
10	No.1 (d_{60} = 2.5, d_{30} = 0.52, d_{10} = 0.05)	1.3	20
11	No.2 (d_{60} = 2.5, d_{30} = 0.52, d_{10} = 0.15)	1.3	20
12	No.3 (d_{60} = 2.5, d_{30} = 0.6, d_{10} = 0.31)	1.3	20
13	No.4 (d_{60} = 2.5, d_{30} = 0.9, d_{10} = 0.6)	1.3	20
14	No.5 (d_{60} = 2.55, d_{30} = 1.5, d_{10} = 1.2)	1.3	20

Test 2: Pore Tortuosity Test			
Group	**Gradation**	**Dry Density (g/cm^3)**	**C (NaCl) (g/L)**
1	0–0.1 mm (100%)	1.3	20
2	0.1–0.25 mm (100)	1.3	20
3	0.25–0.5 mm (100%)	1.3	20
4	0.5–1 mm (100%)	1.3	20
5	1–2 mm (100%)	1.3	20

Test 3: Molecular Diffusion and Mechanical Dispersion Tests				
Group	**Gradation**	**Flow Velocity (cm/s)**	**Dry Density (g/cm^3)**	**C (NaCl) (g/L)**
1	0.25–0.5 mm (100%)	0	1.3	20
2	0.25–0.5 mm (100%)	0	1.3	60
3	0.25–0.5 mm (100%)	1.36×10^{-4}	1.3	60
4	0.25–0.5 mm (100%)	6.61×10^{-4}	1.3	60
5	0.25–0.5 mm (100%)	1.60×10^{-3}	1.3	60
6	0.25–0.5 mm (100%)	6.16×10^{-3}	1.3	60

3. Test Method

3.1. One-Dimensional Dispersion Test

The hydrodynamic dispersion coefficient is a tensor related to the average pore velocity and the characteristics of the porous medium. Studies show that the dispersion is directional even in an isotropic medium and is more complex when the medium is anisotropic [15]. Given this information, only the one-dimensional dispersion characteristics of a solute in coral sands were explored in this study. The test was conducted with a custom-designed one-dimensional dispersion test device, as shown in Figure 3. The complete test device consisted of four parts (denoted by numbers in the figure): dispersion columns (01), a freshwater supply tank (02), a tracer supply tank (03), and a data acquisition system (04). The dispersion columns were composed of several organic glass columns; each column was 8 cm in inner diameter and 50 cm in height. The preparation and tests of the porous medium samples were performed in the dispersion columns, with a 5-cm-thick buffer layer of glass beads placed separately at the top and bottom of the samples. The sensors used in the data acquisition system were CS655 multi-parameter sensors (Campbell Scientific, State of California, America), which can simultaneously measure volumetric moisture content, temperature, conductivity, dielectric constant, signal propagation time, and signal attenuation. The tracer was a 20 g/L NaCl solution and the samples were single-grained coral sands. The test was conducted by continuously injecting the tracer and collecting the data at a fixed location. The samples in the dispersion columns were first saturated with fresh water, then the valve of the freshwater supply tank was shut, followed by opening the valve of the tracer supply tank and starting data acquisition in a synchronous manner. The relative concentration of NaCl was calculated according to $C = (C_{cj} - C_0)/(C_{max} - C_0)$, with C_{cj} denoting the NaCl concentration at time j. C_0 was the NaCl concentration at the initial time, and C_{max} was the maximum concentration (final steady concentration) in the test.

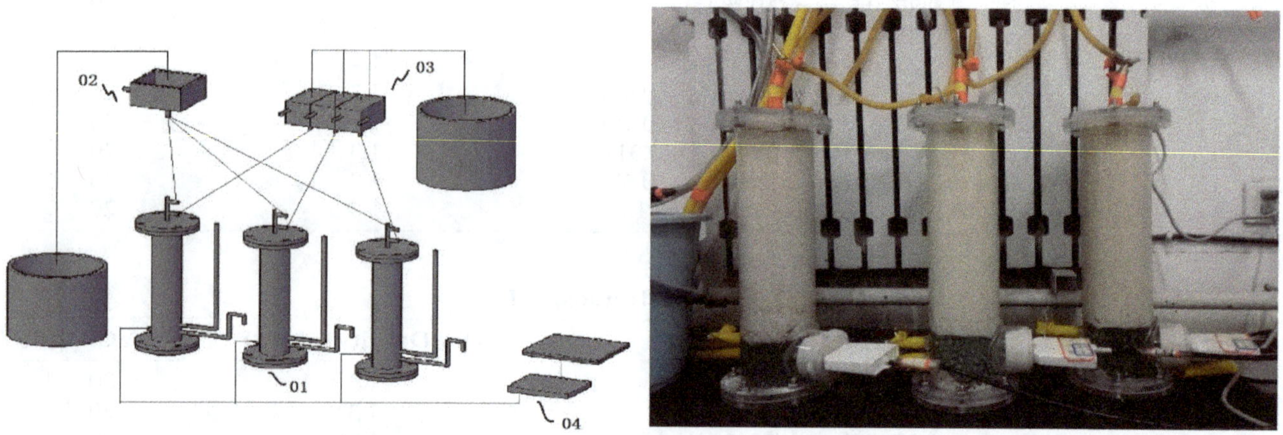

(a) Schematic Diagram (b) Real Product

Figure 3. One-dimensional dispersion device (01: one-dimensional soil column, 02: freshwater supply tank, 03: tracer supply tank, 04: data acquisition system).

The concentration distribution was derived by solving the one-dimensional steady flow problem in a semi-infinite soil column with a constant concentration at one end. With t (s) as the time since the tracer has been turned on, x (cm) as the distance of the multi-parameter sensor from the tracer injection port, $E(x, t)$ (dS/m) as the conductivity measured at time t, E_0 (dS/m) as the conductivity of the tracer, and u (cm/s) as the average flow velocity of the pore fluid in the soil column, the longitudinal dispersion coefficient D_L (cm^2/s) is calculated as given below, and the coordinate system is shown in Figure 4.

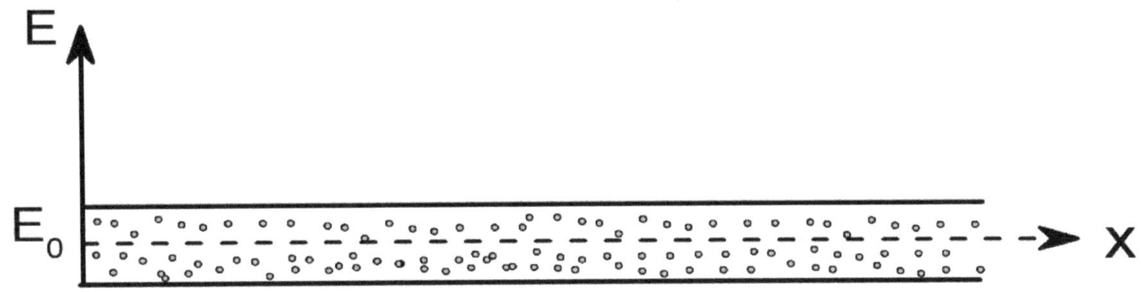

Figure 4. Boundary problem with a constant concentration at one end of the one-dimensional soil column.

The above problem is converted into a mathematical equation as follows:

$$\begin{cases} \frac{\partial E}{\partial t} = D_L \frac{\partial^2 E}{\partial x^2} - u\frac{\partial E}{\partial x} \\ E(x,0) = 0, x \geq 0 \\ E(0,t) = E_0, t > 0 \\ E(\infty,t) = 0, t > 0 \end{cases} \tag{1}$$

The Laplace transform of Equation (1) with respect to t gives

$$\begin{cases} D_L \frac{d^2\overline{E}}{dx^2} - u\frac{dp\overline{E}}{dx} = p\overline{E} \\ \overline{E}(0,p) = \frac{E_0}{p}, \text{t} > 0 \\ \overline{E}(\infty,p) = 0, \text{t} > 0 \end{cases} \tag{2}$$

where $\overline{E}(x,p)$ is a function of x, and p is a parameter. The original problem is converted into the following definite solution problem as an ordinary differential equation:

$$D_L \frac{d^2\overline{E}}{dx^2} - u\frac{dp\overline{E}}{dx} - p\overline{E} = 0 \tag{3}$$

The solution of the second-order homogeneous linear ordinary differential equation is:

$$\frac{E(\text{x},\text{t})}{E_0} = \frac{1}{2}erfc\left[\frac{x-ut}{2\sqrt{D_L t}}\right] + \frac{1}{2}e^{\frac{ux}{D_L}}erfc\left[\frac{x+ut}{2\sqrt{D_L t}}\right] \tag{4}$$

where *erfc* is the complementary error function, and ξ in $erfc(u) = 1 - \frac{2}{\sqrt{\pi}}\int_0^{\text{u}} \text{e}^{-\xi^2 d\xi}$ is the mathematical expectation value.

When x or t is large:

$$\frac{E(\text{x},\text{t})}{E_0} = \frac{1}{2}erfc\left[\frac{x-ut}{2\sqrt{D_L t}}\right] = \frac{1}{\sqrt{\pi}}\int_{\frac{\text{x}-\text{ut}}{2\sqrt{\text{D}_\text{L}\text{t}}}}^{\infty} e^{-\varepsilon^2} d\xi \tag{5}$$

Let $\xi^2 = \eta^2/2$ and $\text{d}\xi = 1/\sqrt{2}d\eta$ (where η is the substitute for the mathematical expectation value ξ and does not have practical meaning), then we have:

$$\frac{E(\text{x},\text{t})}{E_0} = 1 - \frac{1}{\sqrt{2\pi}}\int_{-\infty}^{\frac{\text{x}-\text{ut}}{\sqrt{2\text{D}_\text{L}}}} e^{-\frac{\eta^2}{2}} d\eta = 1 - \text{N}\left[\frac{x-ut}{\sqrt{2D_L t}}\right] \tag{6}$$

Then, according to Equation (6):

$$0.1587 = 1 - \text{N}\left[\frac{x - ut_{0.1587}}{\sqrt{2D_L t_{0.1587}}}\right] 0.8413 = \text{N}\left[\frac{x - ut_{0.1587}}{\sqrt{2D_L t_{0.1587}}}\right]$$

According to the normal distribution table, $N(-1) = 0.1587, N(1) = 0.8413$, thereby leading to:

$$\frac{x - ut_{0.1587}}{\sqrt{2D_L t_{0.1587}}} = 1$$

$$\frac{x - ut_{0.8413}}{\sqrt{2D_L t_{0.8413}}} = -1 \tag{7}$$

Solving this equation leads to the dispersion coefficient as follows:

$$\left[\frac{x - ut_{0.1587}}{\sqrt{2D_L t_{0.1587}}} - \frac{x - ut_{0.8413}}{\sqrt{2D_L t_{0.8413}}}\right]^2 = 4 \tag{8}$$

Here, $t_{0.1587}$ is the time when $(C_{cj} - C_0)/(C_{max} - C_0) = 0.1587$ s and $t_{0.8413}$ is the time when $(C_{cj} - C_0)/(C_{max} - C_0) = 0.8413$ s.

3.2. Molecular Diffusion and Pore Tortuosity Tests

Hydrodynamic dispersion is a process that includes molecular diffusion and mechanical dispersion. Molecular diffusion is the process of molecular transport associated with the stochastic movement of molecules due to a concentration gradient. Mechanical dispersion is the process of mechanical mixing that takes place in porous media as a result of the movement of fluid through the pore space. When the average pore flow velocity is 0, dispersion takes the form of molecular diffusion. In this study, each sample consisted of single-grained coral sands, and there was a total of six test groups, with particle sizes ranging from 0–0.1 mm, 0.1–0.25 mm, 0.25–0.5 mm, 0.5–1 mm, and 1–2 mm, while the dry density of each sample was a fixed value: $\rho_d = 1.3$ g/cm^3. The tracer and sensor types were the same as those in the one-dimensional dispersion test. The test instrument was a custom-designed and built molecular diffusion device, as shown in Figure 5. There was a control partition in the middle of the device, and after the partition was lifted, the fluids in the left and right sample box compartments could flow back and forth between the compartments. The sizes of the left and right sample boxes were both 10 cm × 10 cm × 10 cm, and the left sample was saturated with NaCl solution while the right sample was saturated with fresh water. Timing and data reading were started simultaneously with the lifting of the partition.

Figure 5. Molecular diffusion and pore tortuosity test device.

Let *EN-t* and *EW-t* be the measured conductivity at time *t* on the left and right sides, respectively, and *EM-t* = (*EN-t* + *EW-t*)/2 be the mean value of the two measured conductivities at time *t*.

3.3. Molecular Diffusion and Mechanical Dispersion Tests

When the pore flow velocity is greater than 0, molecular diffusion and mechanical dispersion processes coexist. With an increase in average pore flow velocity, the weight of molecular diffusion versus mechanical dispersion changes. In order to further study the relationship between the weight and the flow velocity, a custom-designed molecular diffusion and mechanical dispersion test device was employed in this study, as shown in Figure 6. The device consisted of three components, namely, a dispersion body, a water supply part, and a data acquisition system. The dispersion body was made up of organic glass tubes and samples, with each tube being 8 cm in inner diameter and 80 cm in length. The samples were coral sands with particle sizes of 0.25–0.5 mm and a dry density of 1.3 g/cm^3. The data acquisition system was equipped with CS655 multi-parameter sensors; the tracer was a 20 g/cm^3 NaCl solution.

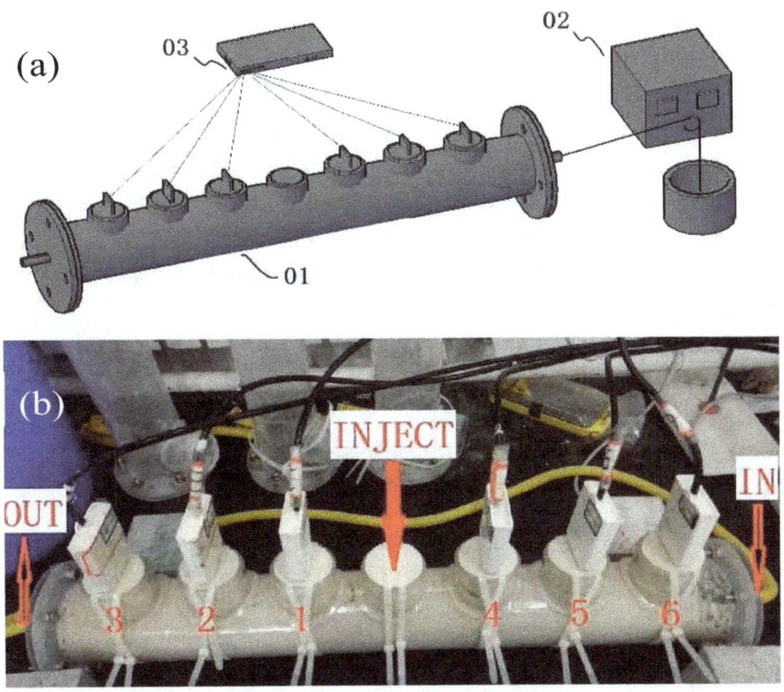

Figure 6. Molecular diffusion and mechanical dispersion test device. (**a**) Schematic Diagram (**b**) Real Product (01: dispersion body, 02: water supply part, 03: data acquisition system).

The required coral sand samples were first prepared in the dispersion tubes. The samples were then saturated with fresh water, after which the inlet and outlet valves of the dispersion tube were shut. The NaCl solution was then injected into the tracer injection port and the injection port was shut. This was completed while simultaneously starting data acquisition, with the pore flow velocity in the device being 0. The above steps were repeated using different concentrations of NaCl solution.

Next, dispersion tests with pore flow velocities greater than 0 were conducted according to the above steps, with the exception that the inlet and outlet valves of the dispersion tube were reopened after the injection of the tracer to allow the pore fluid at a certain flow velocity to pass through the samples. The pore flow velocity was controlled by adjusting the output power of the water pump. By repeating the above test procedure, dispersion tests were performed at several different pore flow velocities.

4. Test Results and Analyses

4.1. Effect of Particle Size on the Dispersion Coefficient

In order to uncover the effect of particle size on the one-dimensional dispersion coefficient, dispersion tests were performed on a total of six groups of single-grained coral sand samples, with the sample groups having a particle sizes of < 0.1 mm, 0.1–0.25 mm, 0.25–0.5 mm, 0.5–1 mm, 1–2 mm, and 2–5 mm; the groups had particle size ranges within the categories of silt, fine sand, medium sand, coarse sand, gravelly sand, and crushed stone (angular gravel), respectively. These groupings were determined according to the soil classification method in the Code for the Investigation of Geotechnical Engineering (GB50021-2001) (Ministry of Construction of the People's Republic of China, 2009) [16]. The dry density of all samples was 1.3 g/cm^3.

Figure 7 presents a one-dimensional dispersion curve for single-grained coral sands, which indicates that when the particle size was 0–0.1 mm, the dispersion process consisted of three stages: a drainage stage, a displacement stage, and a stabilization stage. In the drainage stage, the original saturated fluid in the soil column was discharged outward under the displacement of the tracer, a circumstance in which the discharged fluid had the same concentration as the original saturated fluid, thereby leading to a flat dispersion curve. In the displacement stage, with the continuous injection of the tracer, the tracer underwent diffusion in the original saturated solution, in addition to producing the displacement effect. The diffusion "interface" existed in the form of a concentration transition zone between the original saturated fluid and the tracer. When the edge of the transition zone reached the sensor position, the measured concentration of the discharged fluid began to increase, indicative of the onset of the displacement stage, as manifested by the curve slope starting to increase. In the stabilization stage, when the tracer had completely displaced the original saturated fluid, the discharged fluid had a concentration similar to that of the tracer, with the curve reaching the highest value and tending to flatten.

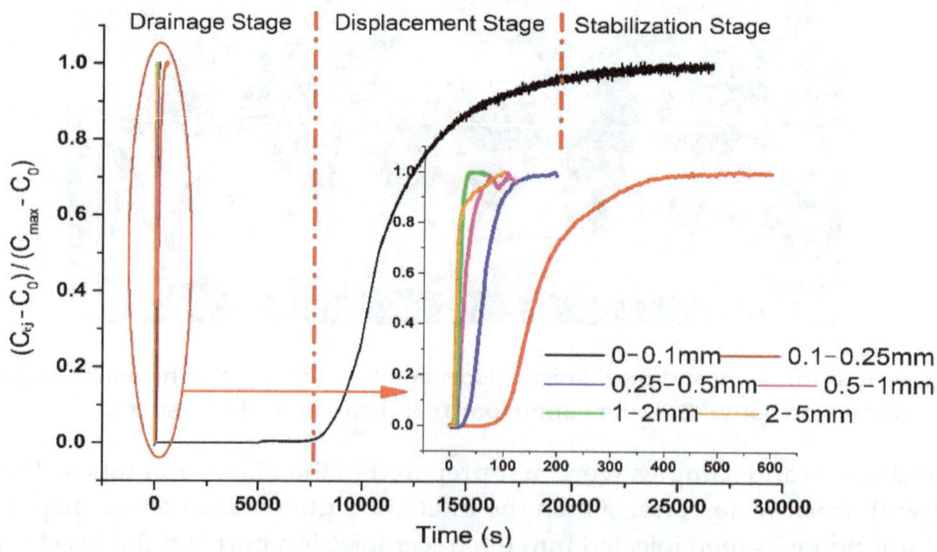

Figure 7. One-dimensional dispersion curves of single-grained coral sands.

As revealed by the above observations, different sample groups of different particle sizes exhibited dispersion pattern similarities, although the drainage stage gradually shortened with increasing particle size. The curve's slope in the displacement stage, i.e., the diffusion rate, increased with increasing particle size, and the time taken for the concentration to reach the stabilization stage gradually shortened with increasing particle size. With the exception of the 0–0.1 mm group, the other three groups were relatively similar to each other in terms of the parameters of the three stages, indicating a significant difference in dispersion characteristics between the particle size 0–0.1 mm and the other particle sizes.

Figure 8 shows the variation pattern of the one-dimensional dispersion coefficient with respect to particle size. The dispersion coefficient of coral sands increased with increasing particle size. When the particle size exceeded 0.25 mm, the dispersion coefficient increased rapidly, with the dispersion coefficient differing by a factor of 202. When the particle size was 0.1–2 mm (i.e., fine sands–gravels), the dispersion coefficient was between 0.152–2.37 cm^2/s, with the difference being a factor of approximately 15. However, when the particle size was greater than 2 mm, the change in the dispersion coefficient with respect to particle size became smaller, with the dispersion coefficient tending to reach a constant value as the particle size increased, finally reaching a value of 2.5 cm^2/s in this test condition. These observations suggest that 0.25 mm and 2 mm could be considered the characteristic particle sizes of coral sands.

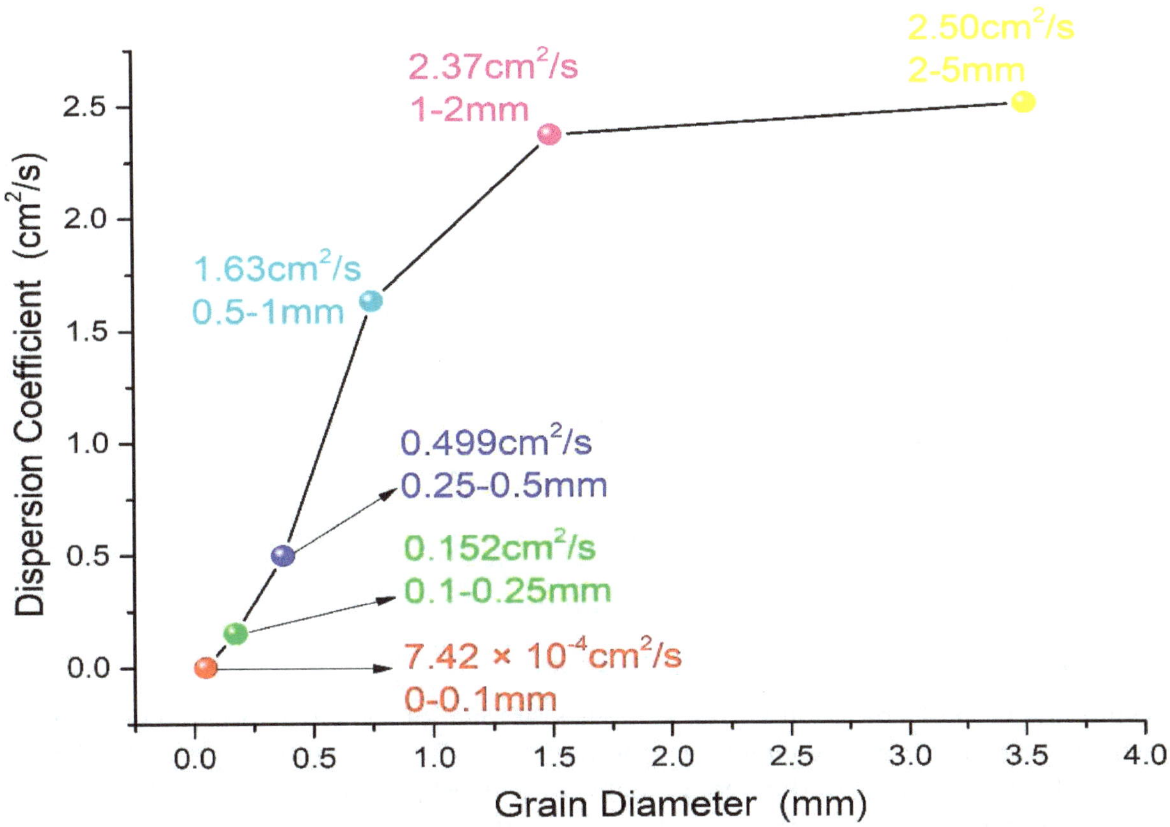

Figure 8. Dispersion coefficients of single-grained soils.

4.2. *Effect of Dry Density on the Dispersion Coefficient*

The effect of the degree of compactness on the one-dimensional dispersion coefficient was investigated using the group of coral sands having a particle size range of 0.25–0.5 mm and dry densities of 1.2 g/cm^3 (relative degree of compactness of 1.021), 1.3 g/cm^3 (relative degree of compactness of 1.212), and 1.4 g/cm^3 (relative degree of compactness of 1.375). The test results, which are shown in Figure 9, indicate that with the increase in density, the displacement stage became longer, the diffusion rate decreased, and the time spent before reaching a steady state increased. Figure 10 represents the variation of the dispersion coefficient with the change in dry density, which reveals that the dispersion coefficient decreased linearly with increasing dry density.

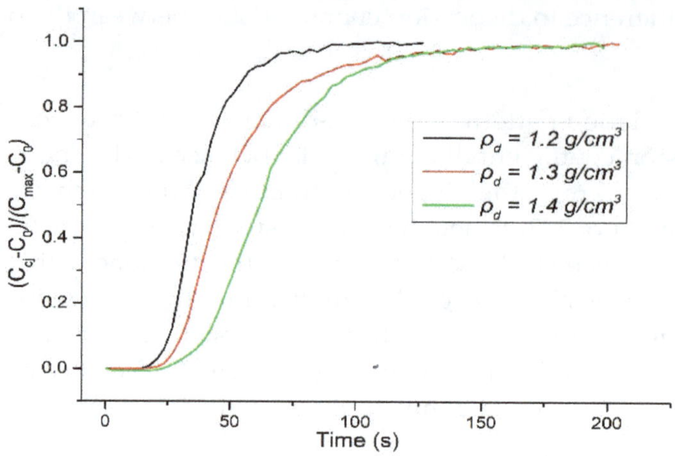

Figure 9. One-dimensional dispersion curves of coral sands with different dry densities.

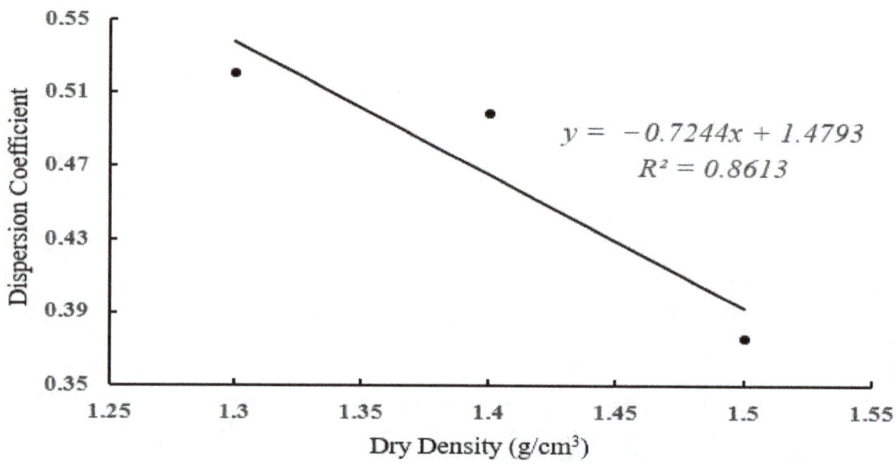

Figure 10. One-dimensional diffusion coefficients of coral sands with different dry densities.

4.3. Effect of Particle Gradation on the Dispersion Coefficient

To simulate natural-gradation sands, stepwise removal of the particles smaller than a certain size was conducted in order to change the gradation [17]. This was followed by one-dimensional dispersion tests on coral sands of different gradations to investigate the effect of particle gradation on the one-dimensional dispersion coefficient. The sample dry density was 1.3 g/cm^3 and the gradation parameters of each sample are listed in Table 2 [18]. As shown by the data, all the samples were poorly graded coral sands, with the exception of the natural gradation sands. The gradation curves of the samples are displayed in Figure 11 and the dispersion coefficient curves are illustrated in Figure 12.

Table 2. Gradation parameters of the samples.

Sample	d_{60}	d_{30}	d_{10}	C_c	C_u
No. 1	2.5	0.52	0.05	2.163	50.000
No. 2	2.5	0.52	0.15	0.721	16.667
No. 3	2.5	0.6	0.31	0.465	8.065
No. 4	2.5	0.9	0.6	0.540	4.167
No. 5	2.55	1.5	1.2	0.735	2.125

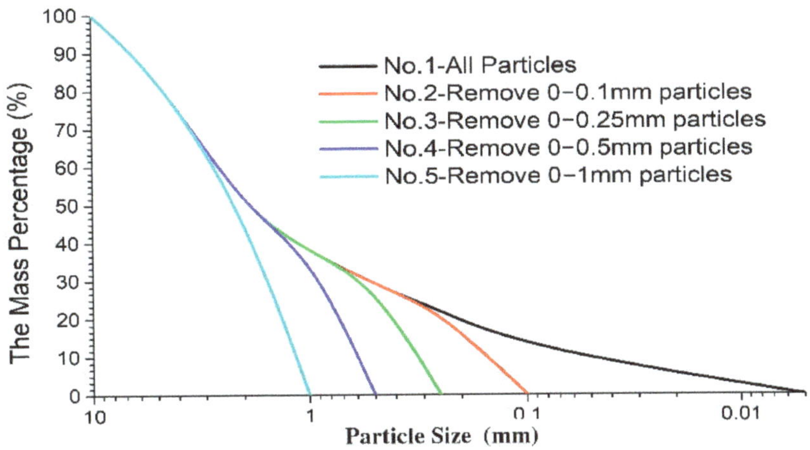

Figure 11. Gradation curves of the samples.

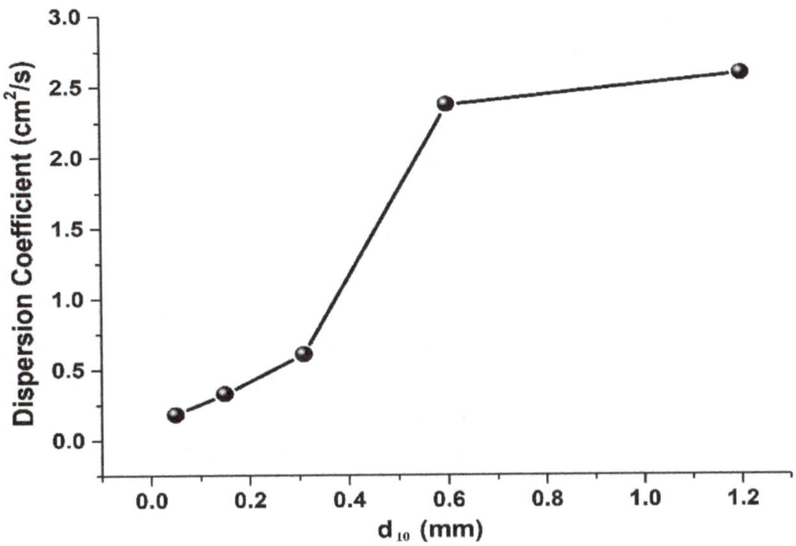

Figure 12. One-dimensional dispersion coefficients of samples with different gradations.

As shown in Figure 12, the dispersion coefficient of the coral sands gradually increased with increasing d_{10}. Following the removal of soil particles smaller than 0.25 mm (d_{10} increased to 0.31 mm from the initial 0.05 mm), the dispersion coefficient of the coral sands increased dramatically. After removal of the soil particles smaller than 0.5 mm (d_{10} exceeded 0.6 mm), the dispersion coefficient of the coral sands tended to reach a constant value. These results suggest that 0.25 mm and 2 mm were the characteristic particle sizes for the dispersion properties of graded coral sands.

5. Analysis of the Dispersion Mechanisms in Coral Sands

There are many significant factors affecting solute transport, such as convection, mechanical dispersion, molecular diffusion, interactions between the solid phase and solute (such as dissolution and adsorption), chemical reactions within the solution, and other source–sink solute interactions (such as attenuation of radioactive elements and absorption of certain solutes by crop roots) [19]. Among these, mechanical dispersion and molecular diffusion are collectively referred to as dispersion. Studies have shown that mechanical dispersion and molecular diffusion generally occur simultaneously during solute transport, although they exhibit different variation patterns with changing average pore flow velocity. When the flow velocity is low, molecular diffusion is stronger than mechanical dispersion. In contrast, when the flow velocity is high, mechanical dispersion is stronger than molecular diffusion. In order to reveal the roles of these two mechanisms in the solute dispersion process of coral sands, pore tortuosity tests, as well as molecular and mechanical dispersion tests, were performed on the coral sands in this study.

5.1. Molecular Diffusion in Coral Sands

Figure 13 depicts a curve showing the diffusion concentrations of solute molecules in coral sands having different particle sizes. E_{NaCl} is the conductivity measured by the left sensor and E_{water} is measured by the right sensor, with $E_{Med} = (E_{NaCl} + E_{water})/2$. When there was a concentration gradient in the saturated porous medium and the pore flow velocity was 0, the NaCl solution concentrations gradually decreased and increased in the left and right samples, respectively, until the two concentrations reached a relative equilibrium. It is noteworthy that the concentrations of the NaCl solution on both sides were not necessarily the same when equilibrium was reached; this was due to the fact that when the concentration gradient was low (defined as the limiting concentration gradient at equilibrium, ranging from 0–1, with larger values representing larger concentration gradients at equilibrium), the "obstruction" effect of the porous medium occurred, and the concentration gradient would not continue to decrease. Porous media with different particle sizes (pore sizes) had different limiting concentration gradients at equilibrium. As shown by Figure 13, when the particle size was larger than 0.1 mm, the time spent prior to reaching the limiting concentration gradient at equilibrium gradually decreased with increasing particle size, concomitant with a gradual decrease in the limiting concentration gradient at equilibrium. When the particle size was larger than 1 mm, the limiting concentration gradient at equilibrium was infinitely close to 0. When the particle size was smaller than 0.1 mm, the molecular driving force generated by the concentration gradient was less than the "obstruction" effect of the porous medium; thus, the limiting concentration gradient at equilibrium was rapidly achieved and relatively large (approaching 1). Figure 14 illustrates how the amount of time taken to reach the limiting concentration gradient at equilibrium varied with particle size. As shown in Figure 14, 0.1–0.25 mm was still the characteristic particle size of coral sands for molecular dispersion, and the group of coral sands with this particle size range took the longest time to reach the limiting concentration gradient at equilibrium. This finding is in agreement with the conclusion drawn from the one-dimensional dispersion test conducted in this study.

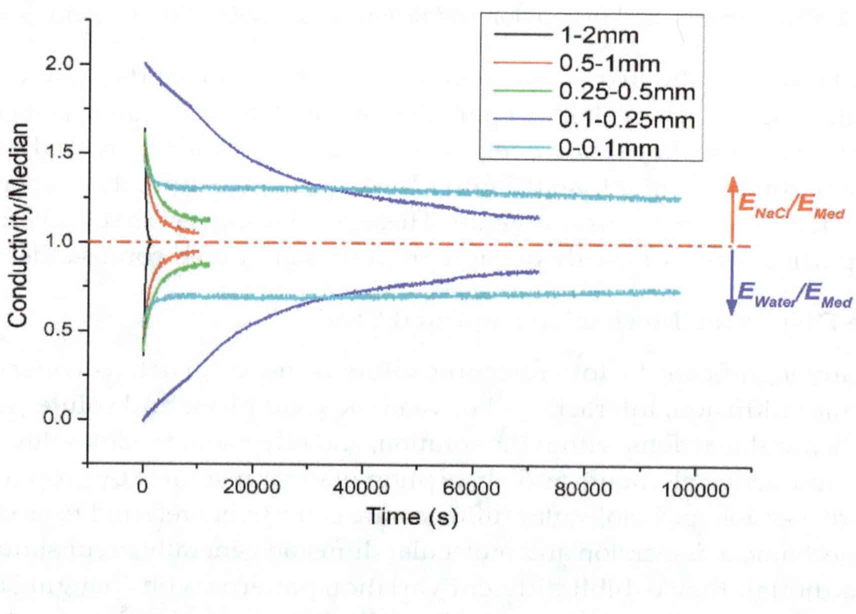

Figure 13. Concentration variation curves during molecular diffusion in coral sands.

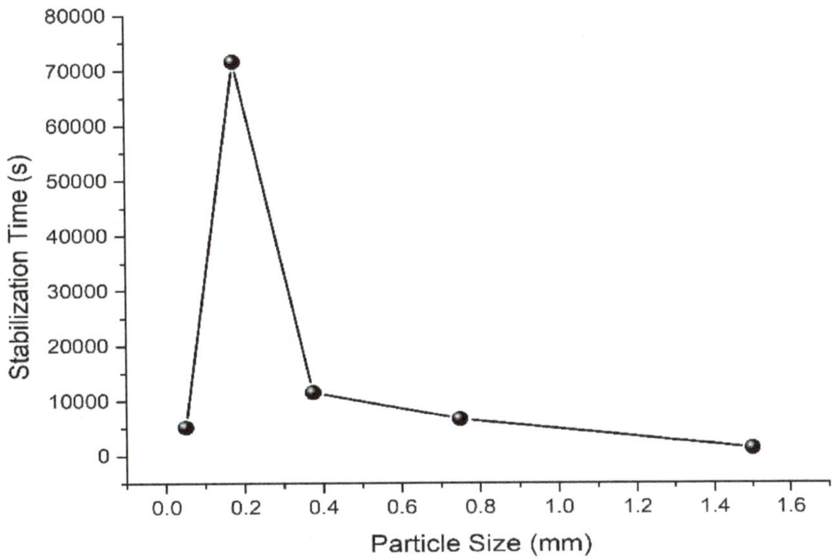

Figure 14. Varying amounts of time required to reach the limiting concentration.

In order to quantitatively characterize the "obstruction" effect of porous media on molecular diffusion, earlier studies introduced the concept of pore tortuosity (θ), i.e., the ratio of the length of the porous medium sample to the actual path traveled by the fluid particles through a sample of that length [20]. Thus, $\theta = D_0/D^*$, where D_0 is the dispersion coefficient in the open water body and D^* is the dispersion coefficient in the porous medium. Given the assumption that the limiting concentration gradient at equilibrium infinitely approaches 0 for molecular diffusion in an open and still water body, the θ of a porous medium should range between 0 and 1. According to the definition of pore tortuosity, $\theta = S/(v \times t)$, S (cm) is the length of the porous medium sample, v (cm/s, taken as scalar without considering direction) is the velocity of fluid particle movement, and t (s) is the time duration of fluid particle movement, namely, the time spent by the fluid particles prior to reaching the limiting concentration gradient at equilibrium. Based on the data in Figure 13, the time t in each particle size group of coral sands could be obtained, with t set to positive infinity in the case of particle sizes less than 0.1 mm. In addition, based on the time t and the known distance of 4.9 cm between the sensor and the central partition, it was possible to obtain the relationship curve of pore tortuosity versus particle size, as shown in Figure 15, where v is the velocity of fluid particle movement at a certain temperature.

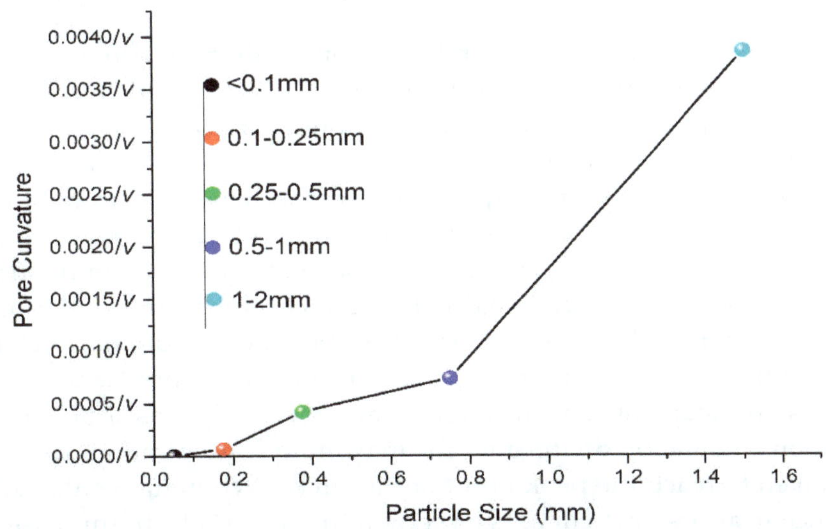

Figure 15. Relationship curve between pore tortuosity and particle size of coral sands.

A previous study on the dispersion characteristics of terrigenous sediments found that θ ranges from 0.01–0.5 [21]. In addition, this parameter has been proposed to be 0.1 for clays and 0.7 for sandy soils [22].

5.2. Mechanical Dispersion in Coral Sands

Dispersion consists of molecular diffusion and mechanical dispersion. When the flow velocity is greater than 0, the two processes usually coexist, albeit with different weights on the overall dispersion coefficient under different conditions (i.e., concentration and flow velocity).

Figure 16 presents the variation curves of relative conductivity at a fixed pore flow velocity of 0 for two different concentrations of tracer NaCl solution (20 g/L versus 60 g/L). Here, E_C is the sensor-measured conductivity, E_0 is the initial conductivity before the injection of the tracer, and E_{max} is the maximum conductivity measured by each sensor. The rate of increase (slope) of relative conductivity was greater for the 60 g/L NaCl solution than for its 20 g/L counterpart, suggesting that the higher the concentration gradient, the greater the molecular diffusion rate.

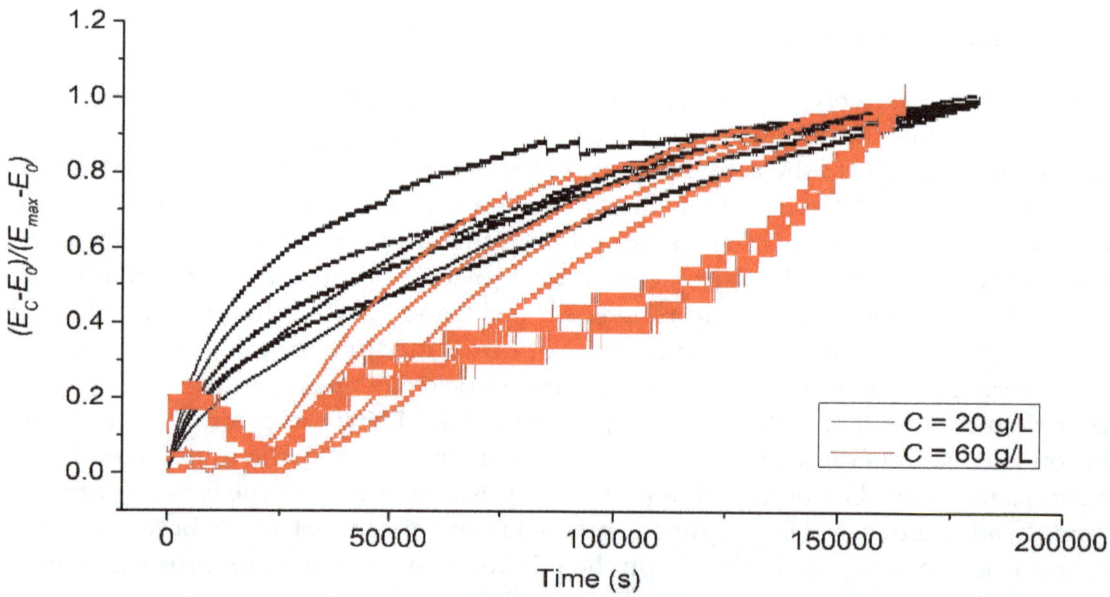

Figure 16. Molecular diffusion curves of different concentration gradients.

Figure 17 shows the variation curves of relative conductivity at different flow velocities, with Figure 17A–D depicting the variation curves of measured conductivities from the sensors EC_1 to EC_3. Figure 17a–d depict the variation curves of measured conductivities from the sensors EC_4 to EC_6. In relation to the tracer injection port, the sensors EC_1 to EC_3 were downstream, while the sensors EC_4 to EC_6 were upstream. The dispersion process downstream from the port would involve both mechanical dispersion and molecular diffusion, with both taking place in the same direction. In contrast, in the dispersion process occurring against the flow direction upstream from the port, mechanical dispersion would take place in a different direction than molecular diffusion. Therefore, the conductivity change detected by the sensor would be a net conductivity change of the solutes with their molecular diffusion overcoming their mechanical dispersion. Generally, when mechanical dispersion dominates, the shape of concentration–time curve will be as shown in Figure 18a. In this situation, concentration attenuation occurs after the maximum concentration has been reached, with t_p denoting the time taken to reach the peak point of the curve. When molecular diffusion dominates, the shape of the concentration–time curve is as shown in Figure 18b. In this case, the concentration remains constant after reaching the maximum value, with t_s denoting the time taken to reach the stable point of the curve.

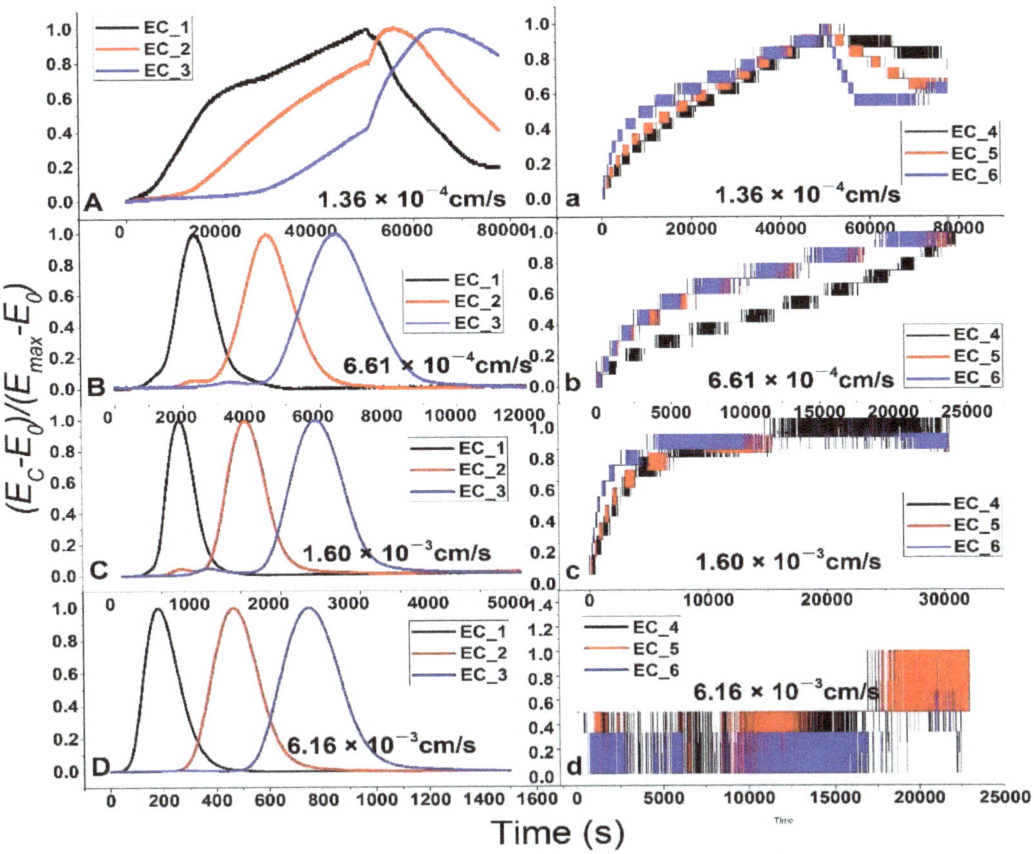

Figure 17. Dispersion mechanisms in coral sands having different pore flow velocities. (Figure 17A–D depicting the variation curves of measured conductivities from the sensors EC_1 to EC_3. Figure 17a–d depict the variation curves of measured conductivities from the sensors EC_4 to EC_6.)

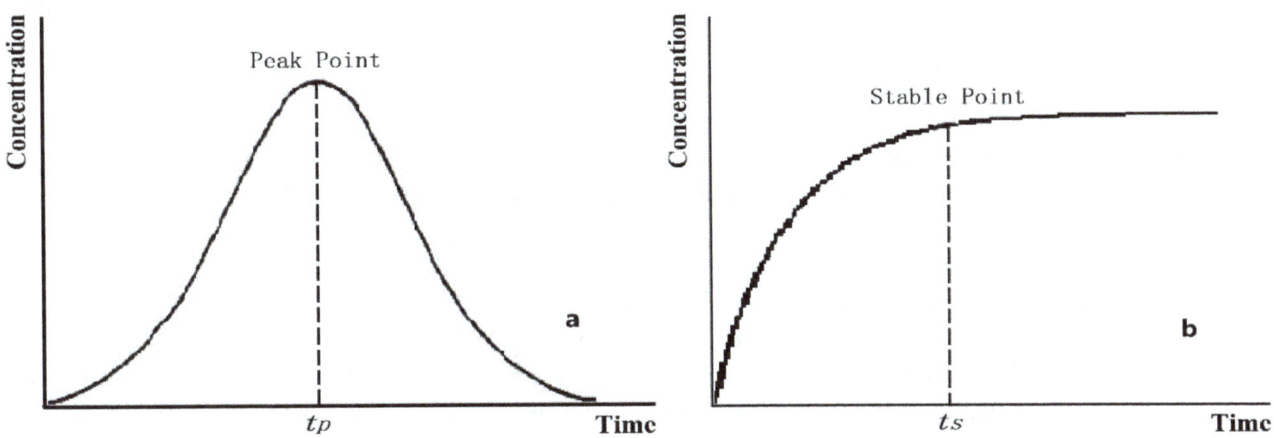

Figure 18. Schematic of the typical curves corresponding to different dispersion mechanisms.

It can be seen that when the flow velocity was 1.36×10^{-4} cm/s (Figures 17A and 17a), the curve characteristics of the left figure fell between those of Figures 18a and 18b, indicating the simultaneous presence of the two mechanisms of molecular diffusion and mechanical dispersion downstream from the port. Data comparison between the left and right figures indicates that the time to reach the maximum conductivity was similar in the two figures, suggesting that the two mechanisms were comparable in terms of the dispersion process under this condition. The relative conductivity decreased after the peak value in the right figure, which was likely the result of the displacement effect of the pore fluids.

With the increase in flow velocity, the curve shape in the left figure (Figure 17A–D) becomes gradually similar to that in Figure 18a, indicating that mechanical dispersion was gradually enhanced. This was manifested by the gradual shortening of the displacement stage and the continuous increase in the diffusion rate. In contrast, the curve shape in the right figure (Figure 17a–d) became increasingly atypical, and the time taken to reach the maximum value gradually increased and was much longer than the corresponding time in the left figure, indicating that molecular diffusion was increasingly subject to the flow velocity. By the time the flow velocity reached 6.16×10^{-3} cm/s, the molecular diffusion upstream from the tracer injection port was basically negligible.

Based on a comprehensive analysis of the above test results, the velocity boundaries of the pore fluids that control the dispersion mechanisms were derived, as shown in Table 3.

Table 3. Velocity threshold value of pore fluids that control the dispersion mechanisms.

Classification	Existing Classification Method [21]	Classification Method in This Study (flow velocity u, cm/s)	Description
I	Very low flow velocity	u approaching 0	Molecular diffusion dominates, while mechanical dispersion is negligible
II	Flow velocity increasing	$0 < u < 1.36 \times 10^{-4}$	The two types of diffusion are comparable
III	Flow velocity continuing to increase	$1.36 \times 10^{-4} < u < 1.60 \times 10^{-3}$	Mechanical dispersion dominates, while molecular diffusion is weak
VI	Higher flow velocity	$1.60 \times 10^{-3} < u < 6.16 \times 10^{-3}$	Mechanical dispersion dominates, while molecular diffusion is negligible
V	Flow velocity too high	$u > 6.16 \times 10^{-3}$	Turbulence and inertial forces weaken mechanical dispersion

6. Conclusions

The solute diffusion coefficient of coral sands is a key factor affecting fresh groundwater conservation, and knowledge of the freshwater dispersion pattern in coral sands will provide both the basic parameters and a theoretical basis for numerical simulation of the formation and evolution of freshwater aquifers under islands, as well as the conservation and utilization of these aquifers. Using coral sands collected from a reef in the South China Sea, a series of tests and theoretical analyses were conducted in this study, from which the following conclusions were drawn:

1) The particle size of coral sands was an important parameter affecting the dispersion coefficient, which increased with increasing particle size. The reason is that when the volume and porosity of the soil are constant, the number of pores decreased and the pore size increased with the increase in particle size. Because the pore fluid was more likely to mechanically flow in larger pores, the diffusion coefficient, therefore, could increase. The diameters of 0.25 mm and 2 mm were discovered to be the characteristic particle sizes of coral sands due to their dispersion characteristics. The dispersion coefficient could vary by a factor of more than 200 between the 0.1–0.25 mm size group and the 0.25–0.5 mm size group, while the dispersion coefficient increased at a relatively small rate in the group with particle sizes larger than 2 mm.
2) Gradation was also an important factor affecting the dispersion coefficient of coral sands, with the diffusion coefficient increasing with increasing d_{10}. When d_{10} was larger than 0.31 mm, the dispersion coefficient increased dramatically. In contrast, when d_{10} was larger than 0.6 mm, the dispersion coefficient tended to be constant. This was because an increase in d_{10} reflected that the filling degree of the fine particles in the soil became worse. Similarly, the pore size and diffusion coefficient increased with an increase in d_{10}. However, when d_{10} was 0.31 mm or less, the diffusion coefficient did not increase with increasing d_{10}. The reason was the increase in pore

size was minimal with respect to the macroscopic scale. Alternatively, the importance of pore size for the macro scale grew with an increase in d_{10}, and the diffusion coefficient also rose. Finally, when the pore size reached a critical value, the influence of pore size on the solute diffusion gradually reduced, and the diffusion coefficient also approached a stable state.

3) The dispersion coefficient of coral sands decreased linearly with an increasing degree of compactness. The dispersion coefficient decreased at a rate of −0.7244 in the group of coral sands with particle sizes of 0.25–0.5 mm. The reason was that when the particle size distribution remained stable, the increase in compactness directly resulted in a decrease in porosity, which could restrict the flow of the pore liquid. The diffusion coefficient, therefore, was reduced. By analyzing the experimental dates, the diffusion coefficient was reduced when the dry density was increased from 1.3 g/cm^3 to 1.5 g/cm^3.
4) Both the concentration of solute in the coral sands and the particle size of the porous medium affected the limiting concentration gradient at equilibrium. With an increase in solute concentration and particle size, the limiting concentration gradient at equilibrium decreased. As particle size increased, the pore tortuosity decreased. This was because when the soil volume and porosity were constant, the number of pores decreased and the pore size increased with an increase in the particle size, and the actual distance that the solution molecules flowed from point A to point B in the soil was shortened. Therefore, the calculated porosity curvature was reduced.
5) The dispersion mechanisms in coral sands could be determined by the flow velocity of the pore fluid. Based on the different weights of molecular diffusion versus mechanical dispersion for different conditions of flow velocity, the dispersion mechanisms could be classified into five types.

Author Contributions: Conceptualization, X.C., C.Z., M.H. and X.W.; Data curation, X.C. and H.L.; Formal analysis, X.C.; Funding acquisition, C.Z., M.H. and X.W.; Project administration, C.Z. and M.H.; Resources, X.W. and H.L.; Writing—original draft, X.C.; Writing—review and editing, C.Z.

References

1. Hu, M.; Cui, X.; Wang, X.-Z.; Liu, H.; Zhang, C. Experimental study on the effect of fine particles on permeability of the calcareous sand. *Rock Soil Mech.* **2019**, *40*, 1–6.
2. Zhu, C.; Zhou, B.; Liu, H. State-of-the-art review of developments of laboratory tests on cemented calcareous soils. *Rock Soil Mech.* **2015**, *36*, 311–319.
3. Zhu, C.; Zhou, B.; Liu, H. Micro-structures and fundamental engineering properties of beach calcarenite from south china sea. *Chin. J. Rock Mech. Eng.* **2015**, *34*, 683–693.
4. Zhu, C.; Chen, H.; Meng, Q.; Wang, R. Microscopic characterization of intra-pore structures of calcareous sands. *Rock Soil Mech.* **2014**, *35*, 1831–1836.
5. Taylor, G. Dispersion of Soluble Matter in Solvent Flowing Slowly through a Tube. *Math. Phys. Sci.* **1953**, *219*, 186–203.
6. Klotz, D.; Moser, H.; Neumaier, F. Dispersivity and velocity relationship from laboratory and field experiments. *J. Hydrol.* **1980**, *45*, 169–184. [CrossRef]
7. De Gennes, P.G. Hydrodynamic dispersion in unsaturated porous media. *J. Fluid Mech.* **1983**, *136*, 189–200. [CrossRef]
8. De Arcangelis, L.; Koplik, J.; Redner, S. Hydrodynamic Dispersion in Network Models of Porous Media. *Phys. Rev. Lett.* **1986**, *57*, 8. [CrossRef] [PubMed]
9. Muhammad Sahimi, I. Fractal and Superdiffusive Transport and Hydrodynamic Dispersion in Heterogeneous Porous Media. *Transp. Porous Media* **1993**, *13*, 3–40. [CrossRef]
10. Lowe, C.P.; Frenkel, D. Do Hydrodynamic Dispersion Coefficients exist. *Phys. Rev. Lett.* **1996**, *77*, 4552–4555. [CrossRef] [PubMed]
11. Zhang, F.; Kang, S.; Pan, Y. Experimental study on hydrodynamic dispersion of adsorption solute in saturated-unsaturated soil. *J. Hydraul. Eng.* **2002**, *3*, 84–90.
12. Li, L.; Zhu, W.; Qu, Y. Experiment study on hydrodynamic dispersion parameters for contaminated soil with low permeability. *Chin. J. Geotech. Eng.* **2011**, *33*, 1308–1312.

13. Shao, A.; Liu, G.; Yang, J. In-lab determination of soil hydrodynamic dispersion coefficient. *Acta Pedol. Sin.* **2002**, *39*, 184–189.
14. Hogh, K.; Jensen, G.; Destouni, M. Advection-dispersion analysis of solute transport in undisturbed soil monoliths. *Ground Water* **1996**, *34*, 1090–1097.
15. Zheng, C.; Gordon, B. *Applied Contaminant Transport Modeling*; Higher Education: Beijing, China, 2009.
16. Ministry of Construction of the People's Republic of China. *GB 55021—2001 Code for Investigation of Geotechnical Engineering*; China Building Industry: Beijing, China, 2009.
17. Xie, D.; Cai, H.; Wei, Y.; Li, W. Scaling principle and method in seepage tests on coarse materials. *Chin. J. Geotech. Eng.* **2015**, *37*, 369–373.
18. Ministry of Water Resources of the People's Republic of China. *GB/T 50123—1999 Standard for Geotechnical Test Methods*; China Planning: Beijing, China, 1999.
19. Qian, H.; Ma, Z. *Hydrogeochemistry*; Geological Publishing House: Beijing, China, 2005.
20. Bear, J. *Dynamics of Fluids in Porous Media*; Elsevier: New York, NY, USA, 1972; 764p.
21. Freeze, R.A.; Cherry, J.A. *Groundwater*; Prentice Hall: Englewood Cliffs, NJ, USA, 1979; 604p.
22. Konikow, L.F.; Bredehoeft, J.D. Ground-water models cannot be validated. *Adv. Water Resour.* **1992**, *15*, 75–83. [CrossRef]

8

Identifying the Frequency Dependent Interactions between Ocean Waves and the Continental Margin on Seismic Noise Recordings

Zhen Guo [1], Yu Huang [1,2,*], Adnan Aydin [3] and Mei Xue [4]

1 College of Civil Engineering, Tongji University, Shanghai 200092, China; zhenguo@tongji.edu.cn
2 Key Laboratory of Geotechnical and Underground Engineering of the Ministry of Education, Tongji University, Shanghai 200092, China
3 Department of Geology and Geological Engineering, The University of Mississippi, University, MS 38677, USA; aaydin@olemiss.edu
4 State Key Laboratory of Marine Geology, Tongji University, Shanghai 200092, China; meixue@tongji.edu.cn
* Correspondence: yhuang@tongji.edu.cn

Abstract: This study presents an exploration into identifying the interactions between ocean waves and the continental margin in the origination of double-frequency (DF, 0.1–0.5 Hz) microseisms recorded at 33 stations across East Coast of USA (ECUSA) during a 10-day period of ordinary ocean wave climate. Daily primary vibration directions are calculated in three frequency bands and projected as great circles passing through each station. In each band, the great circles from all stations exhibit largest spatial density primarily near the continental slope in the western North Atlantic Ocean. Generation mechanisms of three DF microseism events are explored by comparing temporal and spatial variations of the DF microseisms with the migration patterns of ocean wave fronts in Wavewatch III hindcasts. Correlation analyses are conducted by comparing the frequency compositions of and calculating the Pearson correlation coefficients between the DF microseisms and the ocean waves recorded at selected buoys. The observations and analyses lead to a hypothesis that the continental slope causes wave reflection, generating low frequency DF energy and that the continental shelf is where high frequency DF energy is mainly generated in ECUSA. The hypothesis is supported by the primary vibration directions being mainly perpendicular to the strike of the continental slope.

Keywords: ocean waves; double-frequency microseisms; continental margin; continental slope

1. Introduction

Ambient noise (or seismic noise) has been widely used to estimate the seismic site effect parameters (e.g., predominant frequency f_0, sediment thickness, amplification factor, etc.) of a site [1–5] and to characterize both deep (down to the mantle of the Earth) and shallow (within the depth of geological engineering activities) subsurface structures [6–15] for its advantages as a fast, effective, and reliable tool. However, the accuracies and reliabilities of the applications listed above would be strongly affected by the spatial and temporal variations of the ambient noise sources. One example is given in [16] who estimated the amplification factors in Northern Mississippi of United States applying the horizontal-to-vertical spectral ratio (HVSR, or Nakamura) method based on long term ambient noise recordings. As the f_0s in their study region lie in the frequency band of ocean waves induced double-frequency (DF) microseisms (0.1–0.5 Hz), the estimated amplification factors (HVSR values at f_0) fluctuate with time and are strongly correlated with the energy of the DF microseisms as well as the ocean wave height. Many other studies suggest that if the noise sources are not homogeneously distributed, the cross-correlation function cannot be reconstructed causing big errors in subsurface tomography or even failure of subsurface tomography [17–22]. In addition, the ambient noise with frequency greater than 0.1 Hz would be used in shallow subsurface tomography. From this point of view, exploring the source locations, the spatial and temporal characteristics and the generation mechanisms of the ambient noise with frequency greater than 0.1 Hz, especially the DF microseisms, would significantly improve the application of the ambient noise in site effect evaluation and shallow subsurface tomography.

In the spectrum of ambient noise recorded globally, the DF microseisms (or secondary microseisms) manifest themselves as one or more energy peaks in the frequency band of 0.1–0.5 Hz which is roughly twice of ocean waves' frequencies. It is widely accepted that DF microseisms are generated by the nonlinear interaction between ocean waves propagating in opposite directions with similar frequencies (e.g., [23–31]).

Recent studies suggest that two different circumstances may be responsible for generating opposing ocean waves that trigger DF microseisms. The first group considers wave–wave interactions in the open-ocean during strong storms [27,32–34]. However, DF microseisms can be observed worldwide even when there are no strong storms locally or globally. This is explained by the second group of studies who emphasize the role of interactions between the incident and reflected ocean waves at the continental margin [22,35–39]. For example, the authors of [40] observed that Rayleigh waves in a microseism recording at an ocean bottom seismometer in the Pacific Ocean were approaching from California coast during a super-typhoon rather than the location of the typhoon and concluded that the microseisms were generated by interactions of typhoon-induced waves toward and their reflections from the coastal line. In a different set of studies based on correlation analyses between the ocean storms developed close to shorelines and the ambient noise recorded on coastal seafloor or costal land, it was recognized that the long- and short-period DF microseisms (LPDF, 0.1–0.2 Hz and SPDF, 0.2–0.5 Hz, respectively) were excited by swells from distant and local waves, respectively [25,30,31,41,42]. However, regarding the locations where the interactions (reflection) occur, there exist a debate in terms of the water depth (deep or coastal). In [43] the authors summarized the debate and compared theoretical and observed DF characteristics for each case considering the ocean wave frequency composition and velocities. It appears that the relationship between DF microseisms and ocean waves is not yet directly investigated as a function of water depth across the continental margin.

In this study, the continental slope, as a boundary between shallow (continental shelf) and deep (open ocean) water, is explored for its interactions with the ocean waves as well as its role and significance in the generation of DF microseisms. To reach this goal, a total of 10-days (2014/325–334) of ambient noise recordings the WAVEWATCH III® (WWIII) hindcasts of ocean wave energy in Atlantic Ocean, and ocean wave climate parameters are utilized and analyzed.

2. Materials and Methods

2.1. Ambient Noise Data

The selected ambient noise data were recorded at 33 broadband seismic stations (Figure 1, see IRIS Data Management Center for further details of the instruments) of the Transportable Array (TA) network along parts of the middle and southeastern North Atlantic coastal area and the Shenandoah Valley. The original time series of amplitude in vertical (V), north-south (N), and east-west (E) directions archived with a sampling rate of 40 samples-per-second were first parsed into 1-h segments, followed by removing the mean, linear trend, and instrument response in each segment [44]. Then each segment was processed following a 14-step procedure summarized below, to estimate the power spectral density (PSD) in vertical (V) direction (steps 1–4), and the primary vibration direction by the radial-to-transverse spectral ratio (*Ra*) method (steps 5–9) as well as the polarization analysis method based on [45] (steps 10–13) in the three DF bands (DF1, 0.1–0.2 Hz; DF2, 0.2–0.3 Hz; and DF3, 0.3–0.4 Hz). The two methods used to estimate the primary vibration directions are both based on an assumption that DF microseisms propagate dominantly as a fundamental mode Rayleigh wave [34,46–48]. The *Ra* method searches the direction of the largest ratio of radial to transverse components on the horizontal plane which is expected to be along the propagation direction of a Rayleigh wave [29,44]. The polarization analysis method represents the particle motion within a short time range (when the propagation direction is reasonably stable) as an ellipsoid with three axes perpendicular to each other [45], from which the back azimuth of the major axis with highest probability for the whole recording period can be calculated. If the primary energy source is stable and strong enough, and significantly larger than secondary sources, the primary vibration directions obtained by the two methods would agree because the secondary sources do not alter the direction of the major axis but increase the *Ra* values in directions other than the major axis.

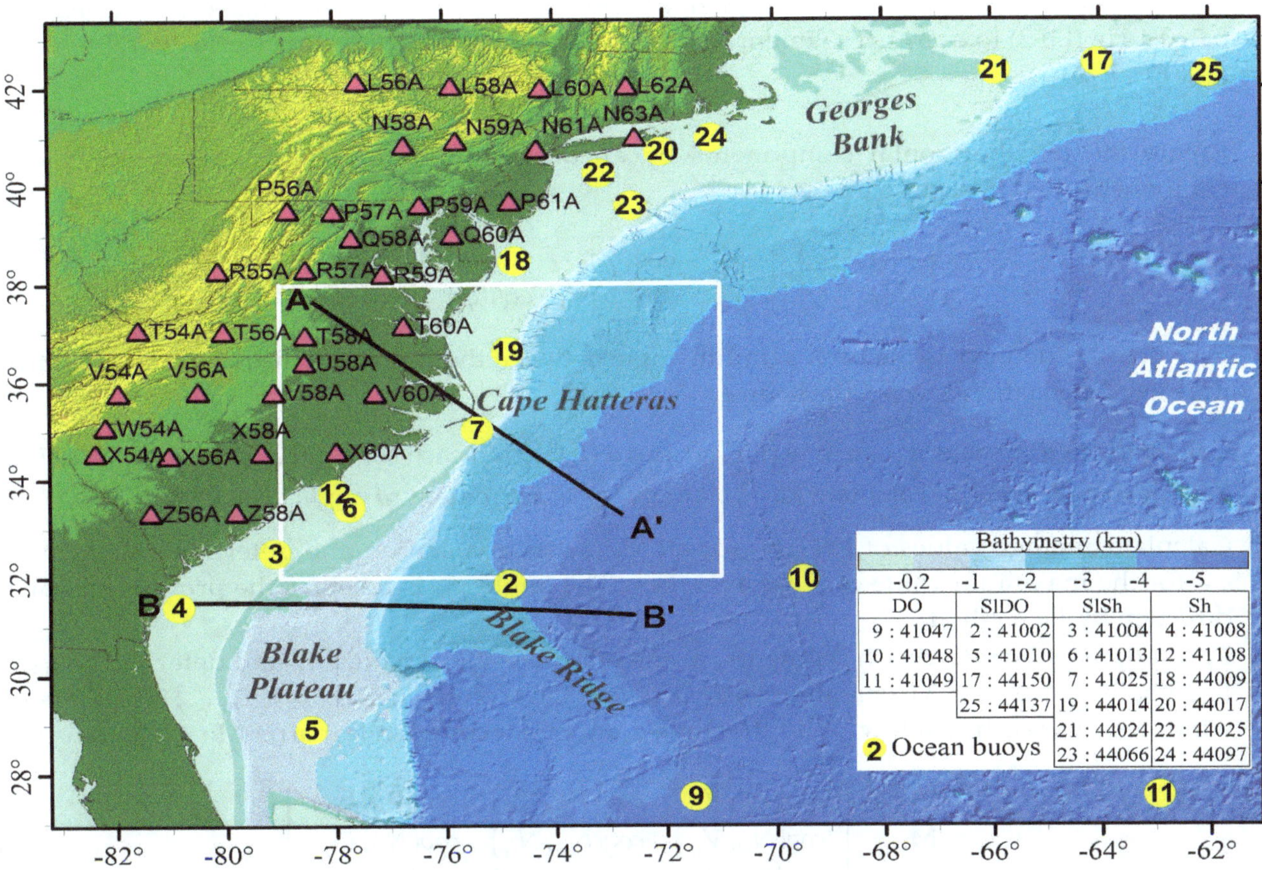

Figure 1. Study area and locations of transportable array (TA) stations (triangles) and buoys (yellow circles) of National Oceanic and Atmospheric Administration, recordings of which are analyzed in this study. The table lists the original numbers and a simplified numbering scheme of the ocean buoys grouped according to their four distinct locations: deep ocean (DO), the continental slope-deep ocean side (SlDO), the continental slope-shelf side (SlSh), and the continental shelf (Sh). The color-encoded relief base map is from [49]. The black lines A-A′ and B-B′ are the transects presented in Figure 9a,b. The white box indicates the area where the underground shear velocity model (Figure 9c,d) was computed based on [15].

The steps of the data analysis are:

(1) Apply an anti-triggering algorithm based on a prescribed range of short (1 s) to long (30 s) term average amplitude ratios (0.2 < STA/LTA < 2.5) to filter each segment for avoiding occasional energy bursts [4,50].

(2) Apply fast Fourier transform with a 10% cosine taper on the filtered segments in three directions to calculate spectra ($V(f), N(f)$, and $E(f)$) and then smooth them using Konno–Ohmachi method with a bandwidth coefficient of 40 [51].

(3) Compute the PSDs in the vertical direction [52] in the unit of $(m/s^2)^2$/Hz dB:

$$PSD(f) = 10\log\left[\frac{1}{0.825}\cdot\frac{2\Delta t}{N}\cdot V(f)^2\right] \quad (1)$$

where Δt is the sample interval (0.01 s); N is the number of samples in each selected time-series segment; the constant 1/0.825 is a scale factor to correct for the 10% cosine taper applied [53].

(4) Plot the PSDs of all segments at each station in time-frequency domain $PSD(t, f)$.

(5) Apply band-pass filter using the frequency bands, F = 0.1–0.2 Hz (DF1), 0.2–0.3 Hz (DF2), and 0.3–0.4 Hz (DF3) to each 1-h segment to produce filtered amplitude-time series $V(t, F), N(t, F)$, and $E(t, F)$.

(6) Rotate the two horizontal components by an angle φ into radial (R) and transverse (T) components [54] in each segment:

$$\begin{bmatrix} R(t,F,\varphi) \\ T(t,F,\varphi) \end{bmatrix} = \begin{bmatrix} -\cos(\varphi) & -\sin(\varphi) \\ -\sin(\varphi) & -\cos(\varphi) \end{bmatrix}\begin{bmatrix} N(t,F) \\ E(t,F) \end{bmatrix} \quad (2)$$

in which φ is defined as the back-azimuth angle between the north and the radial direction from the recording station toward the source.

(7) Calculate the root mean square of $R(t, F, \varphi)$ and $T(t, F, \varphi)$ in each segment and their ratio $Ra(F, \varphi)$.

(8) Repeat steps 6 and 7 to calculate $Ra(F, \varphi)$ at every 1° increment of angle φ in 1°–360°.

(9) Calculate average values of $Ra(F, \varphi)$ for all segments at a station and determine the azimuth φ_m for the maximum value of these averages. The angle φ_m is taken as the primary vibration direction by Ra method.

(10) Further parse the filtered 1 h segments in step 6 into ten 120 s windows and determine V_{ij}, N_{ij}, E_{ij}, where i = 1, 2, and 3 (separating DF1, DF2, and DF3 contents) and j = 1, 2, ... , 30 (number of windows). For each window, build the three-component covariance matrix **M**:

$$\mathbf{M}_{ij} = \begin{bmatrix} cov(V_{ij}, V_{ij}) & cov(V_{ij}, N_{ij}) & cov(V_{ij}, E_{ij}) \\ cov(N_{ij}, V_{ij}) & cov(N_{ij}, N_{ij}) & cov(N_{ij}, E_{ij}) \\ cov(E_{ij}, V_{ij}) & cov(E_{ij}, N_{ij}) & cov(E_{ij}, E_{ij}) \end{bmatrix} \quad (3)$$

(11) Calculate the three eigenvalues λ_{ij} and associated eigenvectors $\vec{x}_{ij}$ of the covariance matrix $\mathbf{M}_{ij}$ by solving:

$$\mathbf{M}_{ij}\vec{x}_{ij} = \lambda_{ij}\vec{x}_{ij} \quad (4)$$

and define the maximum eigenvalue λ_{ij1}, and associated eigenvectors $\left(x_{ij_{11}} \quad x_{ij_{12}} \quad x_{ij_{13}} \right)^{\mathsf{T}}$.

(12) Find the back azimuth angle φ_{ij} corresponding to the major axis of the polarized ellipse:

$$\begin{cases} \varphi_{ij} = \arctan\frac{x_{ij_{13}}}{x_{ij_{12}}} & \text{if } x_{ij_{11}} > 0 \\ \varphi_{ij} = \arctan\frac{x_{ij_{13}}}{x_{ij_{12}}} + 180 & \text{if } x_{ij_{11}} < 0 \end{cases} \quad (5)$$

Compute the probability of the back azimuth angle within 0°–360° range with a 10° bin width. The back azimuth of highest probability is considered as the primary vibration direction by the polarization method.

2.2. Ocean Data

Theoretically, the frequencies of ocean waves that generate DF microseisms should be half of the frequencies of DF peaks. Therefore, daily WWIII hindcast of $E(F/2)$ (log10(m^2/Hz)) distributions within the half frequency bands of corresponding DF peaks are used in this study in order to explore the association between the ocean wave climates in Northern Atlantic Ocean and DF microseisms.

Additionally, a total of 19 ocean buoys in Atlantic Ocean (see Figure 1 for locations) were selected to retrieve recordings of the ocean climate parameters including dominant wave period and significant wave height (detailed descriptions of which can be found at National Data Buoy Center website). In order to clearly display the locations of the buoys on the figure and to facilitate discussion of the observations, these buoys were renumbered and divided into four groups according to their locations (Figure 1): (1) deep ocean (DO) buoys (9, 10, and 11); (2) the continental slope and deep ocean side (SlDO) buoys (2, 5, 17, and 25); (3) the continental slope and shelf side (SlSh) buoys (3, 6, 7, 19, 21, and 23); and (4) the continental shelf (Sh) buoys (the remaining ones).

In order for a direct comparison of frequency compositions of the ocean wave and DF microseisms, the dominant ocean wave frequency of ocean waves at each buoy was simply doubled to determine the double ocean wave frequency (DWF).

2.3. Source Regions of DF Microseisms

2.3.1. Spatial Density of Primary Vibration Directions

The daily primary vibration directions (φ_{m1}) were calculated for all stations and the great circles corresponding to all φ_{m1} were generated in the three DF bands. Any point on a great circle can be considered as a possible energy source of the corresponding station in the corresponding day and DF band. In order to find the possible source areas, spatial density of the points sampled from the great circles with a sampling interval of 100 m were calculated, normalized, and plotted as a color gradient map for each DF band. The relative magnitude of the spatial density reflects the possibility of the area to be an energy source, i.e., the larger the density, the higher the possibility.

2.3.2. Correlation Analyses

As the selected recording period was free of major anomalous ocean activities (e.g., ocean storm, typhoon, hurricane) in Atlantic or Pacific Oceans (according to National Hurricane Center, NHC), the DF microseisms were generated by the interactions of the incoming ocean waves and those reflected at the continental margin. In order to investigate the significance of the continental slope in the excitation of the DF microseisms quantitatively under normal sea states, correlation analyses of time histories between DF microseisms and ocean wave energy were carried out by considering the time-dependent variation of frequency composition and energy levels. A few DF microseism events were identified as their PSD levels are higher than the average PSDs in the whole recording period in the frequency band. Their generation mechanisms were explored by comparing temporal and spatial variations of the PSDs to the migration patterns of ocean wave fronts in WWIII hindcasts in corresponding frequency bands, as well as comparing the frequency band of each event to the DWF of ocean buoys in four groups. For the entire period of microseism recordings, Pearson correlation coefficients (CC) were calculated between the DF microseisms and ocean wave heights recorded at selected buoys in each frequency band.

3. Results

3.1. Power Spectral Density (PSD)

Figure 2a shows the vertical $PSD(t,f)$ plots at the stations selected to cover a wide latitude range (40°N–34°N) (see Figure 1 for locations of the stations). The remaining stations were divided into three groups according to their locations (for their groups, see Figure S1a in the Supplementary Materials to this article) and their $PSD(t,f)$ plots are shown in Figure S1b,d,f. In the recording time period of this study, DF peaks are mainly in the frequency band of 0.15–0.3 Hz at the selected example stations, however, may cover a much wider frequency band at the stations close to the coastline (see Figure S1b in the Supplementary Materials to this article). At all stations, three DF microseism events are identified and labeled with I, II, and III which will be described in Section 3.3.

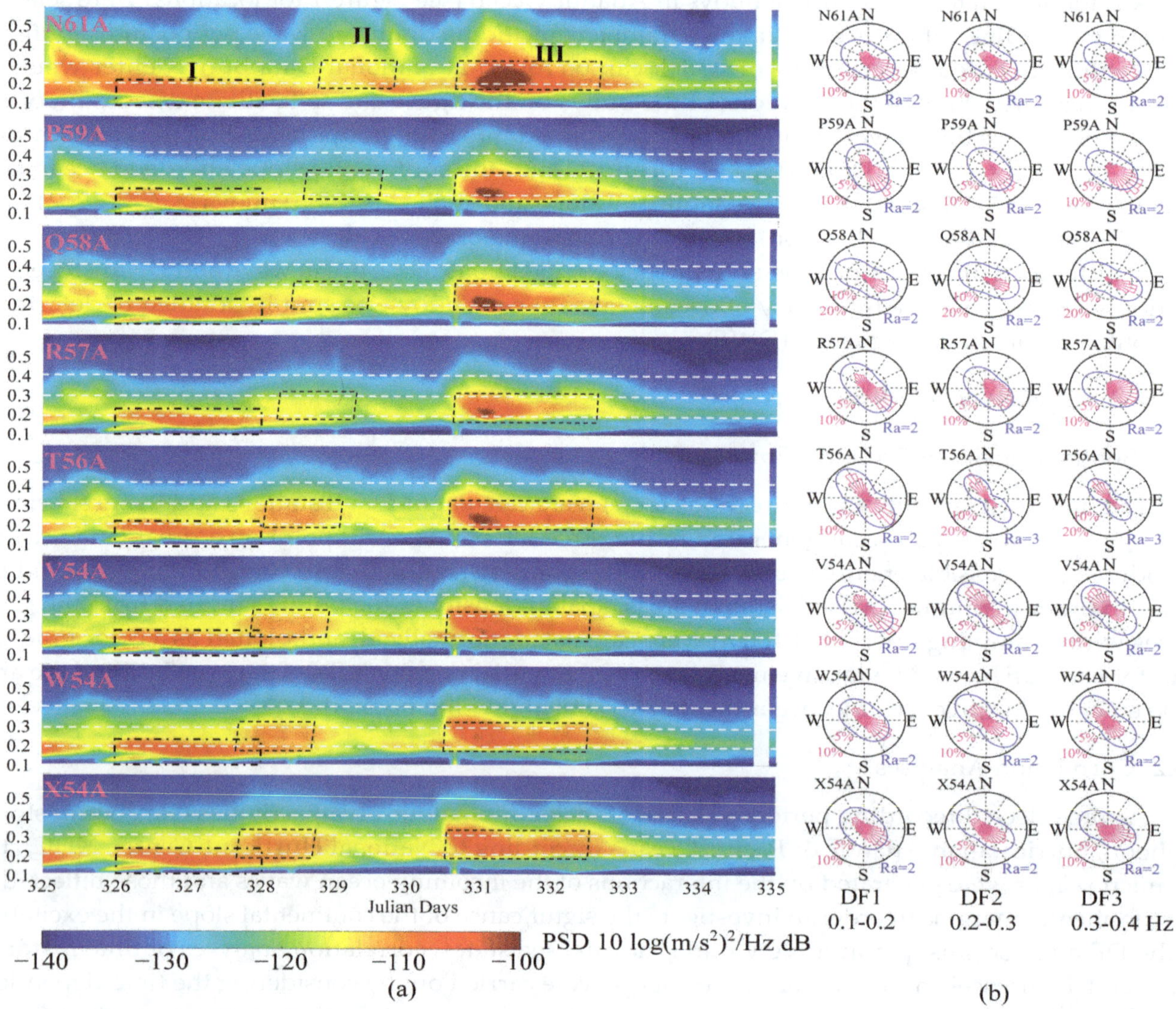

Figure 2. (**a**) Power spectral density (PSD) plots in time-frequency domain of the vertical components at the selected stations. The dashed lines at 0.2 and 0.3 Hz mark the boundaries of the three frequencies ranges. Three relatively strong double-frequency (DF) microseism events are identified and labeled as I, II, and III. (**b**) Polar plots of average $Ra(\varphi)$ values (blue) combined with rose diagrams of back azimuths (calculated by the polarization analysis) (purple) at the three DF bands at the selected stations in the whole recording period. On each plot, the text labels refer to the recording station (e.g., N61A), probability of back azimuth (in purple, e.g., 10%) and the scale of the solid outer circle in multiples of the $Ra(\varphi)$ value (in blue, e.g., $Ra = 2$). The same plots for the remaining stations are shown in Figure S1 in the Supplementary Materials.

3.2. Primary Vibration Directions at DF Peaks

Figure 2b presents the polar plots of average radial-to-transverse spectral ratios $Ra(\varphi)$ (blue outline) and rose diagrams of back azimuths calculated by the polarization analysis (purple) in DF1, DF2, and DF3 bands for the selected stations over the entire recording period of 10 d. The same plots for the remaining stations are given in Figure S1c,e,g in the Supplementary Materials. The longer axis of each $Ra(\varphi)$ outline is identified indicating the average primary vibration direction in 10 days (φ_{m10}), which closely coincide with the major polarized direction.

The φ_{m1}s was calculated as well for all stations, and rose diagrams of them in 10 days were generated for the three DF bands in Figure 3. The main back-azimuth in the three DF bands are shown to be in 110°–150°, which is perfectly consistent with the results in the same area in [55].

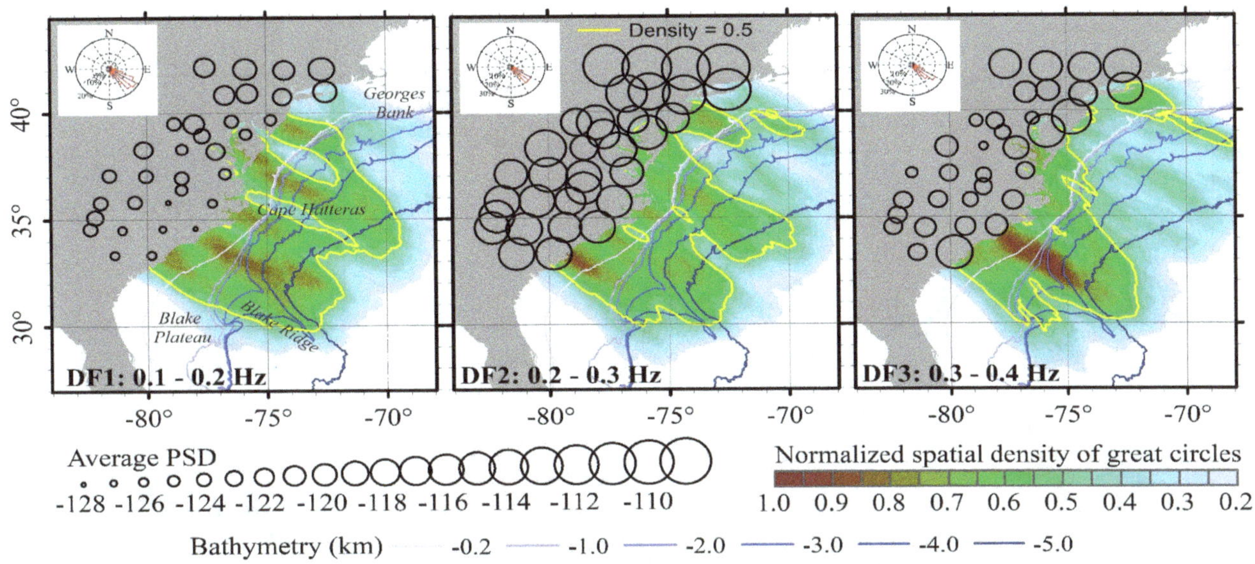

Figure 3. The normalized spatial density (color gradient maps) of the great circles corresponding to the daily primary vibration directions (φ_{m1}) and the vertical PSDs averaged for the entire recording period (scaled circles) in the three DF bands. The yellow lines contour the density of 0.5. The rose diagram shows the probability distribution of all φ_{m1}s.

The spatial density of φ_{m1}s was calculated, normalized, and plotted as a color gradient map for each DF band in Figure 3, as well as the contours at a density of 0.5 and the ocean bathymetries. Comparing the three maps, it can be observed that the areas of high spatial density, e.g., the areas sketched by the density contours of 0.5, shrink towards the continental shelf with increase of frequency, and the spatial density in the area between Blake Ridge and Cape Hatteras is high in all three DF bands.

3.3. *Excitation of the Three Relatively Strong DF Microseism Events*

First, the spectral WWIII hindcasts of ocean wave energy ($E(F/2)$ in log10(m^2/Hz)) in North Atlantic Ocean (see Figure S2 in the Supplementary Materials to this article) were used to explore the temporal and spatial relationships in each frequency band of the energy levels between PSDs and wave energy. This result demonstrates that (1) the primary vibration directions do not point to the areas of high wave energy in open ocean, and (2) variations of ocean wave activities in open ocean of the North Atlantic Ocean have limited influence on the DF microseisms observed in the east coast of the United States.

Based on the spatial density of great circles of φ_{m1}s presented in Figure 3, it can be inferred that the excitations of DF microseisms appear to be associated with the ocean waves in different areas of the continental margin of the western North Atlantic Ocean. Therefore, the excitation mechanisms of the three DF microseism events identified in Figure 2 are explored below with reference to the area of the continental margin as outlined in Figure 4.

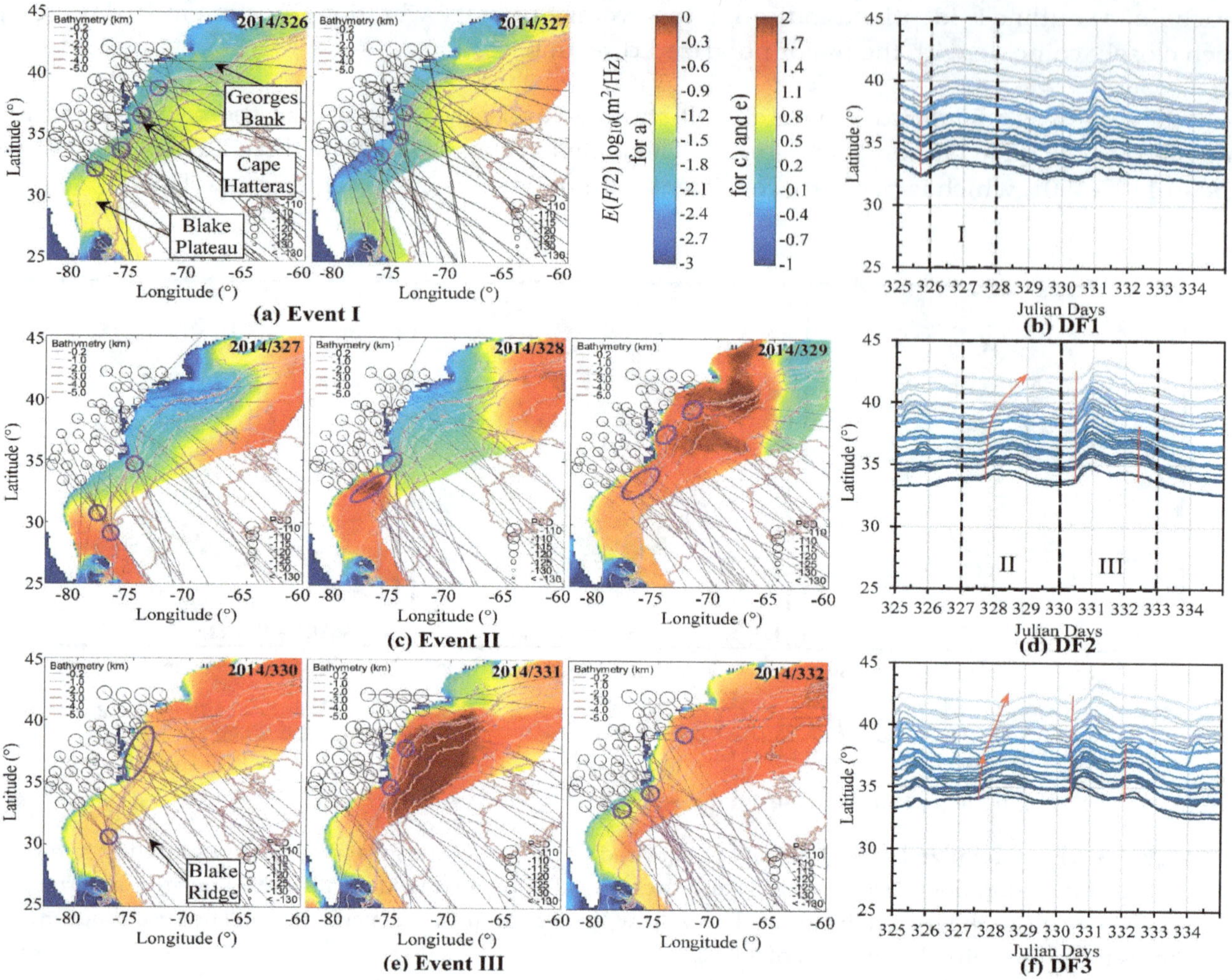

Figure 4. (**a,c,e**) Daily WAVEWATCH III hindcasts of ocean wave spectra ($E(F/2)$ in log10 (m^2/Hz)) in Northern Atlantic Ocean (color gradient maps), PSD levels (small circles with scaled sizes), and primary vibration directions (segments of great circles) at all stations corresponding to the three events (I, II, and III) identified in Figure 2. The purple circles/ellipses delimit the intersections of the great circles; (**b,d,f**) Time history of average PSDs in three frequency bands. In each plot, the curves are stacked by the latitudes of the stations and the relief of each curve shows the change of PSD level. The starting time of the three events are picked and connected to form the red lines, the arrows of which show the impact sequence.

Figure 4a,c,e shows the WWIII hindcasts of $E(F/2)$ in the western Northern Atlantic Ocean (color gradient maps) within the half frequency band of events I, II, and III identified in Figure 2. In the corresponding days and frequency bands of the events, the PSD levels and the primary vibration directions are demonstrated by the small circles with scaled sizes and segments of great circles passing though the corresponding stations. The bathymetries of the western Northern Atlantic Ocean are plotted as well in order to examine the importance of the continental slope in generation of DF microseisms. In Figure 4b,d,f, the arrival times of each event are picked on the PSD-time plots of all stations, and the connection of them form the red lines indicating the sequence of energy impact on stations. In order to facilitate the description of the spatial variations of PSD levels and primary vibration directions, the area where stations are placed are divided into two sections, north and south of Cape Hatteras.

3.3.1. Event I

The energy of event I in DF1 band (LPDF, 0.1–0.2 Hz) starts to appear in late hours of day 2014/325 (red line on day 325 in Figure 4b) and reaches a higher strength on days 2014/326 and 327 (Figures 2 and 4b). The ocean waves in frequency band of 0.05–0.1 Hz in the deep ocean close to the continental slope are uniformly higher on day 326 than day 327, and extend to the continental slope (Figure 4a). In addition, intersections of great circles concentrate at several small areas (purple circles) evenly distributed on the continental slope on day 326. All these observations might explain the occurrence of event I at the same time on day 326. The ocean wave energy decreases in the deep ocean close to the continental slope and increase slightly at the deep ocean area south of Georges Bank on day 327. The primary vibration directions at all stations cover a wider range, and the intersections of great circles are not as concentrated as on day 326. These changes can explain the decrease of PSD levels at most stations except for eight northernmost stations on day 327.

3.3.2. Event II

The event II in DF2 (0.2–0.3 Hz) band occurs on days 2014/327, 328, and 329, and arrives at the stations in south section before north section as indicated by the curved red arrow in Figure 4d. This sequence can be explained by the ocean wave interactions associated with the continental slope (Figure 4c).

On day 327, the wave energy in frequency band of 0.1–0.2 Hz is relatively higher on lower Blake Plateau (water depth 200–2000 m) and many great circles of the stations in the south section intersect on the edge of this area. The PSD levels at the stations in the south section are roughly proportional to their distances to lower Blake Plateau. Then on day 328 the wave front moves north along the continental slope to south of Cape Hatteras while the wave energy increases to the highest. The great circles of almost all stations in south section intersect in this area, and the PSD levels at these stations increase significantly. On day 329, the wave front moves further north to the area between Georges Bank and Cape Hatteras where the strike of the continental slope turns almost 90°, causing the significant increase of wave energy in this area. The great circles at the stations in north section do not intersect in the open ocean where the wave energy is higher, but on the continental slope segments south of Georges Bank and north of Cape Hatteras. The PSD levels at the stations to the north increase but not as significantly as at the southern stations on day 328. The great circles at the southern stations still intersect at the continental slope between Cape Hatteras and Blake Plateau but not as concentrated as on day 328.

3.3.3. Event III

The event III in DF2 (0.2–0.3 Hz) band occurs on days 2014/330, 331, and 332, arrives at the stations almost at the same time, but generates the energy peaks slightly earlier at the stations in the south section (Figure 4d). This sequence can also be explained by the ocean wave interactions associated with the continental slope (Figure 4e).

On day 330, the great circles of most stations point to the area between Georges Bank and Blake Plateau where the ocean waves are roughly the same height, and intersect evenly on this continental slope segment. However, the ocean wave energy on the continental slope of Georges Bank are relatively higher which coincide well with the relatively higher PSD levels at the four northernmost stations. On day 331, the ocean wave energy grows very fast while the wave front moves north quickly to Georges Bank, resulting in high waves on the continental slope segment and deep ocean area between Georges Band and Blake Ridge. Most intersections of the great circles align on this continental slope segment and nearby deep ocean, and the PSD levels are very high at all stations. On day 332, the wave front moves to northeast of Georges Bank and the wave energy decreases moderately. However, the wave energy on the continental slope south of Georges Bank is still relatively high and many great circles of the stations in northern section intersect here, which might be the reason for high PSD levels at the eight northernmost stations. A new high wave front is developed at Blake Ridge where most great circles of stations in southern section intersect, and a new PSD peak appears at the stations in the southern section.

3.4. Correlation between DF Microseisms and Ocean Wave Parameters

The time histories of average PSDs in DF1, DF2, and DF3 bands are shown in Figure 4b,d,f, respectively. The curves are stacked according to the latitudes of the stations and the reliefs show the time-dependent variations of PSD levels. The DWFs at the DO, SlDO, and SlSh buoys are closer to the frequency band of event I, and the waves whose DWFs match events II and III are observed in groups SlDO, SlSh, and Sh (for the time history of DWFs at the buoys, see Figure S3 in the Supplementary Materials to this article). The Pearson correlation coefficients (CC) between the time series of ocean wave height and PSD in the three DF bands are calculated for all pairs of ocean buoy and ambient noise station, and the CC values are then averaged in each buoy group for each DF band at each station, as plotted Figure 5. The right most bar in Figure 5 gives the CC values averaged for all stations. The CC values are normalized within the range [−1.0, 1.0], where positive (negative) values represent the same (opposite) trends of ocean wave height and PSD pairs and the absolute value of CC express the level of consistency in these trends. In the DF1 band, better correlations can be found between the DF microseisms at all stations and the ocean wave heights in the continental slope on deep ocean side (SlDO). In the DF2 band, the DF microseisms at most stations correlate well with the ocean wave heights in SlDO, and some with those in the continental slope on the shelf side (SlSh). In the DF3 band, higher CC values can be found in all SlDO, SlSh, and continental shelf (Sh) groups. Near-zero negative CC values for the deep ocean (DO) buoys implies that deep ocean waves do not exert a positive influence on the DF microseisms.

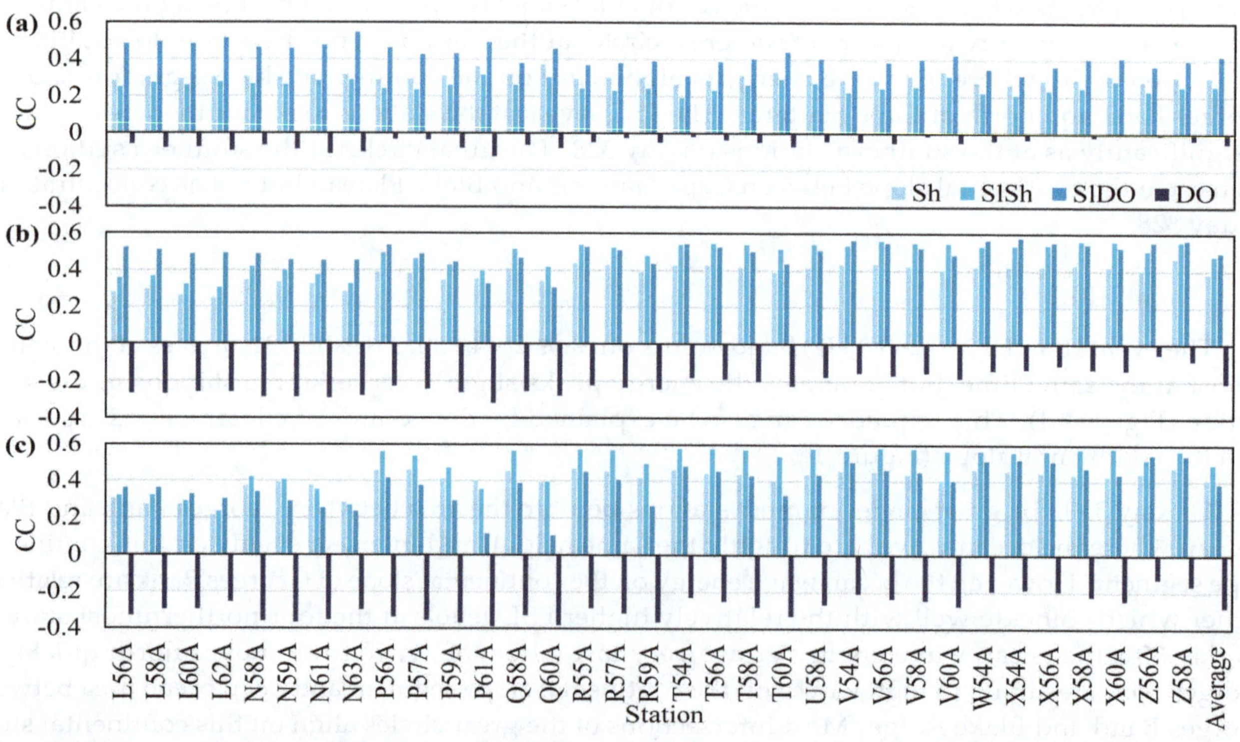

Figure 5. Average Pearson correlation coefficients (CC) between the PSDs in the three DF bands and the ocean wave heights in four groups of ocean buoys.

4. Discussion

4.1. Hypothesis on the Significance of Continental Slope for the Origination of DF Microseisms

The correlation analysis (Figure 5) shows that the DF microseism trends in DF1, DF2, and DF3 bands are most compatible with the ocean wave activities in the continental slope on the deep ocean side (SlDO), the continental slope on both deep ocean and continental shelf sides (SlDO and SlSh), and SlSh, respectively. These domains of wave activities are separated by the continental slope, where waves approaching from the deep ocean zone are reflected back creating nonlinear wave–wave interactions. This naturally leads to a hypothesis that the continental slope plays a significant role in the origination of DF microseisms, and with increasing frequency band, the dominant origination area migrates from SlDO to SlSh in the east coast of the United States. Validity of this hypothesis is explained in detail in the following.

During the time period of event I (Figure 4a), the continental slope is shown to be the boundary of impact from the ocean wave in frequency band of 0.05–0.1 Hz, the increase of wave energy in the deep ocean area south of Georges Bank does not cause an increase of PSD levels at the stations in the north section but the decrease of ocean wave energy on the continental slope do coincide well with the decrease of PSD levels in most stations from day 2014/326 to 327. Comparing events II and III (Figure 4c,e), the PSD levels are much higher on day 331 than day 329, mainly because of the higher wave energy on the continental slope between Georges Bank and Blake Ridge on day 331. Figure 6 shows the differences of the wave energy and PSD levels corresponding to DF2 band between day 331 and 329. A coincidence can be found between the changes of PSD levels and wave energy on the continental slope segment between the Georges Bank and Blake Ridge (outlined by the purple dash-dot line), which supports the hypothesis.

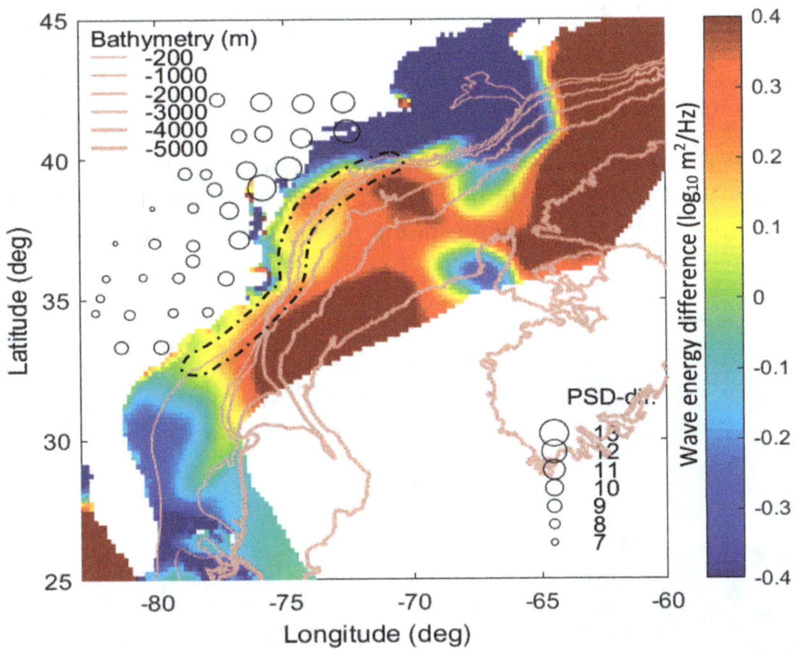

Figure 6. Difference of the ocean wave energy ($E(F/2)$ in $\log_{10}(\mathrm{m}^2/\mathrm{Hz})$) and PSD levels between day 329 and 331 (Figure 4c,e, respectively).

As no hurricane development was reported in Northern Atlantic Ocean during the period of recordings analyzed in this study, most DF microseisms identified from these recordings should be generated mainly by the nonlinear interactions of incoming and reflected ocean waves of similar frequencies. These interactions take place at different intensities and directions as determined by ordinary ocean activities and ocean bottom topography. Theoretically strong reflections leading to strong DF energy can occur only if the incoming waves encounter an obstacle perpendicular to their propagation direction, as shown in Figure 7. The area and energy of constructive interaction is much larger when the incoming wave direction is perpendicular (Figure 7b) rather than nearly parallel (Figure 7a) to the obstacle. Considering that, in a reflection system, at each point of incidence at any angle, the energy normal to the reflector is the largest (Figure 7b), a station receives the strongest signal when the station and incident point is aligned with the line normal to the reflector (Path A in Figure 7b). Such an alignment should therefore correspond to the great circles (lines connecting the wave origination areas to the stations) intersecting the continental slope nearly at orthogonal angles. In order to test this hypothesis, the intersection angles δ between the great circles and the strike of the continental slope are defined as shown in Figure 8a and the frequency histograms of δs in the three DF bands are generated and shown in Figure 8b. The medians of δs in the three DF bands are 72.4°, 74.1°, and 72.5°, respectively, and the largest frequencies of occurrences are within the 81°–90° range, which support the proposed hypothesis.

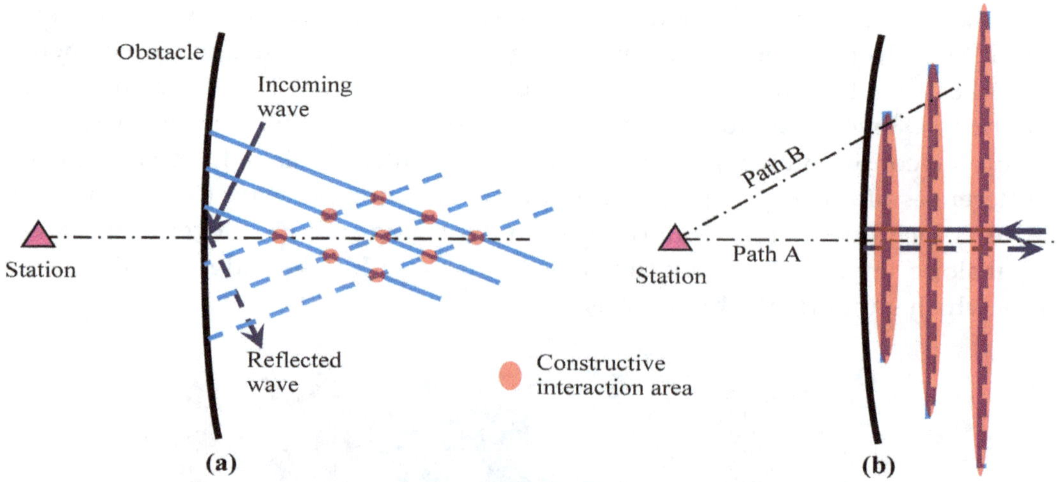

Figure 7. A sketch describing how constructive interactions of ocean waves and their reflections from a barrier (the continental slope or the shoreline) can result in different energies when the angle of incidence is (**a**) large and (**b**) 0°. Note a given station receives the strongest signals along the shortest path A.

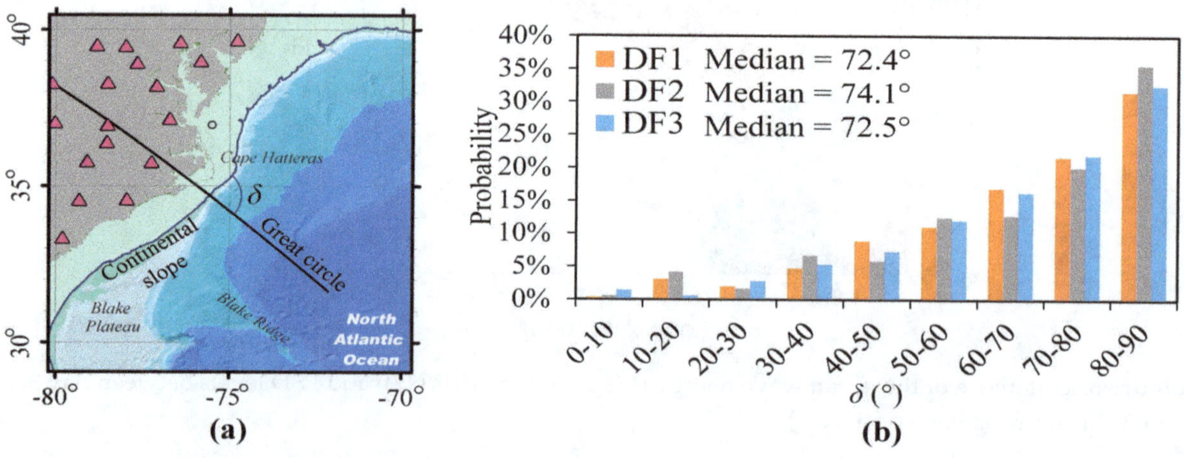

Figure 8. (**a**) Defining the angle δ (<90°) between a great circle and the continental slope's strike. (**b**) Histogram of the angles δs in the three DF bands.

As the waves are reflected from the continental slope, the wave energy should be higher in the continental slope and the nearby deep ocean zone than in the distant deep ocean zone and the continental shelf. To examine this notion, the mean and standard deviation of ocean wave energy were calculated in each of the four buoy groups, 2.19 and 0.44 m in DO, 2.48 and 0.98 m in SlDO, 2.10 and 0.99 m in SlSh, and 1.57 and 0.74 m in Sh. The highest wave energy appears at the SlDO buoys and then DO and SlSh buoys, which coincide well with areas of intersection of the great circle paths (purple dashed ellipses in Figure 4). Similar observations can also be found in [33]. Recalling once more that there was no strong storm in the northern Atlantic Ocean during the microseism recordings, the identified DF microseisms cannot be explained by ocean storms.

The hypothesis is also supported by several ocean bottom observations in shallow and deep waters divided by the continental slope. For example, the authors of [56] concluded that the excitation at DF peaks require some part of the ocean storm to extend over the shallow water based on coherence studies. Comparing ocean waves recorded in shallow waters (100 m deep) in Tasman Sea and microseisms recorded near the shoreline (30 km away from the ocean buoy) of the North Island of New Zealand, the authors of [57] observed that DF peaks were in the > 0.2 Hz band when there was a local wind in the Tasman Sea but lower than 0.2 Hz when the sea was calm and a swell from Southern Ocean arrived across the continental slope. Another example is presented by [31]. Two pressure spectra were obtained from two seafloor stations in the continental shelf (water depth of 0.6 km) and in the continental shelf edge followed by a steep and deep continental slope (water depth of 1 km), respectively, off the coast of southern California (see topographic profiles in [58]). At the shallower site, a high spectral peak with pressure level of around 10^3 Pa^2/Hz was observed at around 0.2 Hz when a storm directly passed overhead, whereas no spectral peak could be clearly identified when there was no passing storm. On the deeper site, two high and sharp pressure spectral peaks appeared at 0.14 and 0.3 Hz with pressure levels of 5×10^3 and 10^4 Pa^2/Hz, respectively. Differences between the effects of shallow and deep ocean on DF peaks' frequencies and energy levels presented in these studies imply that DF microseism in the continental shelf is driven by local weather, whereas that in the deep ocean is excited by the standing waves [35] generated by the interaction between the distant ocean swell and the waves reflected due to the sudden change of water depth at the continental slope.

4.2. Types of Continental Margin

In this study, two types of continental margins are identified as exemplified by the transects A-A′ and B-B′ (profile locations are marked in Figure 1) in Figure 9a,b, respectively. Both types have wide (≈100 km) and shallow (<0.2 km) shelves. Along the first type (A-A′), a high (≈2.5 km) and steep slope sharply changes the profile followed by a long gentle slope (continental rise). In the second type (B-B′), the shelf transition into the slope via a wide and gently dipping plateau (Blake Plateau) followed by a shorter (≈1.5 km) and gentler slope (than the first type), then a wide undulating plateau (continental rise) ending with a relatively steep slope.

These differences between the continental margin profiles appear to modify the mechanism of DF microseism generation as suggested by the consistently high spatial densities in the area covering the Blake Ridge and northern Blake Plateau in all three DF bands (Figure 3). A more gradual transition from the shelf and a shorter continental slope are the most prominent features that can support generation of a relatively stable energy level in these areas. Concavity of the continental slope at the edge of the Blake Plateau potentially causes strong reflections resulting in higher DF energy. In contrast, the continental slope at Cape Hatteras has a convex outline that could cause a diffraction pattern, consequently a lower spatial density in this area as shown in the map of the DF1 band in Figure 3. As the DF3 microseisms are generated in the continental shelf, the rough shoreline at Cape Hatteras may be the reason for higher density observed in the DF3 band in Figure 3.

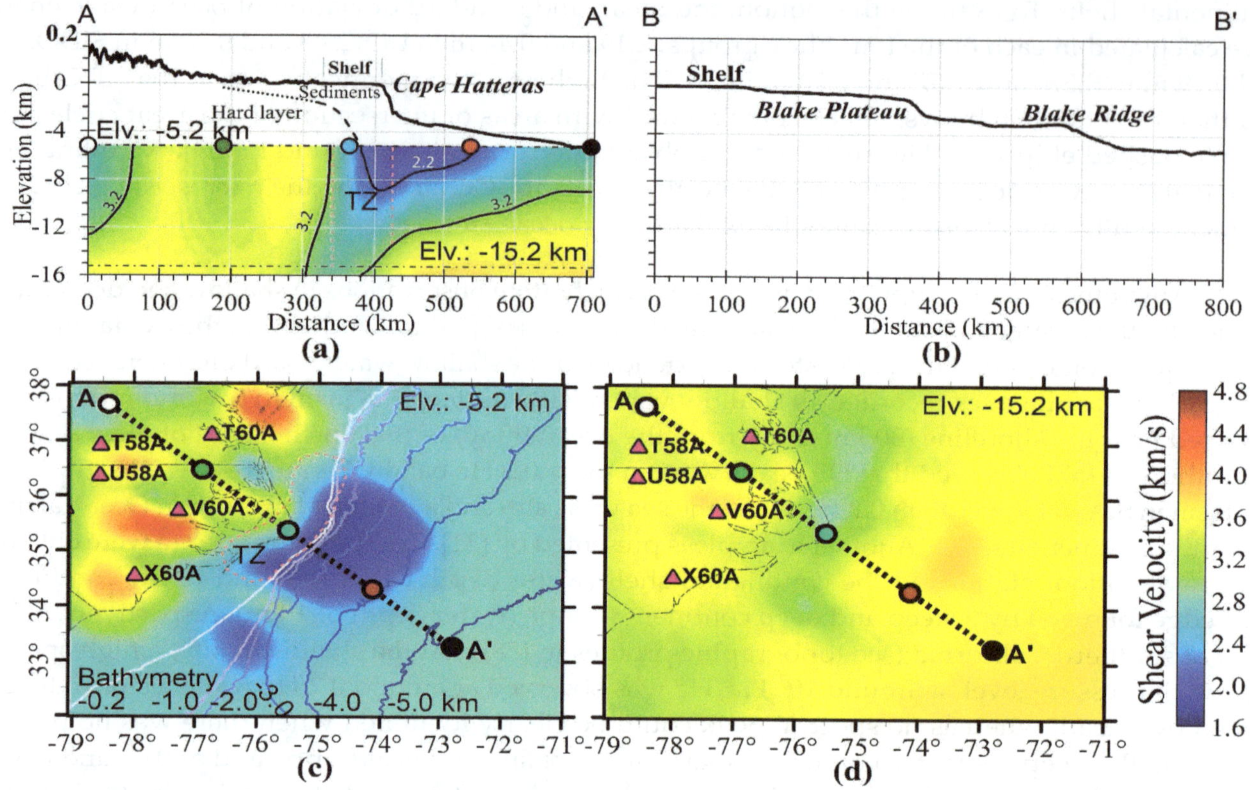

Figure 9. Topographic profiles of the continental margin along (**a**) A-A′ and (**b**) B-B′ (marked in Figure 1) based on General Bathymetric Chart of the Oceans (GEBCO, 2014). The shear velocity profile and contours in (**a**) are reproduced on [15]. The geophysical interface between sediments and bedrock is estimated (black dotted line) by horizontal-to-vertical spectral ratio (HVSR) method and was extended seaward by inference (black dashed line) to connect with the 2.2 km/s shear velocity contour. A transitional zone (TZ) with the largest shear velocity gradient is identified and outlined by the two pink dashed lines in (**a**). The shear velocity models inside the white box outlined in Figure 1 are generated for elevations (Elv.) of −5.2 km (**c**) and −15.2 km (**d**) based on [15].

Recent studies also support the hypothesis about the role of the continental slope and show that submarine ridges act similarly to cause reflection of the waves. In [59], by comparing the seasonal variation of DF microseisms and ocean activities, the authors concluded that the DF microseism in the 0.1–0.2 Hz band on the King George Island (on Antarctic Peninsula) originates from a region of Drake Passage instead of the continental shelf around Antarctic Peninsula even though it is several times wider than that around Cape Hatteras. The ocean at Drake Passage is at least 3 km deep and is delimited by the continental slope of Antarctic Peninsula and an underwater ridge roughly normal to the slope. The authors of [33] showed that the excitation locations of both P- and S-wave microseisms observed by a seismometer array in Japan are distributed along the eastern continental slope of Greenland and Reykjanes Ridge extended from Iceland into the deep ocean. An ocean bottom straight blocked by relief features [34,59] promotes formation of ocean wave reflection at the continental slope.

The hypothesis may appear to fail the test based on the observations in [37], who compared the DF spectra (around 0.15 Hz) at three seismometer stations in the coastal region of Oregon and California with the ocean wave climate parameters' spectra (see their Figure 17). Based on an excellent correlation between DF peak and wave spectra, they concluded that the DF microseism is generated by the wave activities near the shoreline. Significantly different widths of the continental shelves, being much narrower on the western continental margin of North America, and the location of the ocean buoys being on the edge of such a narrow continental shelf (see their Figure 15) can explain the apparent failure. Together with the low frequency (< 0.2 Hz) of their DF peak, it can be argued that the

DF microseism observed by [60] was also a result of the reflections from the nearby continental slope as put forward in the proposed hypothesis.

4.3. Rayleigh Wave Refraction

As mentioned in "Data acquisition and processing" section, the Ra and polarization analysis methods to estimate the primary vibration direction are based on an assumption that DF microseisms propagate dominantly as fundamental mode Rayleigh waves. To verify this assumption, Figure 10 was generated to show the probability distributions of the phase differences between the two orthogonal horizontal components (φ_{HH}) and between the vertical and horizontal components in the primary vibration direction (φ_{VH}) in the three DF bands. The facts that φ_{HH} is dominant in 0° and φ_{VH} is mainly in 65°–80° and −65°–−80° reveal that the energy is propagating as Rayleigh waves dominantly [61], which coincides well with the observation in [55].

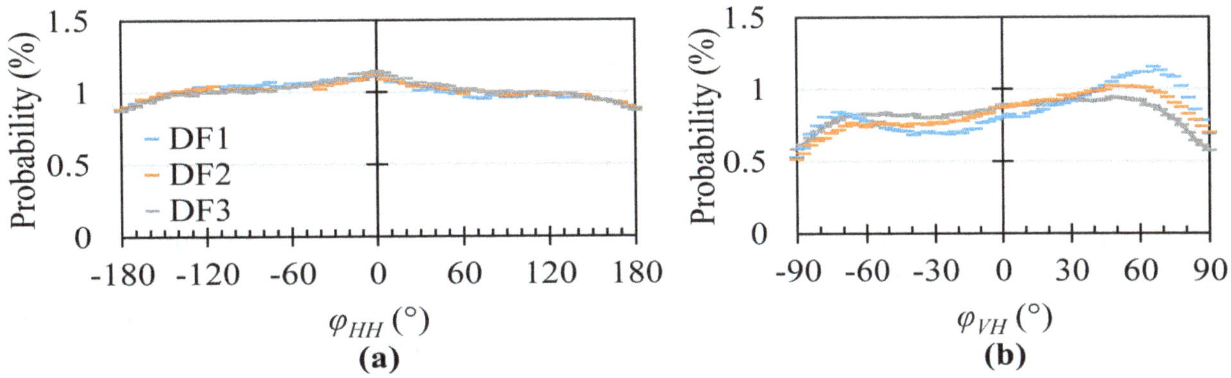

Figure 10. Probability distribution of the phase difference between (**a**) the two horizontal components (φ_{HH}) and (**b**) the vertical and horizontal component in the primary vibration direction (φ_{VH}).

4.3.1. Refraction at the Water–Solid Earth Interface

As explained by [61], when the energy in the DF band is generated from wave–wave interaction in the ocean, it propagates as "pseudo-Rayleigh waves" (p*R*g) in water column and turns to free surface Rayleigh waves (FSRW) when it reaches solid earth. In the water column, due to phase speed difference, the p*R*g exists in different forms: dominantly elastic p*R*g in shallow water with phase speed roughly equal to that of FSRW, and acoustic p*R*g in deep water with phase speed of about 60% of FSRW. Definitions of shallow and deep waters vary with the wave frequencies. According to the analysis in this study, the energy in LPDF (DF1) band is generated around the continental slope where the water depth is generally smaller than 3000 m (Figure 1), especially on the Blake Plateau (≤ 1000 m) and the edge of the Blake Ridge (≈3000 m). According to Figure 14 in [61], with the increase of frequency from 0.1 to 0.2 Hz, the depth in the water column of elastic p*R*g dominance decrease from 3000 to 1500 m, under which the elastic p*R*g transfer to fundamental acoustic p*R*g. Thus, a phase-speed difference might exist at the water–solid earth interface deeper than 1500 m, i.e., areas except the Blake Plateau, however, the DF energy would still propagate vertically in the water column and transfer to solid earth. Even though there is significant energy loss at the interface, the spherical spreading of the DF energy in the solid earth will not change. Therefore, the phase speed difference on the water–solid earth interface is not likely to affect the determination of the source location by great circle. For high-frequency (0.2–0.5 Hz) DF band, the hypothesis put forward in this manuscript claims that the energy is generated in the continental slope and continental shelf where the water depth is smaller than 200 m. The same figure in [61] shows that the energy in this band should also propagate as elastic p*R*g and directly transition to FSRW on the continental shelf. As there is no significant phase speed difference between elastic p*R*g and FSRW, Rayleigh wave refraction at the surface of solid earth is not likely to be significant.

4.3.2. Refraction within the Solid Earth

The Rayleigh wave ray paths are expected to bend also when they travel through the solid earth boundaries with significant impedance contrasts. In order to examine possibility of such boundaries and implications for the validity of the triangulation method, the shear velocity (*Vs*) structure of the study area is explored as described below.

The cross-section A-A′ presented in Figure 9a shows that under the continental slope there exists a layer of material having *Vs* less than 2.2 km/s, which is interpreted as sediments in [15]. Our HVSR survey around the eastern foothills of the Appalachian mountain and near the coastal area suggest that the sediment–hard layer interface lies at a depth as depicted by a dotted line in the A-A′ profile in Figure 9a, and that the average *Vs* of the sediment is about 0.9 km/s [5]. According to [51], the shear velocity contrast at the sediment–hard layer interface should be larger than 2.5 to produce a clear a predominant frequency peak on the HVSR spectrum, therefore, the *Vs* of the hard layer should be around 2.2 km/s. Therefore, this interface is inferred to connect to the 2.2 km/s contour line (depicted as the dashed line in Figure 9a). Shear velocity gradients are calculated along A-A′ at two different elevations by [15] shear velocity model, and a transitional zone (TZ) of largest shear velocity gradient (where *Vs* increases from about 1.7 to 3.2 km/s within a horizontal distance of ≈80 km) is identified and outlined as shown in Figure 9a. The TZ at Cape Hatteras (Figure 9c) extend roughly parallel to the continental slope. The absence of notable shear velocity variations at −15.2 km (Figure 9d) suggests that the transitional zone may not extend to this depth.

As explained above, the Rayleigh waves propagating through the solid earth from the deep ocean near the continental slope (DF1 and DF2 bands) or from the continental shelf (DF3 band) to the inland stations are expected to change propagation direction due to gradual and continuous refraction as they pass through TZ. The total refraction angle ε is defined as shown in Figure 11a only for the great circles passing through TZ which represents the worst scenario of changes. The cumulative probability of ε in the three DF bands is given in Figure 11b. The weighted average (m) of ε in DF1, DF2, and DF3 bands are calculated to be 8.8°, 8.2°, and 8.7°, respectively, and the probabilities of ε values less than these corresponding weighted averages (P < m) are 60.2%, 63.1%, and 59.6%.

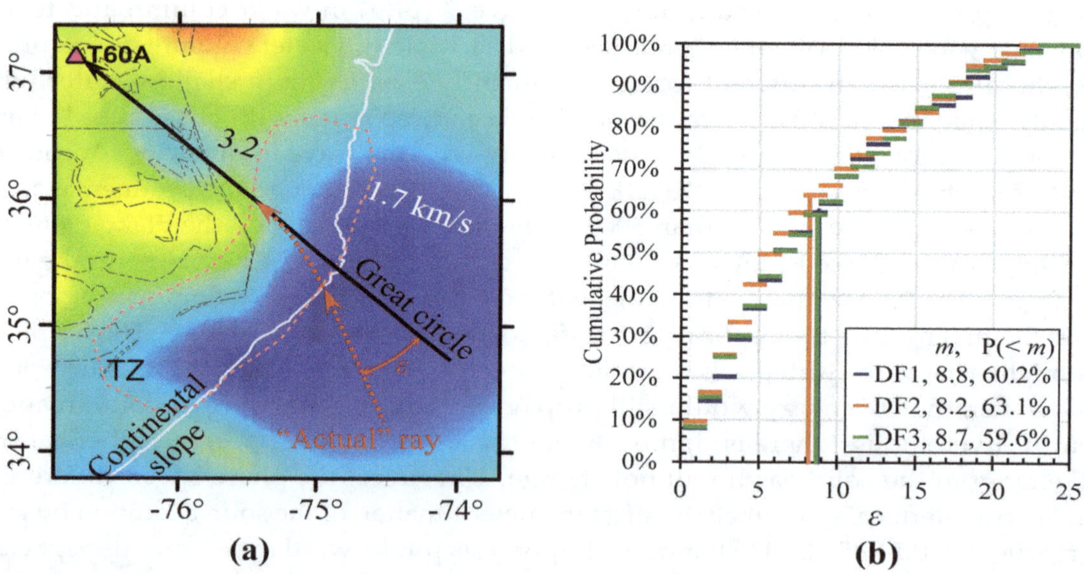

Figure 11. (**a**) Defining angle ε to measure total refraction of Rayleigh wave ray passing through the transitional zone (TZ in Figure 9a,c) taking a ray path to station T60A as an example. The black solid and red dotted lines are the great circle projected in this study and "actual" ray path of Rayleigh wave, respectively. (**b**) Cumulative probability of ε for waves in the three DF bands propagating through TZ.

Another examination of the Rayleigh wave refraction from sediments to bedrock is carried out by tracking hurricane "Sandy" in Atlantic Ocean on 27 October 2012 using 8 h ambient noise recorded on the bedrock in Tishomingo, Mississippi. The DF peak is found at the central frequency of 0.18 Hz with very high energy. The primary vibration direction at this central frequency is calculated by both *Ra* and Polarization analysis methods and projected as great circles which point to the locations of hurricane "Sandy" successfully (see Figure S4 in the Supplementary Materials to this article). Thus, the great circles can be considered as valid projections to the source areas of the DF energy.

4.4. Rayleigh Wave Refraction

According to the hypothesis and the discussion above, the DF microseisms in the LPDF (DF1) band is generated in the deep ocean close to the continental slope and propagate to stations on land. Depending on the location of the stations, the propagation paths are different, i.e., through sediments and bedrock to stations on inland bedrock, and through sediments on coastal stations on sediments, resulting in different attenuations and energy levels at stations. This difference is observed in this study. In Figure 3, the PSD levels in the DF1 band increase from coast (sediments) to inland (Appalachian Mountain) while the distance to source (ocean) increases. The possible explanation is that the attenuation due to spherical spreading is less effective than energy absorption in sediments. A similar trend is observed by [55] in the same study area. In the DF2 band, the PSD levels at coastal stations are slightly larger than or equal to those at inland stations, which might because the attenuation due to spherical spreading and absorption in sediments are equally effective. In the DF3 band, the PSD levels at coastal stations are obviously larger than those at inland stations, as the attenuation is dominantly spherical spreading and absorption effect is equally effective, which reveals that the DF microseisms in this band are generated in the continental shelf and propagate mainly in sediments.

5. Conclusions

This study explored the role and significance of the continental slope in the interactions between the ocean waves and the continental margin as well as the resulting double-frequency (DF) microseisms recorded in ENAM. The primary vibration direction analysis of the ambient noise recordings in the study area shows that these DF microseisms originated in areas of the North Atlantic Ocean, which are generally aligned in the SE direction with the recording stations. The great circles corresponding to these primary vibration directions for different DF peaks intersect at a number of locations enabling delineation of source areas along the continental slope. Correlation analysis between DF microseisms and ocean wave climate by considering the correspondence in their frequency composition and variation in energy levels shows that the DF microseisms in DF1 (0.1–0.2 Hz), DF2 (0.2–0.3 Hz), and DF3 (0.3–0.4 Hz) bands correlate well with the ocean wave activities in the continental slope on the deep ocean side, and in the continental slope on both the deep ocean and shelf sides, and in the continental slope on the shelf side, respectively. These analyses lead to a hypothesis on the frequency dependent interactions of ocean waves with the continental margin and the origination of DF microseisms. The steep continental slope is a key submarine topographic feature which behaves as an obstacle causing reflections of the incoming low frequency (≤ 0.15 Hz) ocean waves and formation of standing waves to generate low frequency (≤ 0.3 Hz) DF microseisms. While the high frequency (≥ 0.15 Hz) ocean waves are reflected at the shallow portion of the continental shelf to excite high frequency (≥ 0.3 Hz) DF microseisms. This hypothesis is also supported by the observations that (1) the great circles corresponding to the primary vibration directions of DF microseisms are mostly normal to the strike of the continental slope; and (2) the ocean wave energy in the continental slope or the nearby deep ocean are higher than in the distant deep ocean and the continental shelf. Additional systematic observations at different parts of the globe will help to determine validity and limits of the proposed hypothesis under all possible climatic and bathymetric conditions.

Understanding the generation mechanisms and locating the source regions of the DF microseisms would improve the reliability of estimating the amplification factor based on ambient noise because the amplification process could be understood by bridging the variations of the input energy levels at the sources and the energy at the site. In addition, DF microseisms recordings at the coastal areas suggests that the sources are not homogeneously distributed. However, surface wave tomography could be carried out only if the noise sources are homogenously distributed. Therefore, locating the source regions could improve the current method of tomography or initiate a new method using non-homogenous ambient noise.

With the frequency dependent interactions between the ocean waves and the continental margin determined, one could further analyze the possible mass wasting on the continental margin caused by the ocean wave energy input and transmission while ocean waves interact with the continental margin.

Supplementary Materials

Figure S1: (a) Map of the USarray transportable array stations and their groups. Vertical power spectral density (PSD) plots in time and frequency domain of (b) coastal stations, (d) Appalachian mountain stations, and (f) south section stations. (c,e,g) Ra outlines and rose diagrams of polarized back azimuth in three DF bands of the stations in the three station groups, Figure S2: Daily WWIII hindcasts of spectral ocean wave height in Northern Atlantic Ocean (color gradient maps), and PSD levels (small circles with scaled sizes) and primary vibration directions (great circles) at all stations in the corresponding DF bands: DF1 in the top two rows, DF2 in the middle two rows, and DF3 in the bottom two rows. The frequency band of ocean wave (half of corresponding DF band) and day are labeled on top of each plot, Figure S3: Time history of double dominant ocean wave frequencies (DWF) and significant ocean wave heights (WH) at the ocean buoys grouped according to their locations, deep ocean (DO), the continental slope on the deep ocean side (SlDO), the continental slope on the shelf side (SlSh), and the continental shelf (Sh) (see Figure 1 for their locations and original names), Figure S4: An example to verify the validity of the great circle projecting to the sources of LPDF microseisms. The great circles at the station T2 on bedrock in Tishomingo, Mississippi project to the sources of DF microseisms induced by hurricane "Sandy" on 27 October 2012 successfully.

Author Contributions: Conceptualization, Z.G.; methodology, Z.G.; formal analysis, Z.G.; investigation, Z.G.; writing—original draft preparation, Z.G.; writing—review and editing, A.A., M.X. and Y.H.; funding acquisition, Y.H. All authors have read and agreed to the published version of the manuscript.

Acknowledgments: The Transportable Array (network TA) data are available on IRIS data management center at http://ds.iris.edu/ds/nodes/dmc/. The ocean buoy data are from the National Data Buoy Center database at http://www.ndbc.gov/ (last accessed Feb. 20, 2019). The WAVEWATCH III®hindcasts of ocean wave energy between 2014/325 and 2014/334 are from https://wwz.ifremer.fr/iowaga/Products. The historical hurricane data in East Pacific Ocean and Atlantic Ocean from National Hurricane Center (NHC) at http://www.nhc.noaa.gov/data/#tcr (last accessed Feb. 20, 2019). The gridded topographic and bathymetric data are downloaded from General Bathymetric Chart of the Oceans (GEBCO) at http://www.gebco.net/data_and_products/gridded_bathymetry_data/ (last accessed Feb. 20, 2019). We wish to thank Colton Lynner for sharing the shear velocity model in the eastern North American margin.

References

1. Nakamura, Y. A method for dynamic characteristics estimation of subsurface using microtremor on the ground surface. *QR of RTRI* **1989**, *30*, 25–33.
2. Bodin, P.; Horton, S. Broadband microtremor observation of basin resonance in the Mississippi embayment, Central US. *Geophys. Res. Lett.* **1999**, *26*, 903–906. [CrossRef]
3. Bodin, P.; Smith, K.; Horton, S.; Hwang, H. Microtremor observations of deep sediment resonance in metropolitan Memphis, Tennessee. *Eng. Geol.* **2001**, *62*, 159–168. [CrossRef]
4. Bard, P.-Y.; SESAME-team. Guideline for the Implementation of the H/V Spectral Ratio Technique on Ambient Vibrations-Measurements, Processing and Interpretations; SESAME European Research Project EVG1-CT-2000-00026, D23.12. 2005. Available online: http://sesame-fp5.obs.ujf-grenoble.fr/index.htm (accessed on 19 February 2020).
5. Guo, Z.; Aydin, A.; Kuszmaul, J. Microtremor recording in northern Mississippi. *Eng. Geol.* **2014**, *179*, 146–157. [CrossRef]
6. Lobkis, O.I.; Weaver, R.L. On the emergence of the Green's function in the correlations of a diffuse field. *J. Acoust. Soc. Am.* **2001**, *110*, 3011–3017. [CrossRef]

7. Snieder, R. Extracting the Green's function from the correlation of coda waves: A derivation based on stationary phase. *Phys. Rev. E* **2001**, *69*, 046610. [CrossRef]
8. Shapiro, N.M.; Campillo, M. Emergence of broadband Rayleigh waves from correlations of the ambient seismic noise. *Geophys. Res. Lett.* **2004**, *31*, L07614. [CrossRef]
9. Shapiro, N.M.; Campillo, M.; Stehly, L.; Ritzwoller, M.H. High resolution surface wave tomography from ambient seismic noise. *Science* **2005**, *307*, 1615–1618. [CrossRef]
10. Sabra, K.G.; Gerstoft, P.; Roux, P.; Kuperman, W.A.; Fehler, M.C. Extracting time-domain Green's function estimates from ambient seismic noise. *Geophys. Res. Lett.* **2005**, *32*, L03310. [CrossRef]
11. Sabra, K.G.; Gerstoft, P.; Roux, P.; Kuperman, W.A.; Fehler, M.C. Surface wave tomography from microseisms in Southern California. *Geophys. Res. Lett.* **2005**, *32*, L14311. [CrossRef]
12. Yang, Y.; Ritzwoller, M. Characteristics of ambient seismic noise as a source for surface wave tomography. *Geochem. Geophys. Geosyst.* **2008**, *9*, Q02008. [CrossRef]
13. Yang, Y.; Ritzwoller, M.; Levshin, A.; Shapiro, N. Ambient noise Rayleigh wave tomography across Europe. *Geophys. J. Int.* **2007**, *168*, 259–274. [CrossRef]
14. Stephenson, W.J.; Hartzell, S.; Frankel, A.D.; Asten, M.; Carver, D.L.; Kim, W.Y. Site characterization for urban seismic hazards in lower Manhattan, New York City, from microtremor array analysis. *Geophys. Res. Lett.* **2009**, *36*, L03301. [CrossRef]
15. Lynner, C.; Porritt, R.W. Crustal structure across the eastern North American margin from ambient noise tomography. *Geophys. Res. Lett.* **2017**, *44*, 6651–6657. [CrossRef]
16. Guo, Z.; Aydin, A. A modified HVSR method to evaluate site effect in Northern Mississippi considering ocean wave climate. *Eng. Geol.* **2016**, *200*, 104–113. [CrossRef]
17. Stehly, L.; Campillo, M.; Shapiro, N.M. A study of the seismic noise from its long-range correlation properties. *J. Geophys. Res.* **2006**, *111*, B10306. [CrossRef]
18. Shapiro, N.M.; Ritzwoller, M.H.; Bensen, G.D. Source location of the 26 sec microseism from cross-correlation of ambient seismic noise. *Geophys. Res. Lett.* **2006**, *33*, L18310. [CrossRef]
19. Tsai, V.C. On establishing the accuracy of noise tomography travel-time measurements in a realistic medium. *Geophys. J. Int.* **2009**, *178*, 1555–1564. [CrossRef]
20. Zeng, X.; Ni, S. A persistent localized microseismic source near the Kyushu Island, Japan. *Geophys. Res. Lett.* **2010**, *37*, L24307. [CrossRef]
21. Ermert, L.; Villaseñor, A.; Fichtner, A. Cross-correlation imaging of ambient noise sources. *Geophys. J. Int.* **2016**, *204*, 347–364. [CrossRef]
22. Guo, Z.; Xue, M.; Aydin, A.; Ma, Z. Exploring source regions of single- and double-frequency microseisms recorded in eastern North American margin (ENAM) by cross-correlation. *Geophys. J. Int.* **2020**, *220*, 1352–1367. [CrossRef]
23. Longuet-Higgins, M.S. A theory for the generation of microseisms. *Philos. Trans. R. Soc. Lond.* **1950**, *243*, 1–35.
24. Hasselmann, K. A statistical analysis of the generation of microseisms. *Rev. Geophys. Space Phys.* **1963**, *1*, 177–210. [CrossRef]
25. Bromirski, P.D.; Duennebier, F.K.; Stephen, R.A. Mid-ocean microseisms. *Geochem. Geophys. Geosyst.* **2005**, *6*, Q04009. [CrossRef]
26. Essen, H.-H.; Krüger, F.; Dahm, T.; Grevemeyer, I. On the generation of secondary microseisms observed in northern and central Europe. *J. Geophys. Res.* **2003**, *108*, 2506. [CrossRef]
27. Obrebski, M.J.; Ardhuin, F.; Stutzmann, E.; Schimmel, M. How moderate sea states can generate loud seismic noise in the deep ocean. *Geophys. Res. Lett.* **2012**, *39*, L11601. [CrossRef]
28. Rhie, J.; Romanowicz, B. A study of the relation between ocean storms and the Earth's hum. *Geochem. Geophys. Geosyst.* **2006**, *7*, Q10004. [CrossRef]
29. Schimmel, M.; Stutzmann, E.; Ardhuin, F.; Gallart, J. Polarized Earth's ambient microseismic noise. *Geochem. Geophys. Geosyst.* **2011**, *12*, Q07014. [CrossRef]
30. Stephen, R.A.; Spiess, F.N.; Collins, J.A.; Hildebrand, J.A.; Orcutt, J.A.; Peal, K.R.; Vernon, F.L.; Wooding, F.B. Ocean seismic network pilot experiment. *Geochem. Geophys. Geosyst.* **2003**, *4*, 1092. [CrossRef]
31. Webb, S.C. Broadband seismology and noise under the ocean. *Rev. Geophys.* **1998**, *36*, 105–142.
32. Traer, J.; Gerstoft, P.; Bromirski, P.D.; Shearer, P.M. Microseisms and hum from ocean surface gravity waves. *J. Geophys. Res.* **2012**, *117*, B11307. [CrossRef]
33. Nishida, K.; Takagi, R. Teleseismic S wave microseism. *Science* **2016**, *353*, 919–921. [CrossRef] [PubMed]

34. Gerstoft, P.; Bromirski, P. "Weather bomb" induced seismic signals. *Science* **2016**, *353*, 869–870. [CrossRef] [PubMed]
35. Elgar, S.; Herbers, T.H.C.; Guza, R.T. Reflection of ocean surface gravity waves from a natural beach. *J. Phys. Oceanogr.* **1994**, *24*, 1503–1511. [CrossRef]
36. Ardhuin, F.; Stutzmann, E.; Schimmel, M.; Mangeney, A. Ocean wave sources of seismic noise. *J. Geophys. Res.* **2011**, *116*, C09004. [CrossRef]
37. Bromirski, P.D.; Duennebier, F.K. The near-coastal microseism spectrum: Spatial and temporal wave climate relationships. *J. Geophys. Res.* **2002**, *107*, B8. [CrossRef]
38. Ardhuin, F.; Roland, A. Coastal wave reflection, directional spread, and seismoacoustic noise sources. *J. Geophys. Res.* **2012**, *117*, C00J20. [CrossRef]
39. Gualtieri, L.; Stutzmann, E.; Capdeville, F.; Farra, V.; Mangeney, A.; Morelli, A. On the shaping factors of the secondary microseismic wavefield. *J. Geophys. Res.* **2015**, *120*. [CrossRef]
40. Sutton, G.; Barstow, N. Ocean bottom microseisms from a distant supertyphoon. *Geophys. Res. Lett.* **1996**, *23*, 499–502. [CrossRef]
41. Bromirski, P.D.; Flick, R.E.; Graham, N. Ocean wave height determined from inland seismometer data: Implications for investigating wave climate change in the NE Pacific. *J. Geophys. Res.* **1999**, *104*, 20753–20766. [CrossRef]
42. Dorman, L.M.; Schreiner, A.E.; Bibee, L.D.; Hildebrand, J.A. *Deep-Water Seafloor Array Observations of Seismo-Acoustic Noise in the Eastern Pacific and Comparisons with Wind and Swell, in Natural Physical Source of Underwater Sound*; Kerman, B., Ed.; Springer: New York, NY, USA, 1993; pp. 165–174.
43. Ebeling, C.W. Chapter One–inferring ocean storm characteristics from ambient seismic noise: A historical perspective. *Adv. Geophys.* **2012**, *53*, 1–33.
44. Guo, Z.; Aydin, A. Double-frequency microseisms in ambient noise recorded in Mississippi. *Bull. Seismol. Soc. Am.* **2015**, *105*, 1691–1710. [CrossRef]
45. Park, J.; Verono III, F.L.; Lindberg, C.R. Frequency dependent polarization analysis of high-frequency seismograms. *J. Geophys. Res.* **1987**, *92*, 12664–12674. [CrossRef]
46. Cessaro, R.K. Sources of primary and secondary microseisms. *Bull. Seismol. Soc. Am.* **1994**, *84*, 142–148.
47. Friedrich, A.; Krüger, F.; Klinge, K. Ocean-generated microseismic noise located with the Gräfenberg array. *J. Seismol.* **1998**, *2*, 47–64. [CrossRef]
48. Nishida, K.; Kawakatsu, H.; Fukao, Y.; Obara, K. Background Love and Rayleigh waves simultaneously generated at the Pacific Ocean floors. *Geophys. Res. Lett.* **2008**, *35*, L16307. [CrossRef]
49. Paskevich, V. *Srtm30plus-na_pctshade.tif—SRTM30PLUS color-encoded shaded relief image of North America (approximately 1km)—GeoTIFF image: Open-File Report 2005-1001, U.S. Geological Survey, Coastal and Marine Geology Program*; Woods Hole Science Center: Woods Hole, MA, USA, 2005; 30p.
50. Bard, P.-Y. Microtremor measurements: A tool for site effect estimation? In Proceedings of the 2nd International Symposium on the Effects of Surface Geology on Seismic Motion, Yokohama, Japan, 1–3 December 1998; pp. 1251–1279.
51. Konno, K.; Ohmachi, T. Ground-motion characteristics estimated from spectral ratio between horizontal and vertical components of microtremor. *Bull. Seismol. Soc. Am.* **1998**, *88*, 28–241.
52. McNamara, D.E.; Buland, R.P. Ambient noise levels in the continental United States. *Bull. Seismol. Soc. Am.* **2004**, *94*, 1517–1527. [CrossRef]
53. Bendat, J.S.; Piersol, A.G. *Random Data: Analysis and Measurement Procedures*; Wiley-Interscience: New York, NY, USA, 1971.
54. Havskov, J.; Ottemoller, L. *Routine Data Processing in Earthquake Seismology*; Springer: New York, NY, USA, 2010; 347p.
55. Koper, K.D.; Burlacu, R. The fine structure of double-frequency microseisms recorded by seismometers in North America. *J. Geophys. Res.* **2015**, 120. [CrossRef]
56. Babcock, J.M.; Kirkendall, B.A.; Orcutt, J.A. Relationship between ocean bottom noise and the environment. *Bull. Seismol. Soc. Am.* **1994**, *84*, 1991–2007.
57. Kibblewhite, A.C.; Ewans, K.C. Wave-wave interactions, microseism, and infrasonic ambient noise in the ocean. *J. Acoust. Soc. Am.* **1985**, *78*, 981–994. [CrossRef]
58. Schreiner, A.E.; Dorman, L.M. Coherence lengths of seafloor noise: Effect of ocean bottom structure. *J. Acoust. Soc. Am.* **1990**, *88*, 1503–1514. [CrossRef]

59. Lee, W.S.; Sheen, D.H.; Yun, S.; Seo, K.W. The origin of double-frequency microseism and its seasonal variability at King Sejong Station, Antarctica. *Bull. Seismol. Soc. Am.* **2011**, *106*, 1446–1451. [CrossRef]
60. Koper, K.D.; Hawley, V.L. Frequency dependent polarization analysis of ambient seismic noise recorded at a broadband seismometer in the central United States. *Earthq. Sci.* **2010**, *23*, 439–447. [CrossRef]
61. Bromirski, P.D.; Stephen, R.A.; Gerstoft, P. Are deep-ocean-generated surface-wave microseisms observed on land? *J. Geophys. Res.* **2013**, *118*, 3610–3629. [CrossRef]

9

Features of Earthquake-Induced Seabed Liquefaction and Mitigation Strategies of Novel Marine Structures

Yu Huang [1,2,*] and Xu Han [1]

1 Department of Geotechnical Engineering, College of Civil Engineering, Tongji University, Shanghai 200092, China; hanxu@tongji.edu.cn
2 Key Laboratory of Geotechnical and Underground Engineering of the Ministry of Education, Tongji University, Shanghai 200092, China
* Correspondence: yhuang@tongji.edu.cn

Abstract: With the accelerated development of marine engineering, a growing number of marine structures are being constructed (e.g., seabed pipelines, drilling platforms, oil platforms, wind turbines). However, seismic field investigations over recent decades have shown that many marine structures were damaged or destroyed due to liquefaction. Seismic liquefaction in marine engineering can have huge financial repercussions as well as a devastating effect on the marine environment, which merits our great attention. As the effects of seawater and the gas component in the seabed layers are not negligible, the seabed soil layers are more prone to liquefaction than onshore soil layers, and the liquefied area may be larger than when liquefaction occurs on land. To mitigate the impact of liquefaction events on marine engineering structures, some novel liquefaction-resistant marine structures have been proposed in recent years. This paper reviews the features of earthquake-induced liquefaction and the mitigation strategies for marine structures to meet the future requirements of marine engineering.

Keywords: marine engineering; seismic liquefaction; novel liquefaction-resistant structures; mitigation strategies

1. Introduction

An increasing number of structures are being constructed in offshore areas; these include wharfs, cross-sea bridges, seabed tunnels, wind turbines and oil platforms. An important challenge of our times is to develop eco-friendly and renewable energy sources in marine areas [1,2]. Thus, offshore engineering has greatly developed in various countries and the pace of marine resource development is gradually accelerating [3]. The United Nations pointed out that the 21st century is the century of the oceans.

However, marine geohazards occur frequently because of the complex and harsh marine environment [4,5]. Under cyclic loading, such as storms, sea ices, waves and earthquakes, the strength and stiffness of marine soft clay will decrease and liquefaction may occur [6]. Seabed liquefaction can lead to catastrophic consequences, such as the creation of a submarine slope, pile foundation instability, flotation of buried pipelines, and overturning of wind turbines [7–9]. For example, Christian et al. reported a flotation accident on a 3.05-m diameter steel pipeline in Lake Ontario in 1974, which was induced by seabed liquefaction [10]. In 2010, huge waves caused liquefaction of the seabed soil in some areas of the Yellow River Delta in China, and an offshore platform capsized, causing two deaths and economic losses of 5.92 million RMB [11]. In the 2011 Tohoku Earthquake, Kamisu and Hiyama wind farms located 300 kilometers away from the epicenter survived without major damage because the wind turbine system (~3 s) is designed to have a dominant period of ~3 s, which is considerably different from that of the seismic motions at the farm sites (0.07–1.0 s). However, one wind turbine with a monopile foundation tilted due to the seismic seabed liquefaction [12]. These liquefaction-induced accidents had huge financial impacts and severely affected the environment. Due to the serious consequences, researchers have made great efforts in the studies of seabed liquefaction induced by various types of excitation. For instance, Jia and Ye carried out systematic wave flume experiments and numerical simulation works respectively, which well explained the hydrodynamic behaviors (liquefaction and re-suspension) of marine deposits under the sea wave loads [13,14]. Sui et al. considered distribution gradient terms of soil properties and analyzed liquefaction of an inhomogeneous seabed caused by waves [15]. Huang et al. comprehensively reviewed the mechanisms of wave-induced liquefaction and relevant remedial measures [16]. Additionally, in some high-altitude areas, ice-induced vibration needs to be considered, which may also cause liquefaction around marine structures [17,18]. The duration of wave and ice loads is much longer than that of earthquakes. However, earthquakes can produce more energy in a short time compared to waves or ice sheets, so marine structures located in the earthquake zone will be at great risk due to seismic liquefaction. Unfortunately, most earthquakes occur on the seafloor, especially in offshore areas [19]. In Japan, a large earthquake occurs off the coast every three to four years on average, with potential to cause severe damage to marine structures [20]. Therefore, it is important to understand the effect of seismic seabed liquefaction on marine engineering structures.

Significant advances have been made in the study of onshore seismic liquefaction and anti-liquefaction measures [21,22]. However, the ocean environment is more complicated than the onshore one. Earthquake-induced seabed liquefaction has some unique features. For example, the dynamic response of the seawater during an earthquake event can also cause liquefaction in the seabed [23]. Furthermore, the biggest feature and most important development trend which ocean engineering faces is moving from shallow to deep sea. In marine engineering, especially in the abyssal environment, reinforcing the seabed soil skeleton or improving the pore water to prevent liquefaction are not always applicable because of the difficulty and cost. In recent years, scientists have been extensively studying earthquake-induced seabed liquefaction and damage mitigation related to the design of new marine structures. For example, Groot et al. systematically summarized the physical principles of various triggering mechanisms for liquefaction affecting ocean construction [8]. Esfeh et al. used an advanced liquefaction model with FLAC3D and successfully analyzed the liquefaction effect on floating structures [24]. Through dynamic centrifugal tests, Yu et al. studied the dynamic behaviors of different types of foundation (mono-pile and gravity) under seismic loadings that caused liquefaction [25]. Wang et al. presented a comprehensive review of research on mainstream wind turbine foundations and new suction bucket foundations based on both experimental and numerical methods [26].

However, the characteristics of marine seismic liquefaction and the latest marine structures proposed for reducing liquefaction damage have not been reviewed systematically. This article summarizes previous studies and outlines specific issues of seismic liquefaction in marine engineering. Moreover, perspectives on novel liquefaction-resistant marine structures are presented to help cope with the future trends and challenges of ocean engineering. This paper can help readers understand the problems of marine engineers in designing liquefaction-resistant marine structures, and provide useful guidelines on the subject.

2. Seismic Field Investigations in Marine Engineering

A large number of earthquakes occur in highly populated coastal areas such as the Pacific Rim earthquake zone and the Mediterranean earthquake zone. Therefore, earthquake damage investigations such as the seismic survey of the Grand Banks submarine landslide were conducted as early as 1929 [27]. In 1964, after the Alaska earthquake, investigations on seabed liquefaction and submarine landslides were also carried out [28,29]. Due to the difficulties of underwater surveying, there are a limited number of seismic damage investigations on the sea floor compared with those on land. However, from the existing cases, we can still conclude that earthquake-induced seabed liquefaction has caused serious damage to marine engineering structures in the past. Recently, there has been growing interest in seismic field investigations in marine engineering. Sumer et al. summarized seismic liquefaction around ocean engineering structures in Japan and Turkey [20]. Kardogan et al. reported on historical cases of earthquake-induced liquefaction of pile-support wharf structures [30]. This article supplements these studies and summarizes earthquake-induced liquefaction field investigations in marine engineering (Table 1) to provide readers with a systematic understanding of historical cases over the last three decades.

Table 1. Major historical cases of seismic liquefaction in marine engineering.

Date	Earthquakes	Magnitude	Details	References
17 October 1989	1989 Loma Prieta Earthquake	6.9	Monterey Bay Aquarium Research Institute's pier subsided approximately 30 cm led by liquefaction, evidence of seabed liquefaction extending seaward over 600 m, a large number of pipelines failed, some fuel tanks tilted at the dock	[30,31]
17 January 1995	1995 Hyogoken-nanbu Earthquake	7.2	All 240 berths in Kobe Port suffered at least some damages, quay walls moved laterally seaward	[32]
17 August 1999/12 November 1999	1999 Kocaeli, earthquake/Duzce earthquake	7.4/7.1	Almost all the backfill and sheet-piled structures were liquefied behind dock walls, some structures were displaced seaward, seabed settled and some marine structures collapsed	[20,33]
26 January 2001	2001 Bhuj Earthquake	7.7	Dams built on alluavia badly damaged, intake tower titled induced by liquefaction, differential settlement and lateral spreading occurred	[34]
26 December 2004	2004 Great Sumatra Earthquake	9.0+	Liquefaction-induced coastal structures and embankment failures occurred	[35]
12 January 2010	2010 Haiti Earthquake	7.0	A piece of coastal land disappeared, large delta area liquefied	[36]
27 February 2010	2010 Chile Earthquake	8.8	Liquefied zone covered an area with a length of almost 1000km in the north-south direction, several piles-supported facilities were damaged because of liquefaction-induced lateral spreading, some tanks at gas facility titled	[37,38]
11 March 2011	2011 off the Pacific coast of Tohoku Earthquake	9.0	Liquefaction occurred in the river delta area, offshore ground failed, uplift of pipelines and fuel tanks occurred, sand boiled on quay wall, dike collapsed for liquefaction at the bottom, a wind turbine tilted	[39]
4 September 2010 (start on)	2010–2011 Canterbury Earthquake Sequence (CES)	7.1 (mainshock)	Severe seismic liquefaction damage to infrastructures happened, recurrent and large-area liquefaction in offshore area	[40,41]
6 February 2012	2012 Negros Earthquake	6.7	Columns titled and spans of bridge dismembered, induced by liquefaction, large area settlement of coastal roadbed	[42]
14 November 2016	2016 Kaikōura Earthquake	7.8	Gravel and sand ejected near the entrance to the harbor, the pier settled below the surface of water, foundation connection failed and wharves damaged	[43,44]
28 September 2018	2018 Indonesia Sulawesi Earthquake	7.5	Extensive liquefaction happened in offshore areas, floatation of pipelines was observed, a piece of coastal land disappeared, devastating tsunami took place caused by liquefaction	[45]

3. Features of Earthquake-Induced Seabed Liquefaction

Both submarine seismic liquefaction and onshore seismic liquefaction can be explained using the principle of effective stress. However, the amount of seismic damage in marine areas indicates that seabed seismic liquefaction has many characteristics that differ from those of onshore seismic liquefaction.

3.1. Marine Deposits Layer

Based on extensive soil liquefaction cases, particle size distribution curve boundaries for the possibility of liquefaction can be drawn, as shown in Figure 1 [20]. Generally, if the curve of the seabed soil samples falls within the range defined by the two blue boundaries, it is necessary for us to consider the risks of soil liquefaction in engineering design. Well-sorted aeolian sands are widespread in offshore areas, which is inclined to liquefaction easily [46]. For example, the offshore areas of China are mainly layered soils composed of sand, silt and clay (as shown in Figure 2) [47,48], and the submarine soil layers in the North Sea of Europe are dominated by sands [49].

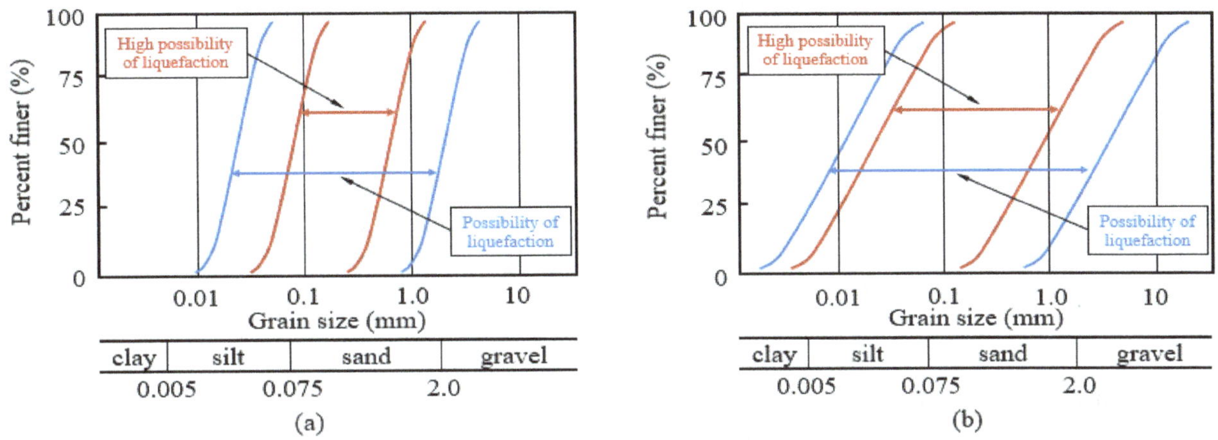

Figure 1. Grain size distribution curve boundaries for the possibility of liquefaction: (**a**) soil with low coefficient of uniformity; (**b**) soil with high coefficient of uniformity (modified from [20]).

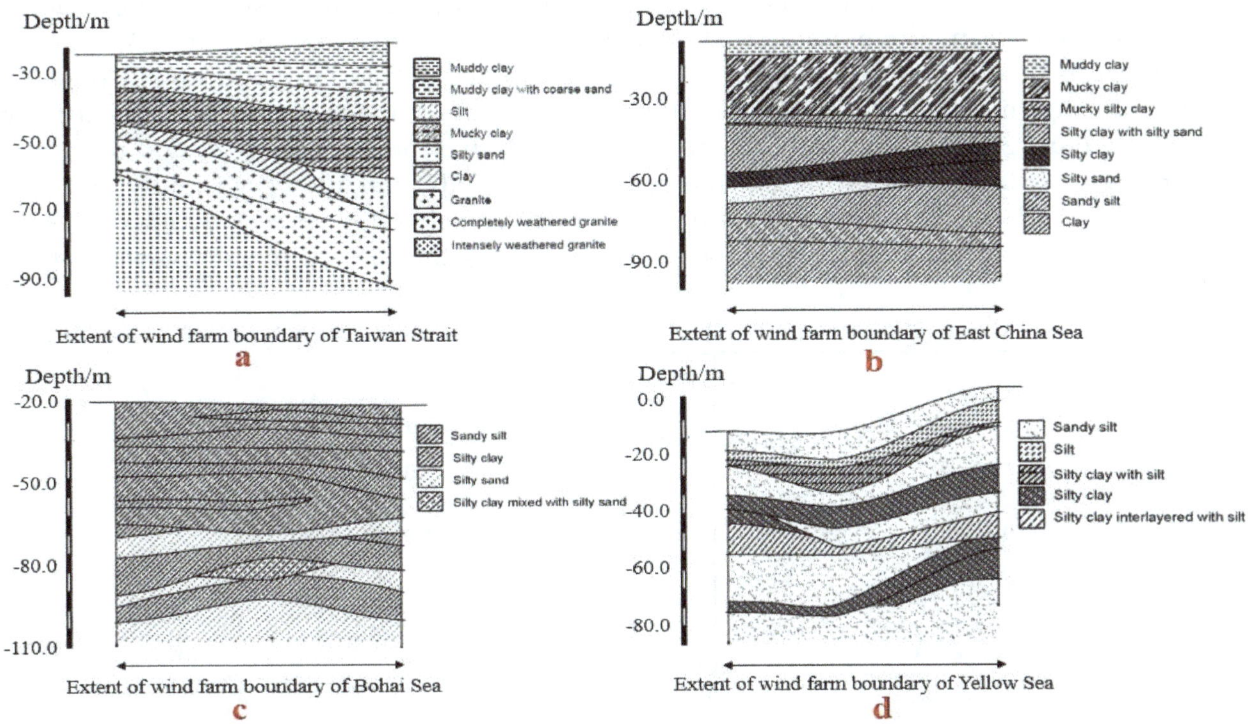

Figure 2. Typical ground profiles in Chinese waters: (**a**) Taiwan Strait; (**b**) East China Sea; (**c**) Bohai Sea; (**d**) Yellow Sea (modified from [48]).

In addition, another important feature of the marine deposits layer is the presence of calcareous sands. It worth noting that calcareous sands are widely distributed in the South China Sea, the Gulf of Mexico, the coasts of Australia, etc. Calcareous sands may have higher resistance to liquefaction than siliceous sands [50]; however, they are also at a great risk of liquefaction [51]. The liquefaction mechanisms of calcareous sands are not very clear yet due to their unique structural characteristics, such as crushability, high content of angular particles and mineralogy surface roughness [52]. Studies on the seismic liquefaction behavior of calcareous sands are of great significance in marine engineering and need to be further carried out.

3.2. *Influence of Sea Water*

When analyzing seismic earthquake forces in marine areas, it is necessary to consider the increase in pore-water pressure on the seabed caused by earthquake-induced water waves acting in offshore areas. Thus, the external excitation that triggers the liquefaction of the soil is not only seismic action but also wave action. Waves can cause two types of seafloor liquefaction: instantaneous liquefaction and residual liquefaction [16].

Moreover, after the seafloor is liquefied, the soil is liable to form mud flows due to the action of waves and seawater; the suspended flow can diffuse over a long distance, which results in lateral spread over a larger area compared with land liquefaction. When liquefaction occurs in soil layers below the seabed surface, the pore-water pressure dissipates much more slowly than on land, and the strength recovery is slower [53].

3.3. *Influence of Submarine Gas Composition*

Gas is always present in gas-charged sediments which are widespread in marine or offshore environments [54]. Under normal conditions, methane is the dominant gas component [55]. As there are many differences between the behavior of unsaturated soils and typical saturated soils under seismic loading [56], it is necessary to clarify and summarize the differences in their liquefaction characteristics.

Firstly, seismic cyclic loading is likely to cause the discharge of shallow seabed gas, which will accelerate the increase in pore pressure and make liquefaction more likely to occur [53]. Moreover, research suggests that small amount of tiny gas bubbles suppress the accumulation of soil pore-water pressure, but may increase the instantaneous liquefaction risk under waves or vertical seismic motion [57]. Figure 3 shows the change in pore pressure with depth for saturated and unsaturated soils. If the soil contains some air or gas, the pore pressure will dissipate very rapidly with depth. In unsaturated soil, the pore pressure gradient can be extremely large, especially near the seabed surface, which means considerable lift can be generated during the passage of a wave trough [58].

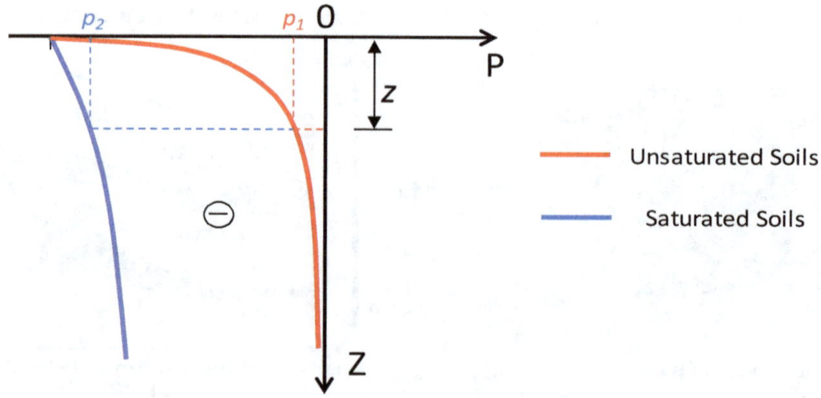

Figure 3. Typical pore pressure distributions in saturated and unsaturated soils during the passage of a wave trough (modified from [58]).

Additionally, natural gas hydrates are widely distributed in marine sediments. Under standard conditions, 1 unit volume of hydrate can release about 164 units of methane [59]. When a large amount of gas migrates upward, it may be trapped under the low-permeability soil layer, which can reduce the effective stress to zero and cause potential instability [60]. An earthquake can trigger dissociation of a large amount of gas hydrate. Moreover, Xu et al. studied the shear behavior of dissociated gas hydrate in undrained conditions using DEM and found that the dissociation of gas hydrate produced significant excess pore pressure and volume expansion, and occasionally static liquefaction [61].

In conclusion, under the influence of sea water and trapped gas, seabed soil layers are more prone to liquefaction than onshore soil layers, and the liquefied area may be larger.

4. Seismic Liquefaction Mitigation Strategies of Novel Marine Structures

4.1. Conventional Liquefaction-Resistance Measures

Generally, for seabed liquefiable foundation soils, it is imperative to lower the risks of liquefaction. Measures to reduce liquefaction damage can be summarized into three categories [16]: (i) reinforcement of seabed soil; (ii) improvement of pore water; and (iii) improvement of structures. In the design stages, the advantages and disadvantages of various remedial measures are compared, and the most suitable and economical method is selected. Sometimes, a combination of two or more countermeasures leads to better results.

4.2. Liquefaction-Resistance Measures of New Marine Structures

In recent years, novel liquefaction-resistant marine structures to prevent liquefaction have been widely researched. This is because traditional measures are difficult and costly to take on the ocean floor. It is worth noting that marine structures are divided into two types in this paper: (1) non-supported structures, such as submarine pipelines and cables; and (2) foundation-supported structures, such as wind turbines, drilling platforms and oil platforms.

4.2.1. Non-Supported Structures

This section provides a brief introduction to pipelines. Submarine pipelines are an important part of the marine oil and gas extraction system, and are currently the most convenient and economical tool for transporting oil and gas. Seismic liquefaction is one of the main causes of damage to submarine pipelines, mostly through the following two failure modes: (1) due to the difference in soil gravity between the pipe and the liquefied seabed, the pipe will rise or sink; (2) seabed sliding causes lateral movement of the pipe. In conventional mitigation measures, the pipelines are buried deeper; however, it is difficult to do so in a marine environment. Ren et al. proposed a new measure for liquefaction damage prevention by reinforcing pipelines with wing plates, and verified the feasibility through shaking-table tests [62]. Yang proposed a simple portal frame to limit the displacement of pipelines. Compared with general anchoring reinforcements, the portal frame allows a certain upward displacement of the pipeline, which greatly reduces the stress of the pipeline and improves the safety when liquefaction occurs [63].

4.2.2. Foundation-Supported Structures

As listed in Table 1, many marine facilities experienced strong earthquakes and were damaged to a certain extent. Among various foundation-supported structures (offshore drilling platforms, oil platforms, wind turbines, cross-sea bridges, etc.), offshore wind turbines are gradually becoming the focus of attention. This is mainly because of the trend of developing clean and eco-friendly energies. Wind energy as a representative has aroused great research interest. Europe is a pioneer of offshore wind turbine (OWT) construction [64]. A 2019 report on European offshore wind turbines showed that OWTs are moving towards the deeper sea (Figure 4). As OWTs are deployed in deeper water, the OWT foundations are being modified, as shown in Figure 5.

Meanwhile, the wind turbine is a slender structure, which has a larger length/width ratio compared to other marine structures, so it is very sensitive to lateral loads [65]. Under the combined effect of winds, waves, and possible seismic loads, the structure–soil interaction will become quite complicated. The soils around the foundations of wind turbines may be greatly disturbed and have a great potential of liquefaction. Thus, in this section, we take the wind turbine as a typical marine structure and highlight research on new foundation structures of liquefaction resistance as applied to it.

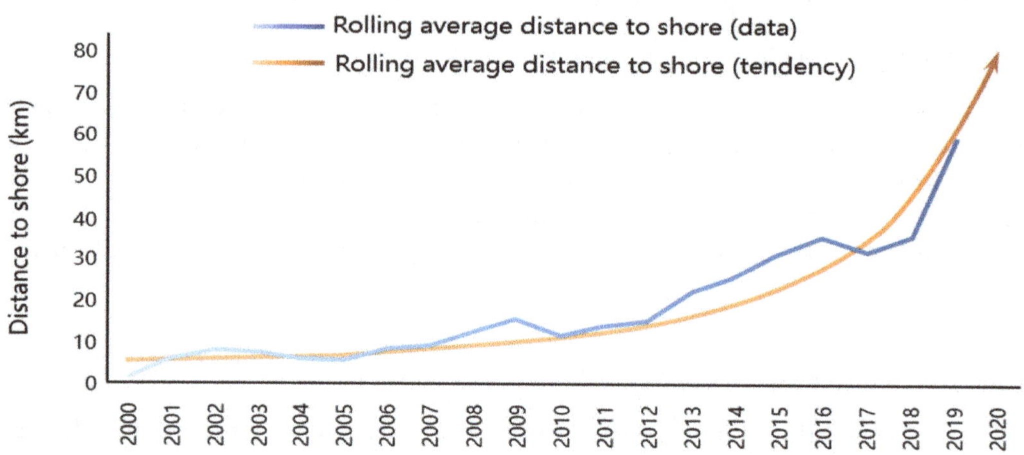

Figure 4. Average distance to shore of offshore wind turbine (OWTs) in Europe (modified from [66]).

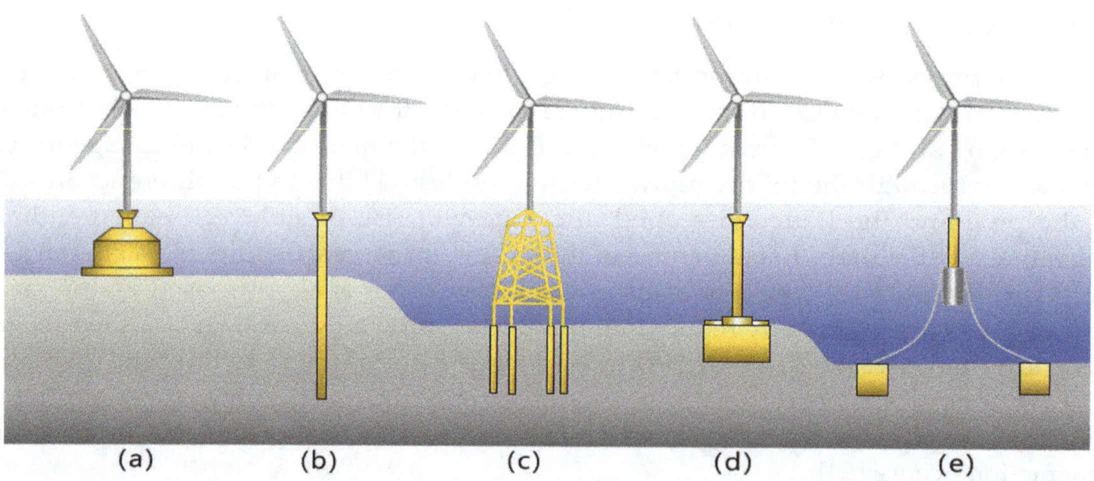

Figure 5. Major foundation types used in OWT design: (**a**) gravity foundation; (**b**) mono-pile; (**c**) jacket foundation piles; (**d**) suction bucket (mono-pod); (**e**) floating wind turbine with anchors (modified from [67]).

At present, there are five main types of foundation structure for OWTs: gravity, monopile, jacket, suction bucket and floating foundation. Many researchers have studied the damage-mitigation performance of the above foundation forms, mainly by numerical methods and dynamic centrifuge experiments. To meet the various marine engineering challenges in the future and improve the liquefaction-resistant performance of the foundations, many innovative structure improvements have been proposed (Table 2).

Table 2. Main OWT foundation types and novel liquefaction-resistant structure improvements (based on [2,26,68–71]).

Foundation Type	Application Scape	Descriptions	Novel Anti-Liquefaction Structure Improvements
Gravity	Shallow water (0~10 m)	Simple structure, long construction period and low cost, compaction effect on soil body	Cross-shaped structure [72]
Monopile	Shallow water (0~30 m)	Industrialization, large disturbance to soil, high cost, scour effect, poor resistance to liquefaction	Hybrid monopile foundation [73,74], tripod foundation [75]
Jacket	Intermediate water (10~50 m)	Applicable to various geological conditions, difficult installation and high cost	—
Suction Bucket	Intermediate water (5~60 m)	Fast construction, reusable, most applicable for soft clay, low cost, good resistance to liquefaction	Umbrella suction anchor foundation [64,68], large-scale prestressed concrete bucket foundation [76,77], tripod suction bucket foundations [78], modified suction caisson with external skirt [79,80], modified suction buckets with honeycomb compartment [81]
Floating	Deep water (>50 m)	Flexible installation, unstable foundation and a little high cost	Anchor piles and suction anchors [24]

The monopile foundation is the main type of OWT foundation currently in use. A new adaptation is the multi-pile foundation (to some extent, the jacket foundation can also be classified as a multi-pile foundation). Hao et al. carried out dynamic centrifugal model tests on the tripod foundation and found that it has better resistance to liquefaction than the common monopile one [75]. Wang et al. proposed a new hybrid monopile foundation, also based on centrifugal tests, and found that the mixed foundation has smaller lateral displacement and enhanced liquefaction resistance than ordinary monopile foundations [73,74]. General views of these two alternatives are illustrated in Figure 6. In fact, the concept of a hybrid monopile can be used to strengthen existing structures.

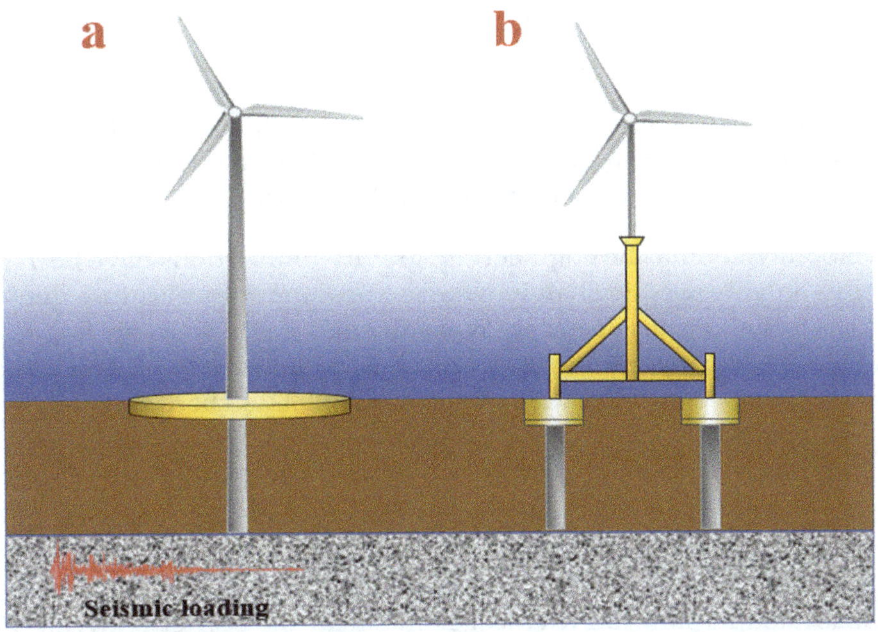

Figure 6. New structures of monopile foundation to increase liquefaction resistance: (**a**). hybrid monopile; (**b**). tripod foundation (based on [74,75]).

New anti-liquefaction jacket foundations have not been proposed in published articles to our knowledge. However, Ju et al. used the finite element method to analyze the seismic response of NREL 5-MW jacket-type OWT under combined loads (earthquakes, waves and winds), and found that the first-mode tuned mass dampers are necessary, which can effectively reduce the vibration induced by combined loads when liquefaction occurred [82].

As shown in Table 2, the suction bucket foundation is a hot topic currently. Many modifications of the suction bucket foundation have been proposed and implemented, such as the large-scale prestressed concrete bucket foundation, with certain success in real engineering by mitigating liquefaction damage (details in Section 4.2.3). Many other new liquefaction resistant structures are still in the research stage of model testing and numerical calculations; these include suction buckets with honeycomb compartments, a modified suction caisson with an external skirt, an umbrella suction anchor foundation, and so on (Figure 7). Experiments by Wang et al. showed that the honeycomb-compartment bucket can reduce soil settlement by about 50% according to experimental data [81]. Li et al. found that the external skirt provides the modified suction caisson with a higher lateral capacity [79,80]. Liu et al. studied a new umbrella suction anchor foundation with anchor branches that closely fit the seafloor; this system improves the anti-overturning ability of the master cylinder and the anti-scouring ability of the surrounding seabed soil [64,68]. Compared with conventional foundations, these new structures show good liquefaction-resistance performance, and have broad application prospects in practical marine engineering.

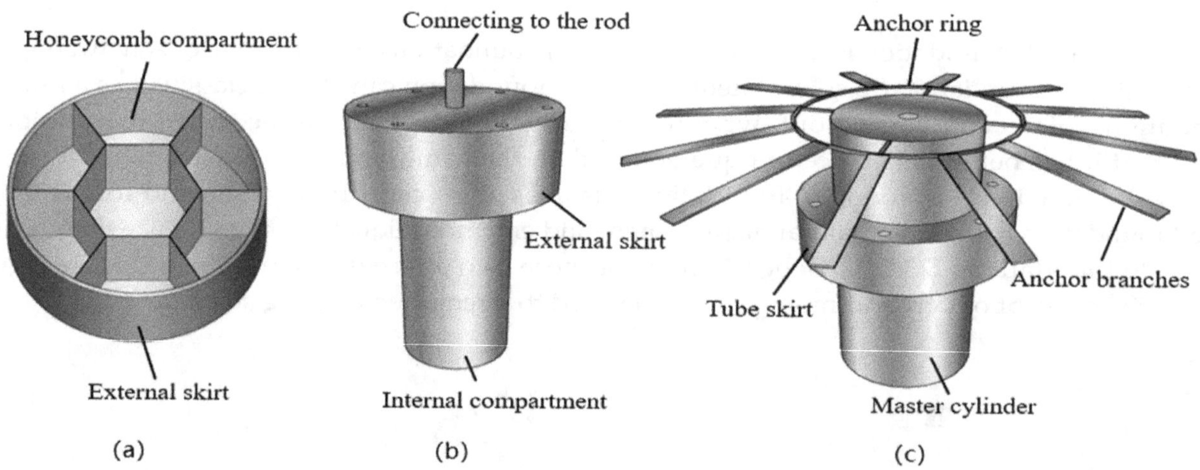

Figure 7. New structures of suction bucket foundations with increased liquefaction resistance: (**a**) modified suction buckets with honeycomb compartment; (**b**) modified suction caisson with external skirt; (**c**) umbrella suction anchor foundation (based on [64,79,81]).

With the development of marine engineering, gravity foundations have been gradually phased out because they can only be used in shallow waters and cannot meet future demands. In contrast, floating foundations are suitable for deep-sea environments. It is foreseeable that research on floating foundations will increase in the coming years, and new liquefaction-resistant structures of floating foundations may be developed and applied in the field, which will become the next research hotspot.

4.2.3. An Example on Site

In this section, we enter a concrete example in reference to the field of wind turbines (large-scale prestressed concrete bucket foundation in Qidong Sea), in order to make readers understand the issues discussed in this article more clearly.

Qidong Sea is located in Jiangsu Province, China, near the border between the East China Sea and the Yellow Sea. In October 2010, the first large-scale prestressed concrete bucket foundation (diameter 30 m, buried depth 7 m) was constructed in this area. As shown in Figure 2, the ground conditions China's four major marine areas are soft and layered. In this wind farm, the geological survey showed that the soil properties from 0 to 33.5 m below the seabed are mainly silty sand and sandy silt, and the soil properties change into dense silty fine sand with the buried depth greater than 33.5 m [83]. These soils are liable to liquefy under strong seismic motion. Therefore, the effect of soil liquefaction needs to be considered in the design process of the wind turbines.

According to the detailed geological surveys and the seismic fortification intensity of this site (7 degrees), Zhang et al. used the ADINA program to analyze the liquefaction-resistance ability of soils below and inside this foundation and showed improvements due to the overburden pressure of the foundation and the constraint effect of the skirt [76,77]. They found that the concrete bucket foundation could still work after soil liquefaction. However, they only added the design ultimate wind loads to the structure, and the dynamic effects of seismic waves combined with the winds were not considered.

Many other new structures have not been constructed in real engineering, but some related research works are also based on site geological conditions. For example, the model tests of modified suction buckets with honeycomb compartment were also carried out in Jiangsu Province [84], and the umbrella suction anchor foundation has been designed for the Yellow River Delta area in the future [64].

In addition, although there are no actual engineering cases of new measures, many scholars have studied earthquake-induced liquefaction in Taiwan, Mexico and other sites. Kuo et al. focused on Changbin offshore wind farm in Taiwan Strait, and evaluated the liquefaction potential based on the typical ground profile of this site [85]. Mardfekri et al. proposed a probabilistic framework to evaluate the vulnerability of wind turbines in the Gulf of Mexico [86]. Martín del Campo et al. used numerical methods to analyze a wind turbine in Mexico under combined loads of earthquakes and winds, and made the fragility analysis [87].

In the above research, we can see that there are not many examples of new liquefaction-resistant structures that have been built. Works of this topic are still dominated by model tests and numerical simulations. The advanced numerical models are generally consistent with the results of the dynamic centrifugal tests. Numerical calculation has the advantages of being efficient and low cost while being able to evaluate many parameters and provide insight into the entire process of liquefaction-induced failure of structures. However, in view of the complexity of the marine environment, pore-pressure models and soil-structure interactions need to be further studied. Thus, numerical analysis of seismic seabed liquefaction will still be a focus of future research.

5. Conclusions

The features of onshore seismic liquefaction are quite different from seabed liquefaction, which is more complicated and requires great attention and extensive research. In addition, mitigation strategies of novel marine foundation structures were reviewed considering their resistance of liquefaction. Based on the reviewed studies, the following conclusions can be drawn.

(1) This article summarizes seismic liquefaction field investigations in marine engineering to provide a systematic understanding of the historical cases published over recent decades. These cases show that seismic-induced liquefaction has a huge impact on marine structures and should be taken into account when designing in the future.

(2) Seabed seismic liquefaction has different characteristics to those seen in land seismic liquefaction. The effect of seawater and trapped or escaping gas on seismic liquefaction is not negligible; seabed soil layers are more prone to liquefaction than onshore soil layers, and the liquefied area may be larger than on land.

(3) Many novel improvements of foundation structures that reduce liquefaction damage in marine engineering have been proposed in recent years; these include the hybrid monopile foundation, umbrella suction anchor foundation, and anchor piles with suction for floating foundations, etc. Experimental and numerical analyses show that these new marine structures have better

liquefaction-resistance performance than traditional structures and need to be further promoted in engineering design.

(4) Having the advantages of low cost, fast construction and reusability, the suction bucket modification used in OWTs is the most widely studied concept nowadays. However, the monopile is the main foundation type for OWTs in current use. The hybrid monopile concept can be used to strengthen existing monopile structures to increase their liquefaction resistance. In addition, it is foreseeable that the research on floating foundations is likely to expand in the coming years, and new liquefaction-resistant structures with floating foundations may become the next research hotspot.

(5) Many other marine structures have been designed while taking into account seismic liquefaction. However, the prevention of submarine seismic liquefaction damage is still facing many difficulties and challenges. Thus, we should give priority to marine geological disaster prevention in project site selection and design to minimize the damage caused by seismic liquefaction around marine structures.

Author Contributions: Conceptualization, Y.H. and X.H.; methodology, Y.H.; investigation, X.H.; writing—original draft preparation, X.H.; writing—review and editing, Y.H. and X.H.; supervision, Y.H.; funding acquisition, Y.H. All authors have read and agreed to the published version of the manuscript.

Acknowledgments: The authors appreciate the assistant of our group members in the discussion.

References

1. Willis, D.J.; Niezrecki, C.; Kuchma, D.; Hines, E.; Arwade, S.R.; Barthelmie, R.J.; DiPaola, M.; Drane, P.J.; Hansen, C.J.; Inalpolat, M.; et al. Wind energy research: State-of-the-art and future research directions. *Renew. Energy* **2018**, *125*, 133–154. [CrossRef]
2. Wu, X.; Hu, Y.; Li, Y.; Yang, J.; Duan, L.; Wang, T.; Adcock, T.; Jiang, Z.; Gao, Z.; Lin, Z.; et al. Foundations of offshore wind turbines: A review. *Renew. Sustain. Energy Rev.* **2019**, *104*, 379–393. [CrossRef]
3. Zhang, M.; Zhang, W.; Huang, Y.; Cai, Y.; Shen, S. Failure mechanism of submarine slopes based on the wave flume test. *Nat. Hazards* **2019**, *96*, 1249–1262. [CrossRef]
4. Jia, Y.; Zhu, C.; Liu, L.; Wang, D. Marine Geohazards: Review and Future Perspective. *Acta Geol. Sin.-Engl. Ed.* **2016**, *90*, 1455–1470. [CrossRef]
5. Huang, Y.; Jin, P. Impact of human interventions on coastal and marine geological hazards: A review. *Bull. Eng. Geol. Environ.* **2018**, *77*, 1081–1090. [CrossRef]
6. Yang, Q.; Ren, Y.; Niu, J.; Cheng, K.; Hu, Y.; Wang, Y. Characteristics of soft marine clay under cyclic loading: A review. *Bull. Eng. Geol. Environ.* **2018**, *77*, 1027–1046. [CrossRef]
7. Zhang, W.; Askarinejad, A. Centrifuge modelling of submarine landslides due to static liquefaction. *Landslides* **2019**, *16*, 1921–1938. [CrossRef]
8. de Groot, M.B.; Bolton, M.D.; Foray, P.; Meijers, P.; Palmer, A.C.; Sandven, R.; Sawicki, A.; Teh, T.C. Physics of liquefaction phenomena around marine structures. *J. Waterw. Port Coast. Ocean Eng.* **2006**, *132*, 227–243. [CrossRef]
9. Jeng, D.-S.; Chen, L.; Liao, C.; Tong, D. A numerical approach to determine wave (current)-induced residual responses in a layered seabed. *J. Coast. Res.* **2019**, *35*, 1271–1284. [CrossRef]
10. Christian, J.T.; Taylor, P.K.; Yen, J.K.C.; Erali, D.R. Large diameter underwater pipe line for nuclear power plant designed against soil liquefaction. In Proceedings of the Offshore Technology Conference, Houston, TX, USA, 6–8 May 1974. [CrossRef]
11. Du, F. Research on Geological Causes of Shengli Well Workover Platform III Overturning Accident. Master's Thesis, Ocean University of China, Qingdao, China, 30 May 2013.

12. Bhattacharya, S.; Goda, K. Use of offshore wind farms to increase seismic resilience of Nuclear Power Plants. *Soil Dyn. Earthq. Eng.* **2016**, *80*, 65–68. [CrossRef]
13. Jia, Y.; Zhang, L.; Zheng, J.; Liu, X.; Jeng, D.-S.; Shan, H. Effects of wave-induced seabed liquefaction on sediment re-suspension in the Yellow River Delta. *Ocean Eng.* **2014**, *89*, 146–156. [CrossRef]
14. Ye, J.; Wang, G. Numerical simulation of the seismic liquefaction mechanism in an offshore loosely deposited seabed. *Bull. Eng. Geol. Environ.* **2016**, *75*, 1183–1197. [CrossRef]
15. Sui, T.; Jin, Y.; Wang, Z.; Zhang, C.; Shi, J. Effects of the Soil Property Distribution Gradient on the Wave-Induced Response of a Non-Homogeneous Seabed. *J. Mar. Sci. Eng.* **2019**, *7*, 281. [CrossRef]
16. Huang, Y.; Bao, Y.; Zhang, M.; Liu, C.; Lu, P. Analysis of the mechanism of seabed liquefaction induced by waves and related seabed protection. *Nat. Hazards* **2015**, *79*, 1399–1408. [CrossRef]
17. Ding, H.Y.; Qi, L.; Du, X.Z. Estimating soil liquefaction in ice-induced vibration of bucket foundation. *J. Cold Reg. Eng.* **2003**, *17*, 60–67. [CrossRef]
18. Chunning, J.I.; Hai, Z.; Lei, G.U.O. Study of soil dynamic liquefaction in ice-induced vibration of offshore platform. *Mar. Sci. Bull.* **2008**, *27*, 81–87.
19. Momma, H.; Fujiwara, N.; Suzuki, S. Deep-sea monitoring system for submarine earthquakes, environment. *Sea Technol.* **1998**, *39*, 72–76.
20. Sumer, B.M.; Ansal, A.; Cetin, K.O.; Damgaard, J.; Gunbak, A.R.; Hansen, N.-E.O.; Sawicki, A.; Synolakis, C.E.; Yalciner, A.C.; Yuksel, Y.; et al. Earthquake-induced liquefaction around marine structures. *J. Waterw. Port Coast.* **2007**, *133*, 55–82. [CrossRef]
21. Bao, X.; Jin, Z.; Cui, H.; Chen, X.; Xie, X. Soil liquefaction mitigation in geotechnical engineering: An overview of recently developed methods. *Soil Dyn. Earthq. Eng.* **2019**, *120*, 273–291. [CrossRef]
22. Huang, Y.; Wen, Z.; Wang, L.; Zhu, C. Centrifuge testing of liquefaction mitigation effectiveness on sand foundations treated with nanoparticles. *Eng. Geol.* **2019**, *249*, 249–256. [CrossRef]
23. Rodriguez-Castellanos, A.; Martinez-Calzada, V.; Efrain Rodriguez-Sanchez, J.; Orozco-del-Castillo, M.; Carbajal-Romero, M. Induced water pressure profiles due to seismic motions. *Appl. Ocean Res.* **2014**, *47*, 9–16. [CrossRef]
24. Esfeh, P.K.; Kaynia, A.M. Numerical modeling of liquefaction and its impact on anchor piles for floating offshore structures. *Soil Dyn. Earthq. Eng.* **2019**, *127*, 105839. [CrossRef]
25. Yu, H.; Zeng, X.; Li, B.; Lian, J. Centrifuge modeling of offshore wind foundations under earthquake loading. *Soil Dyn. Earthq. Eng.* **2015**, *77*, 402–415. [CrossRef]
26. Wang, X.; Zeng, X.; Li, J.; Yang, X.; Wang, H. A review on recent advancements of substructures for offshore wind turbines. *Energy Convers. Manag.* **2018**, *158*, 103–119. [CrossRef]
27. Nof, D. Rotational turbidity flows and the 1929 Grand Banks earthquake. *Deep-Sea Res. Part I-Oceanogr. Res. Pap.* **1996**, *43*, 1143–1163. [CrossRef]
28. Parsons, T.; Geist, E.L.; Ryan, H.F.; Lee, H.J.; Haeussler, P.J.; Lynett, P.; Hart, P.E.; Sliter, R.; Roland, E. Source and progression of a submarine landslide and tsunami: The 1964 Great Alaska earthquake at Valdez. *J. Geophys. Res.-Solid Earth* **2014**, *119*, 8502–8516. [CrossRef]
29. Landen, D. Alaska Earthquake, 27 March 1964. *Science* **1964**, *145*, 74–76. [CrossRef]
30. Kardogan, P.S.O.; Bhattacharya, S. Review of Liquefaction Around Marine and Pile-Supported Wharf Structures. In *Proceedings of 3rd International Sustainable Buildings Symposium*; Firat, S., Kinuthia, J., AbuTair, A., Eds.; Springer International Publishing: Berlin, Germany, 2018; Volume 6, pp. 893–903.
31. Greene, H.G.; Gardnertaggart, J.; Ledbetter, M.T.; Barminski, R.; Chase, T.E.; Hicks, K.R.; Baxter, C. Offshore and onshore liquefaction at moss landing SPIT, central California-result of the October 17, 1989, Loma-Prieta Earthquake. *Geology* **1991**, *19*, 945–949. [CrossRef]
32. Werner, S.D.; Dickenson, S.E.; Taylor, C.E. Seismic risk reduction at ports: Case studies and acceptable risk evaluation. *J. Waterw. Port Coast. Ocean Eng.-Asce* **1997**, *123*, 337–346. [CrossRef]
33. Sumer, B.M.; Kaya, A.; Hansen, N.E.O. Impact of liquefaction on coastal structures in the 1999 Kocaeli, Turkey earthquake. In *International Offshore and Polar Engineering Conference Proceedings, 12th International Offshore and Polar Engineering Conference (ISOPE-2002), Kyushu, Japan, 26-31 May 2002*; Chung, J.S., Matsui, T., Chen, J., Kyozuka, Y., Eds.; International Society Offshore & Polar Engineers: Cupertino, CA, USA, 2002; pp. 504–511.
34. Krinitzsky, E.L.; Hynes, M.E. The Bhuj, India, earthquake: Lessons learned for earthquake safety of dams on alluvium. *Eng. Geol.* **2002**, *66*, 163–196. [CrossRef]

35. Sengara, I.W.; Puspito, N.; Kertapati, E. Survey of geotechnical engineering aspects of the December 2004 great sumatra earthquake and indian ocean tsunami and the March 2005 nias-simeulue earthquake. *Earthq. Spectra* **2006**, *22*, S495–S509. [CrossRef]
36. Hornbach, M.J.; Braudy, N.; Briggs, R.W.; Cormier, M.-H.; Davis, M.B.; Diebold, J.B.; Dieudonne, N.; Douilly, R.; Frohlich, C.; Gulick, S.P.S.; et al. High tsunami frequency as a result of combined strike-slip faulting and coastal landslides. *Nat. Geosci.* **2010**, *3*, 783–788. [CrossRef]
37. Verdugo, R.; Gonzalez, J. Liquefaction-induced ground damages during the 2010 Chile earthquake. *Soil Dyn.Earthq. Eng.* **2015**, *79*, 280–295. [CrossRef]
38. Bray, J.; Rollins, K.; Hutchinson, T.; Verdugo, R.; Ledezma, C.; Mylonakis, G.; Assimaki, D.; Montalva, G.; Arduino, P.; Olson, S.M.; et al. Effects of Ground Failure on Buildings, Ports, and Industrial Facilities. *Earthq. Spectra* **2012**, *28*, S97–S118. [CrossRef]
39. Yamaguchi, A.; Mori, T.; Kazama, M.; Yoshida, N. Liquefaction in Tohoku district during the 2011 off the Pacific Coast of Tohoku Earthquake. *Soils Found.* **2012**, *52*, 811–829. [CrossRef]
40. Bastin, S.H.; Bassett, K.; Quigley, M.C.; Maurer, B.; Green, R.A.; Bradley, B.; Jacobson, D. Late Holocene Liquefaction at Sites of Contemporary Liquefaction during the 2010–2011 Canterbury Earthquake Sequence, New Zealand. *Bull. Seismol. Soc. Am.* **2016**, *106*, 881–903. [CrossRef]
41. Leite, J.; Lourenco, P.B.; Ingham, J.M. Statistical Assessment of Damage to Churches Affected by the 2010–2011 Canterbury (New Zealand) Earthquake Sequence. *J. Earthq. Eng.* **2013**, *17*, 73–97. [CrossRef]
42. Aurelio, M.A.; Dianala, J.D.B.; Taguibao, K.J.L.; Pastoriza, L.R.; Reyes, K.; Sarande, R.; Lucero, A., Jr. Seismotectonics of the 6 February 2012 Mw 6.7 Negros Earthquake, central Philippines. *J. Asian Earth Sci.* **2017**, *142*, 93–108. [CrossRef]
43. Cubrinovski, M.; Bray, J.D.; de la Torre, C.; Olsen, M.; Bradley, B.; Chiaro, G.; Stocks, E.; Wotherspoon, L.; Krall, T. Liquefaction-Induced Damage and CPT Characterization of the Reclamations at CentrePort, Wellington. *Bull. Seismol. Soc. Am.* **2018**, *108*, 1695–1708. [CrossRef]
44. Orense, R.P.; Mirjafari, Y.; Asadi, S.; Naghibi, M.; Chen, X.; Altaf, O.; Asadi, B. Ground performance in Wellington waterfront area following the 2016 Kaikōura Earthquake. *Bull. N. Z. Soc. Earthq. Eng.* **2017**, *50*, 142–151. [CrossRef]
45. Sassa, S.; Takagawa, T. Liquefied gravity flow-induced tsunami: First evidence and comparison from the 2018 Indonesia Sulawesi earthquake and tsunami disasters. *Landslides* **2019**, *16*, 195–200. [CrossRef]
46. Bucci, M.G.; Almond, P.C.; Villamor, P.; Tuttle, M.P.; Stringer, M.; Smith, C.M.S.; Ries, W.; Bourgeois, J.; Loame, R.; Howarth, J.; et al. Controls on patterns of liquefaction in a coastal dune environment, Christchurch, New Zealand. *Sediment. Geol.* **2018**, *377*, 17–33. [CrossRef]
47. Lu, X.; Zheng, Z.; Zhang, J. Progress in the study on the bucket foundations of offshore platform. *Adv. Mech.* **2003**, *33*, 27–40.
48. Bhattacharya, S.; Wang, L.; Liu, J.; Hong, Y.J.W.E.E. Chapter 13—Civil Engineering Challenges Associated with Design of Offshore Wind Turbines with Special Reference to China. In *Wind Energy Engineering*; Academic Press: Cambridge, MA, USA, 2017; pp. 243–273. [CrossRef]
49. Bhattacharya, S.; Carrington, T.; Aldridge, T. Observed increases in offshore pile driving resistance. *Proc. Inst. Civil Eng.-Geotech. Eng.* **2009**, *162*, 71–80. [CrossRef]
50. Salem, M.; Elmamlouk, H.; Agaiby, S. Static and cyclic behavior of North Coast calcareous sand in Egypt. *Soil Dyn. Earthq. Eng.* **2013**, *55*, 83–91. [CrossRef]
51. Wang, Y.; Qiu, Y.; Ma, L.; Li, Z. Experimental study on the cyclic response of Nanhai Sea calcareous sand in China. *Arab. J. Geosci.* **2019**, *12*, 677. [CrossRef]
52. Sandoval, E.A.; Pando, M.A. Experimental assessment of the liquefaction resistance of calcareous biogenous sands. *Earth Sci. Res. J.* **2012**, *16*, 55–63. [CrossRef]
53. Guangbiao, S.; Qimin, F. Review of studies on earthquake liquefaction failure of submarine soil layer. *J. Nat. Disasters* **2007**, *16*, 70–75. [CrossRef]
54. Fleischer, P.; Orsi, T.H.; Richardson, M.D.; Anderson, A.L. Distribution of free gas in marine sediments: A global overview. *Geo-Mar. Lett.* **2001**, *21*, 103–122. [CrossRef]
55. Sills, G.C.; Wheeler, S.J. The significance of gas for offshore operations. *Continent. Shelf Res.* **1992**, *12*, 1239. [CrossRef]
56. Sobkowicz, J.C.; Morgenstern, N.R. The Undrained Equilibrium Behavior of Gassy Sediments. *Can. Geotech. J.* **1984**, *21*, 439–448. [CrossRef]

57. Sumer, B.M.; Truelsen, C.; Fredsoe, J. Liquefaction around pipelines under waves. *J. Waterw. Port Coast. Ocean Eng.* **2006**, *132*, 266–275. [CrossRef]
58. Sumer, B.M. Introduction and Physics of Liquefaction. In *Liquefaction Around Marine Structures*; Liu, P.L.-F., Ed.; World Scientific Publishing: Singapore, 2014; Volume 39, pp. 1–16. [CrossRef]
59. Max, M.D.; Clifford, S.M. The state, potential distribution, and biological implications of methane in the Martian crust. *J. Geophys. Res.-Planets* **2000**, *105*, 4165–4171. [CrossRef]
60. van Paassen, L.A.; Vinh, P.; Mahabadi, N.; Hall, C.; Stallings, E.; Kavazanjian, E., Jr. Desaturation via Biogenic Gas Formation as a Ground Improvement Technique. In *Panam Unsaturated Soils 2017: Plenary Papers, 2nd Pan-American Conference on Unsaturated Soils ((PanAm-UNSAT), Dallas, USA, 12–15 November 2017*; Hoyos, L.R., McCartney, J.S., Houston, S.L., Likos, W.J., Eds.; AMER Soc Civil Engineers United Engineering Center: New York, NY, USA, 2018; pp. 244–256.
61. Xu, M.; Song, E.; Jiang, H.; Hong, J. DEM simulation of the undrained shear behavior of sand containing dissociated gas hydrate. *Granul. Matter* **2016**, *18*, 79. [CrossRef]
62. Ren, C.; Li, Z. Research of Anti-liquefaction of buried pipelines by experimentation. In Proceedings of the 2007 Ocean Engineering Conference, Guiyang, China, 1 November 2007; Wu, Y., Ed.; Shipbuilding of China: Shanghai, China, 2007; pp. 622–627.
63. Yang, Y. Submarine Pipeline Buckling on Uneven Seabed and the Stability of the Submarine Pipeline in Liquefied Soil. Master's Thesis, Tianjin University, Tianjin, China, November 2014.
64. Li, H.; Liu, H.; Liu, S. Dynamic analysis of umbrella suction anchor foundation embedded in seabed for offshore wind turbines. *Geomech. Energy Environ.* **2017**, *10*, 12–20. [CrossRef]
65. Katsanos, E.I.; Thons, S.; Georgakis, C.T. Wind turbines and seismic hazard: A state-of-the-art review. *Wind Energy* **2016**, *19*, 2113–2133. [CrossRef]
66. Walsh, C. *Offshore Wind in Europe–Key Trends and Statistics 2019*; Wind Europe: Brussels, Belgium, 2020.
67. Kaynia, A.M. Seismic considerations in design of offshore wind turbines. *Soil Dyn. Earthq. Eng.* **2019**, *124*, 399–407. [CrossRef]
68. Liu, H.; Li, H. A New Suction Anchor Foundation of the Yellow River Delta Offshore Wind Power. *Period. Ocean Univ. China* **2014**, *44*, 71–76.
69. Oh, K.-Y.; Nam, W.; Ryu, M.S.; Kim, J.-Y.; Epureanu, B.I. A review of foundations of offshore wind energy convertors: Current status and future perspectives. *Renew. Sustain. Energy Rev.* **2018**, *88*, 16–36. [CrossRef]
70. Perez-Collazo, C.; Greaves, D.; Iglesias, G. A review of combined wave and offshore wind energy. *Renew. Sustain. Energy Rev.* **2015**, *42*, 141–153. [CrossRef]
71. Zhang, P.; Han, Y.; Ding, H.; Zhang, S. Field experiments on wet tows of an integrated transportation and installation vessel with two bucket foundations for offshore wind turbines. *Ocean Eng.* **2015**, *108*, 769–777. [CrossRef]
72. Sturm, H. Geotechnical performance of a novel gravity base type shallow foundation for offshore wind turbines. *Geotechnik* **2011**, *34*, 85–96. [CrossRef]
73. Wang, X.; Zeng, X.; Li, X.; Li, J. Liquefaction characteristics of offshore wind turbine with hybrid monopile foundation via centrifuge modelling. *Renew. Energy* **2020**, *145*, 2358–2372. [CrossRef]
74. Wang, X.; Zeng, X.; Yang, X.; Li, J. Seismic response of offshore wind turbine with hybrid monopile foundation based on centrifuge modelling. *Appl. Energy* **2019**, *235*, 1335–1350. [CrossRef]
75. Hao, Y.; Zeng, X.; Wang, X. Seismic centrifuge modelling of offshore wind turbine with tripod foundation. In Proceedings of the 2013 IEEE Energytech, Cleveland, OH, USA, 21–23 May 2013.
76. Zhang, P.; Ding, H.; Le, C. Seismic response of large-scale prestressed concrete bucket foundation for offshore wind turbines. *J. Renew. Sustain. Energy* **2014**, *6*, 013127. [CrossRef]
77. Zhang, P.; Xiong, K.; Ding, H.; Le, C. Anti-liquefaction characteristics of composite bucket foundations for offshore wind turbines. *J. Renew. Sustain. Energy* **2014**, *6*, 053102. [CrossRef]
78. Kim, D.-J.; Choo, Y.W.; Kim, J.-H.; Kim, S.; Kim, D.-S. Investigation of Monotonic and Cyclic Behavior of Tripod Suction Bucket Foundations for Offshore Wind Towers Using Centrifuge Modeling. *J. Geotech. Geoenviron.* **2014**, *140*, 04014008. [CrossRef]
79. Li, D.; Zhang, Y.; Feng, L.; Gao, Y. Capacity of modified suction caissons in marine sand under static horizontal loading. *Ocean Eng.* **2015**, *102*, 1–16. [CrossRef]
80. Li, D.; Feng, L.; Zhang, Y. Model tests of modified suction caissons in marine sand under monotonic lateral combined loading. *Appl. Ocean Rese.* **2014**, *48*, 137–147. [CrossRef]

81. Wang, X.; Yang, X.; Zeng, X. Seismic centrifuge modelling of suction bucket foundation for offshore wind turbine. *Renew. Energy* **2017**, *114*, 1013–1022. [CrossRef]
82. Ju, S.-H.; Huang, Y.-C. Analyses of offshore wind turbine structures with soil-structure interaction under earthquakes. *Ocean Eng.* **2019**, *187*, 106190. [CrossRef]
83. Ji, J.; Sun, L.; Zhang, J. Bearing Capacity and Technical Advantages of Composite, Bucket Foundation of Offshore Wind Turbines. *Trans. Tianjin Univ.* **2011**, *17*, 132–137.
84. Zhang, P.; Guo, Y.; Liu, Y.; Ding, H. Experimental study on installation of hybrid bucket foundations for offshore wind turbines in silty clay. *Ocean Eng.* **2016**, *114*, 87–100. [CrossRef]
85. Kuo, Y.-S.; Lin, C.-S.; Chai, J.-F.; Chang, Y.-W.; Tseng, Y.-H. Case study of the ground motion analyses and seabed soil liquefaction potential of Changbin offshore wind farm. *J. Mar. Sci. Technol.-Taiwan* **2019**, *27*, 448–462. [CrossRef]
86. Mardfekri, M.; Gardoni, P. Multi-hazard reliability assessment of offshore wind turbines. *Wind Energy* **2015**, *18*, 1433–1450. [CrossRef]
87. Martin del Campo, J.O.; Pozos-Estrada, A. Multi-hazard fragility analysis for a wind turbine support structure: An application to the Southwest of Mexico. *Eng. Struct.* **2020**, *209*. [CrossRef]

10

Finite Element Analysis and Parametric Study of Spudcan Footing Geometries Penetrating Clay Near Existing Footprints

Long Yu *, Heyue Zhang, Jing Li and Xian Wang

State Key Laboratory of Coastal and Offshore Engineering, Dalian University of Technology, Dalian 116024, China; zhangheyue@mail.dlut.edu.cn (H.Z.); ljj28@mail.dlut.edu.cn (J.L.); ljj2828@mail.dlut.edu.cn (X.W.)

* Correspondence: longyu@dlut.edu.cn

Abstract: Most existing research on the stability of spudcans during reinstallation nearing footprints is based on centrifuge tests and theoretical analyses. In this study, the reinstallation of the flat base footing, fusimform spudcan footing and skirted footing near existing footprints are simulated using the coupled Eulerian–Lagrangian (CEL) method. The effects of footprints' geometry, reinstallation eccentricity (0.25*D*–2.0*D*) and the roughness between spudcan and soil on the profiles of the vertical force, horizontal force and bending moment are discussed. The results show that the friction condition of the soil–footing interface has a significant effect on *H* profile but much less effect on *M* profile. The eccentricity ratio is a key factor to evaluate the *H* and *M*. The results show that the geometry shape of the footing also has certain effects on the *V*, *H*, and *M* profiles. The flat base footing gives the lowest peak value in *H* but largest in *M*, and the performances of the fusiform spudcan footing and the skirted footing are similar. From the view of the resultant forces, the skirted footing shows a certain potential in resisting the damage during reinstallation near existing footprints by comparing with commonly used fusiform spudcan footings. The bending moments on the leg–hull connection section of different leg length at certain offset distances are discussed.

Keywords: spudcan; skirted footing; footprint; jack-up; clay; large deformation analysis

---

1. Introduction

1.1. Background

Jack-up units are self-elevating mobile platforms which are used extensively in the offshore oil and gas industry. A typical jack-up consists of a floatable hull and three independent retractable legs. The legs rest on spudcan footings that are usually circular or polygonal in plan and with an inverse cone underneath. Once a jack-up unit is towed to site, its installation begins by lowering the legs to the seabed and pushing the spudcans into the soil and then rising the hull over the water. Then pre-loading can be achieved by pumping water into the hull. The pre-loading makes the spudcan penetrate deeper to provide more resistance. After pre-loading, the water is pumped out and the spudcan's bearing capacity has some reservation. After all the work of the jack-up has finished, it is removed from the site by retracting the legs from the seabed. The processes of installation and extraction of the spudcan leave a permanent seabed depression at each footing site, which is referred to as a "footprint".

The footprint changes the seabed in two ways, as shown in Figure 1: An inclined seabed surface and a varying soil strength profile within the footprint (normally decreasing soil strength due to remolding). Both of them result in additional horizontal forces and bending moments compared with the initial installation. The spudcan–footprint interaction problem is significant as it can lead to significant time loss, cost implications, risks to adjacent structures and potential injury to personnel. Dier et al.

concluded from industry practice data that incidents caused by uneven seabed/scour/footprint are at a rate of 15% of the total [1]. This rate has increased obviously due to increasing demands of jack-up operation close to previous sites in recent years [2].

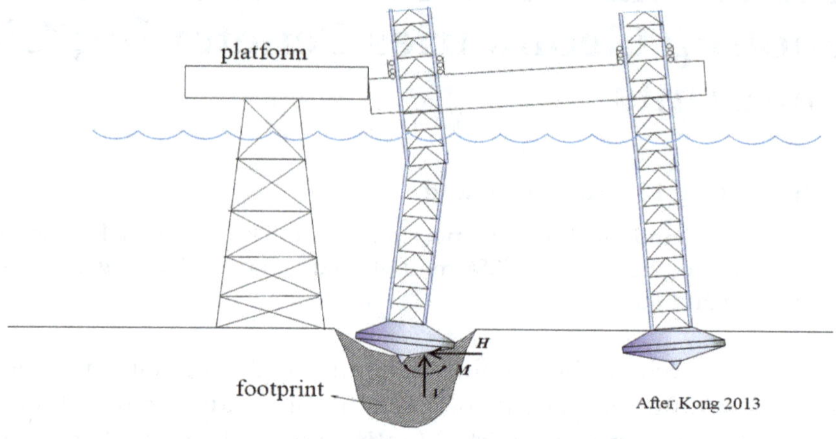

Figure 1. The failure mechanism of spudcan reinstallation near a footprint.

1.2. Previous Work

The problem of jack-up reinstallation near the existing footprints attracted more attention in the recent 10–15 years. In some studies [3,4], footprints resemble an inverted conical shape cavity. The installation, operation, and removal of the spudcan can also remold the surrounding soil, resulting in highly variable shear strength profiles in the vicinity of the footprints [4–8].

Hartono et al. [9] used an experimental method (centrifuge tests) and numerical analysis (simulated with ABAQUS/CEL) respectively to investigate the efficacy of reaming technique in mitigating the footprint hazards. He found that the numerical results demonstrate good agreement with experiment results and reaming can be a viable option to mitigate spudcan–footprint interaction. He strongly suggested making numerical modeling as a viable tool for site-specific assessment of spudcan–footprint interaction problem. Like Hartono, the CEL large deformation method is adopted in this study to investigate the reinstallation behaviors of flat base footing, fusiform spudcan footing, and skirted footing.

Spudcans are the most common footings used for jack-up units. Along with the improvement of technology and the increasing demands of operating on the very soft soils, the footings become larger in diameter and flatter at the base. The geometries of typical fusiform spudcan footings are shown in Figure 2. The investigations from some research shows that, by comparing with fusiform spudcan footing, skirted footing may have a higher bearing capacity [7] and have some potential in mitigating punch-through failure [10,11].

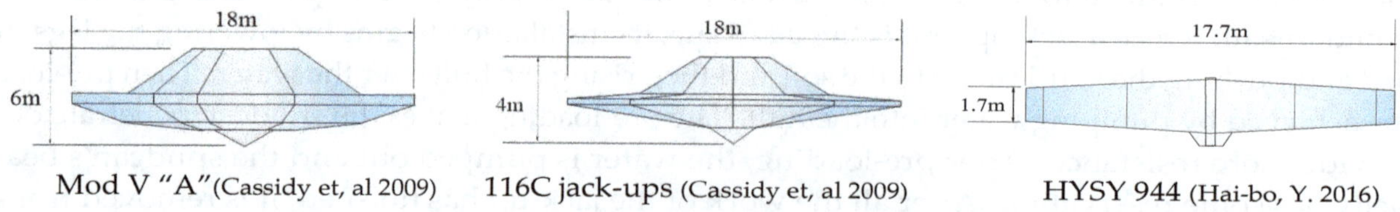

Figure 2. Three typical spudcans [3,12].

Cassidy et al. [3] used a 1:250 scale model of current Mod 'V 'jack-up in their centrifuge tests to simulate the interaction between real spudcan and soil. Kong et al. [13,14] replaced real spudcan with flat base footing in their centrifuge tests to eliminate the variables related to spudcan geometry. Zhang et al. [15] simulated spudcan with flat base footing in his numerical study to make sure that the touchdown level of footing could be identified clearly.

Gan et al. [4,6,8] studied the spudcan–footprint interaction considering the 'real' initial penetration. Their research showed that the soil is obviously disturbed during the initial penetration and will recover with time. To simplify the problem, many of the following studies assumed an artificial footprint, such as Kong [13], Zhang et al. [15], Jun et al. [16–19]. The assumption of an artificial reverse cone footprint may respond to a fully recovered 'real' footprint after a long period from the initial penetration. Therefore, the three idealized footprints TA, TB, and TC following Kong [13] are adopted in this study to simplify the numerical model.

The distance from the central line of the footprint to the reinstalled spudcan center was termed as offset distance or reinstallation eccentricity (β), which was proved to be a key issue to the profiles of bending moment (M) and horizontal force (H). Stewart [20] carried out centrifuge model tests and the results showed that both M and H increased to an obvious value when $\beta/D = 0.5$ to 1.0, where D is the diameter of the reinstalled spudcan, and H reached to the maximum value when $\beta/D = 0.75$. Cassidy et al. [3] founded that M and H were most obvious when $\beta/D = 0.5$ and became very small when $\beta/D > 1.5$. Carrington [21] carried out large deformation numerical analyses to simulate the reinstallation processes with $\beta/D = 0.167$ to 0.407, and obtained a most critical case at $\beta/D = 0.29$. Kong et al. [13,14]. Investigated the effect of footprints with various size and slope angles. In their study the critical case was $\beta/D = 1.0$.

Some research showed that the fixity condition at the leg–hull connecting point has a significant effect on the reinstallation behavior near a footprint [3,5,22–25]. It can be concluded that harder fixity tends to increase the maximum value of M and H but reducing the lateral movement of the spudcan during reinstallation.

1.3. Motivation of Present Study

Most existing research on the stability of spudcans during reinstallation nearing footprints is from centrifuge tests and theoretical analyses. In this study, the reinstallation of flat base footing, fusiform spudcan footing and skirted footing near existing footprints are simulated using the coupled Eulerian–Lagrangian (CEL) method. The effects of footprints' geometry, reinstallation eccentricity and the roughness between spudcan and soil on the profiles of vertical force, horizontal force and bending moment are discussed. One purpose of this study is to reveal the mechanisms of those factors which affect V, H, and M profiles during reinstallation, by presenting the soil flow mechanisms of selected cases.

The other purpose is to discuss the effect of footing geometry shape during reinstallation near existing footprints. Flat base footings, fusiform spudcan footings, and skirted footings are investigated in this paper. Fusiform spudcan footings have been widely used in practice. Skirted footings have been proved to have some potential in bearing capacity and mitigating punch-through failure, but its behavior in mitigating footprint hazards is still not very clear. Flat base footings have advantages in eliminating the uncertainty when discussing the soil flow mechanism, by comparing with fusiform spudcan footings of which the reverse cone initially touches the seabed. Besides, some large footings in practical engineering have a relatively flat base, such as HYSY 944, as shown in Figure 2.

2. Materials and Methods

2.1. Modeling of Footings

The sign convention and definition of terminology in this study are plotted in Figure 3 and the numerical models of footings investigated in this study are shown in Figure 4.

The diameter of all the footings is D = 15 m and the height of the max area is H_t = 1.75 m. The geometry of the spudcan follows Liu et al. [26] and Yu et al. [27]. The geometry of the skirted footing is $H_s = 0.25D = 3.75$ m and $T_s = 1.75$ m; where H_s is the height of the skirt and T_s is the thickness of the skirt. The geometry of the fusiform spudcan footing is D = 15 m, H_1 = 2.5 m and H_2 = 3.3 m. The distance from the center of Section 1.1 to the reference point (RP) for flat base footing is H_a = 1.75 m, for fusiform spudcan footing is $H_a = 7.55$ m, for skirted footing is $H_a = 5.5$ m. To simplify the problem, the leg and footing are constrained as rigid.

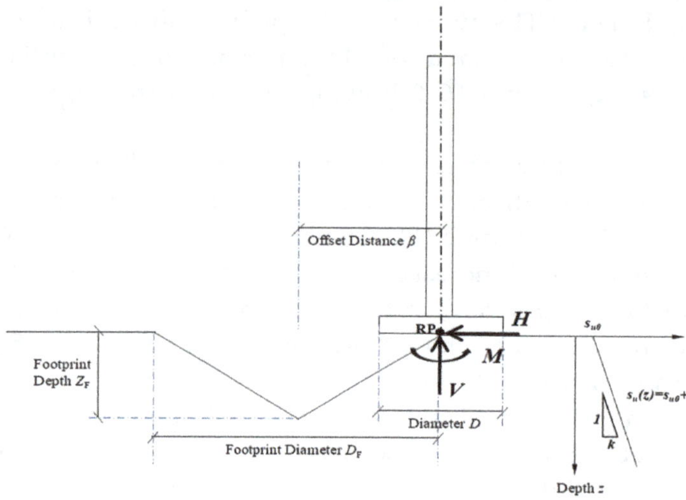

Figure 3. Sign convention and definition of terminology.

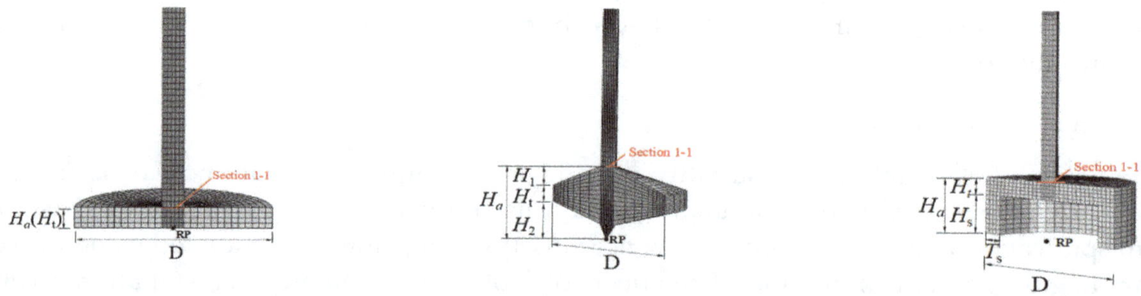

Figure 4. The dimensions of footings: flat base footing, fusiform spudcan footing and skirted footing.

2.2. Model of Soil

To simplify the problem, the footprint in this study is idealized as a reverse conical cave on the soil surface. The ideal elasto-plastic model is used to describe the stress–strain relationship of the soil, obeying the Mohr–Coulomb strength criterion. The undrained shear strength profile is $s_u = 7.5 + 0.92z$ kPa, where z is the soil depth from the mudline. The Poisson's ratio is $\nu = 0.49$. The elastic modulus is $E = 500s_u$. The effective unit weight is $\gamma' = 6.82$ kN/m^3. The internal friction angle and the dilation angle are $\varphi = \Psi = 0°$. The load is achieved by displacement control at a rate of $v = 0.5$ m/s, which is a compromise between the accuracy and the efficiency.

The principle of universal contact is used to simulate the contact property between footing and soil. In tangential direction, the penalty function is selected to model the friction condition, thus different frictions can be tested. In normal direction, "hard" contact is set to simulate the interface, which can transfer positive pressure without limitation but separate under tension.

A half model is modeled because of the symmetry. Both a cuboid soil domain and a cylindrical one have been tested. The results show that the former is more efficient and easier to mesh without the loss of accuracy. Thus, the cuboid soil domain, as shown in Figure 5, is used in this study. The soil is modeled by EC3D8R element (three-dimensional, eight-node linear brick, multimaterial, reduced integration with hourglass control) and the footings are modeled by C3D8R (three-dimensional, eight-node linear brick, reduced integration with hourglass control) element in ABAQUS/Explicit. In order to eliminate the influence of boundary effect, the width, depth, and thickness of the soil are 8*D*, 4*D*, and 4*D*, respectively. In addition, there is an empty element layer, 4 m thick, at the top of the soil to heave up during reinstallation. The mesh close to the footing penetrating path is refined.

The minimum element size is $d_{min}/D = 1/30 = 0.5$ m. The mesh density and soil domain have been proved to be with acceptable accuracy.

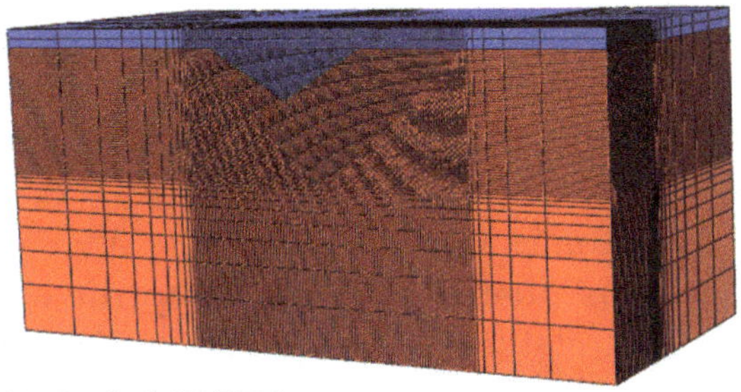

Figure 5. The mesh of soil.

2.3. Numerical Cases

In this study, the reinstallation process of three 15 m diameter footings with offset distances of $\beta/D = 0.25$, 0.5, 0.75, 1.0, 1.25, 1.5, and 2.0 are simulated respectively. Four types of footprints (one of them is a flat surface field) are simulated, as listed in Figure 6. TA, TB, and TC are three footprints with various depths, and FS means flat surface (no footprint). The naming rule of each case is similar to that of Kong et al. [13]. For example, TB-2*D*-0.25*D* means the footprint is TB type with a diameter of $D_F = 2D$, and the eccentricity of reinstallation is $\beta = 0.25D$. All the cases investigated in the paper are listed in Table 1.

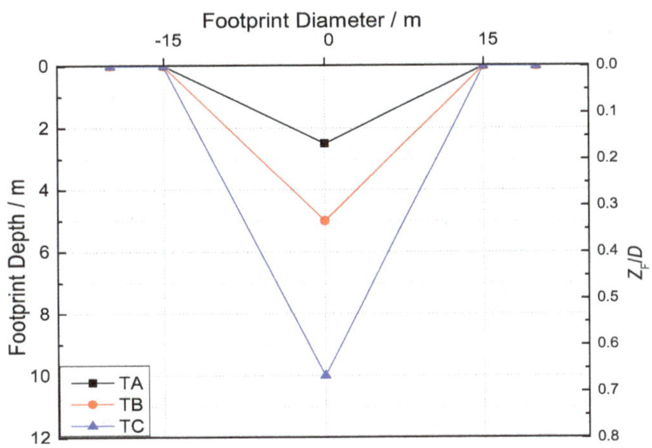

Figure 6. Dimensions of three types of footprint.

Table 1. Numerical cases.

Footprint Type	Prototype (m)		(°)	β (m)	Case Name
	D_F	Z_F			
TA	30	2.5	9.5	0.25*D*, 0.5*D*,	TB-2*D*-0.25*D*(μ)
TB	30	5	18.4	0.75*D*, 1.0*D*,	(μ is friction
TC	30	10	33.7	1.25*D*, 1.5*D*,	coefficient)
Flat surface	-	-	-	2.0*D*	FS

3. Results

3.1. Effect of An Existing Footprint

At first, the flat base footing is taken as an example to show how an existing footprint affects the resistance profile during installation and to show the soil flow mechanism. Zero depth is defined as the maximum cross-section area of footing touching seabed level. *V*, *H*, and *M* denote the vertical force, horizontal force, and bending moment acting on the reference point, respectively.

The *V*, *H*, and *M* profiles are plotted in Figure 7 and the soil flow mechanisms are shown in Figure 8. The positive *H* value means a horizontal force towards the footprint. The positive *M* value means an anti-clockwise bending moment acting on the reference point (RP).

In this case, the horizontal force comes from two parts: (1) When $z/D < 0.15$, with the penetration goes by, the soil under the footing is pushed into the footprint while the footing is constrained without horizontal movements. The relative motion provides a friction force towards the footprint (as the yellow arrow shown in Figure 8a. When $z/D < 0.15$, the maintained zero H/AS_u value of case TB-2*D*-1.0*D* (smooth) as shown in Figure 7a confirms this conclusion. (2) As the footing penetrates deeper ($z/D > 0.15$), the soil on the footing's right side is compressed, which provides a leftward earth pressure (as the green arrows shown in Figure 8c. When the footing reaches the toe of the footprint, 0.33*D*, the soil on the footing's left side heaves up and provides a rightward earth pressure. After that, the total horizontal force reduces. When the penetration depth reaches ~0.8*D*, the soil on both the left and right sides show a symmetric fully flowing back mechanism, as shown in Figure 8d. The symmetric soil flow mechanism results in that the *H* and *M* values are much smaller.

M peaks as soon as the footing touches the seabed (around $z/D = 0.02$). This is because that at this depth the eccentric distance of the resultant vertical resistance force is very large, as shown in Figure 8a, although the vertical resistance is far from the peak value at this depth. As the penetration depth increases, the eccentric distance of vertical force reduces and as a result the corresponding bending moment obviously reduces.

Compared with the centrifuge test results from Kong [25], the horizontal force profile of Kong's lies between the smooth and rough cases of this study (Figure 7a) because the friction characteristic of the centrifuge test on the interface of aluminum footing and the soil is between rough and smooth. It can be seen that the numerical results of this study and the centrifuge test results from Kong [25] have very similar *H* and *M* profile trends.

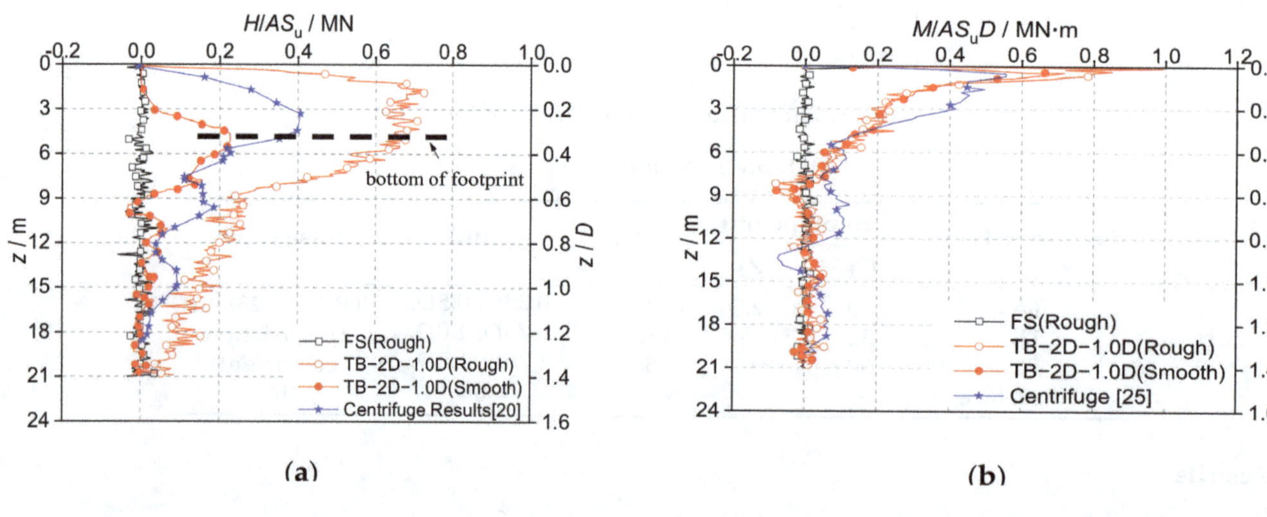

Figure 7. Effect of existing footprint on the *V*, *H*, and *M* responses of the flat base footing.

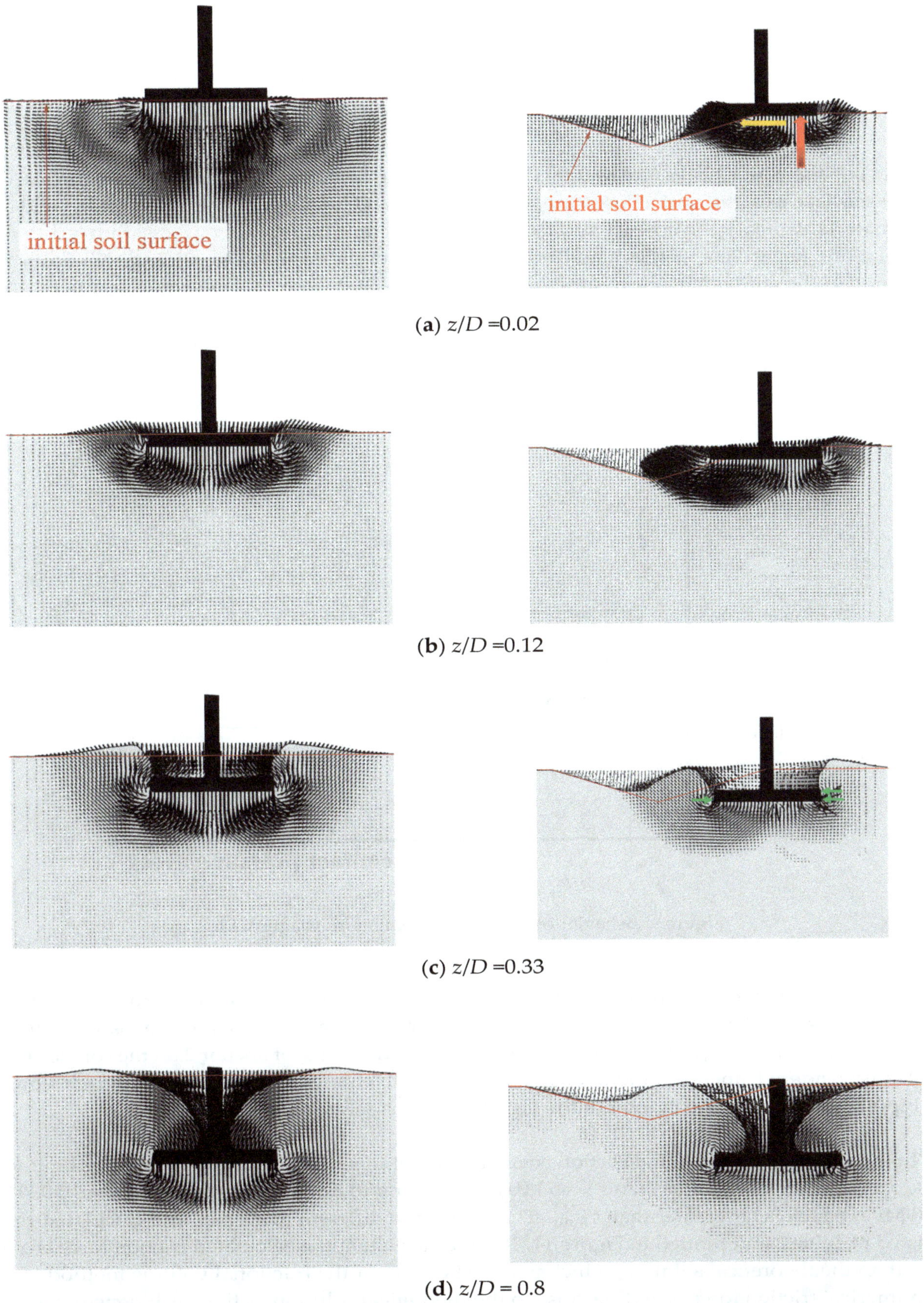

Figure 8. Soil flow mechanism of cases FS (**left**) and TB-2D-1.0D (**right**) for flat base footing.

3.2. *Effect of An Existing Footprint*

In Abaqus/CEL, the basic coulomb friction model defines the maximum allowable friction (shear) stress across an interface to the contact pressure stress, τ_{max}, as a function of the contact pressure:

$$\tau_{max} = \mu\, p \tag{1}$$

in which p is the contact pressure and μ is a friction coefficient that can be any non-negative value. In some special cases, the contact pressure p might be so large that $\tau_{max} = \mu\, p$ exceeds the yield stress in the material beneath the contact surface, thus a shearing limit value, τ_{limit}, is adopted to avoid this situation. Regardless of the magnitude of the contact pressure stress, sliding will occur if the magnitude of the equivalent shear stress reaches τ_{limit}. When both τ_{max} and τ_{limit} exceed s_u (yield stress), the maximum allowable friction (shear) stress equals s_u. All in all, the μ value only affects the friction force before the contact pressure reaches $p_2 = s_u/\mu$. After that, the friction force would be equal to ~s_u due to the yielding of clay. The relationship between equivalent shear stress and the contact pressure is plotted in Figure 9.

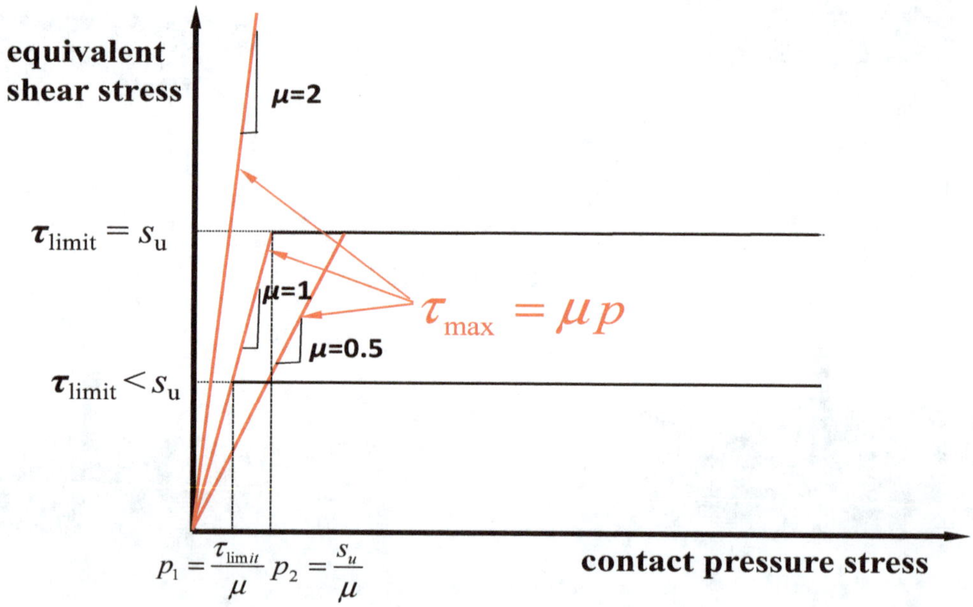

Figure 9. Behavior of the contact element in Abaqus/CEL.

A simple test model, as shown in Figure 10, is created to verify the accuracy of calculating friction force by Abaqus. All the parameters are detailed in Figure 10. The three anchors are disconnected and go right at a speed of 0.1 m/s at the same time. Only the friction force of anchor 2 on the contact surface surf 2 are considered, anchor 1 and anchor 3 are created to eliminate the influence of back-flow soil. An empty element layer of 5 m thick at the top of the soil is set to allow soil heaving.

In the simple test model, the friction coefficient is set to μ = 10,000 and the shearing limit value is set to τ_{limit} = 5.5 kPa (larger than s_u = 5 kPa). According to Equation (1), $\tau_{max} = \mu p$ = 10,000 × 70 = 700 MPa, which is far greater than τ_{limit}. The cases with different mesh size and calculated results are listed in Table 2 and plotted in Figure 11. It can be seen that the calculated friction force is a little lower than the theoretical solution, which may be because of the fractional volume method in CEL. The numerical friction force is getting close to the theoretical solution as the mesh density increases. When the minimum element size is b_{min}/B = 1/30, the calculation error is 6%, which is selected in the following analyses.

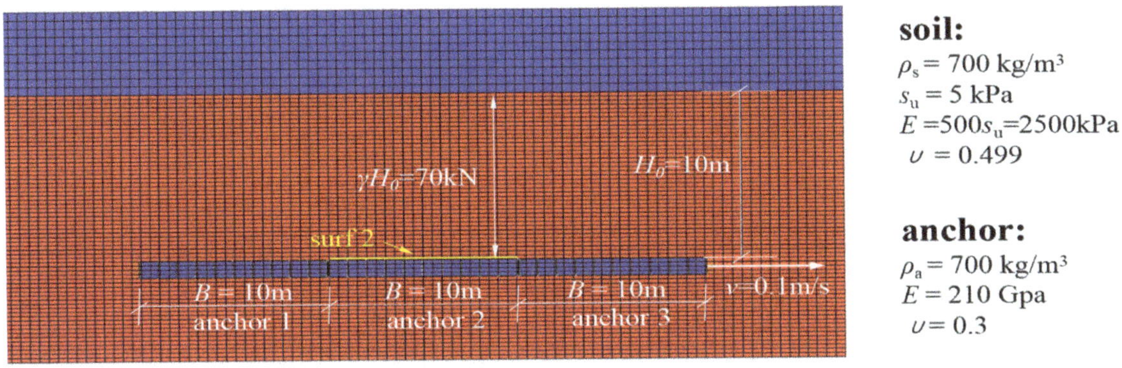

Figure 10. The test model.

Table 2. Numerical cases and results.

The minimum Element Size b_{min}/B	Numerical Friction Force (kPa)	Theoretical Friction Force (kPa)	Calculation Error (%)
1/20	4.55	5	9
1/30	4.7	5	6
1/40	4.75	5	5
1/80	4.85	5	3

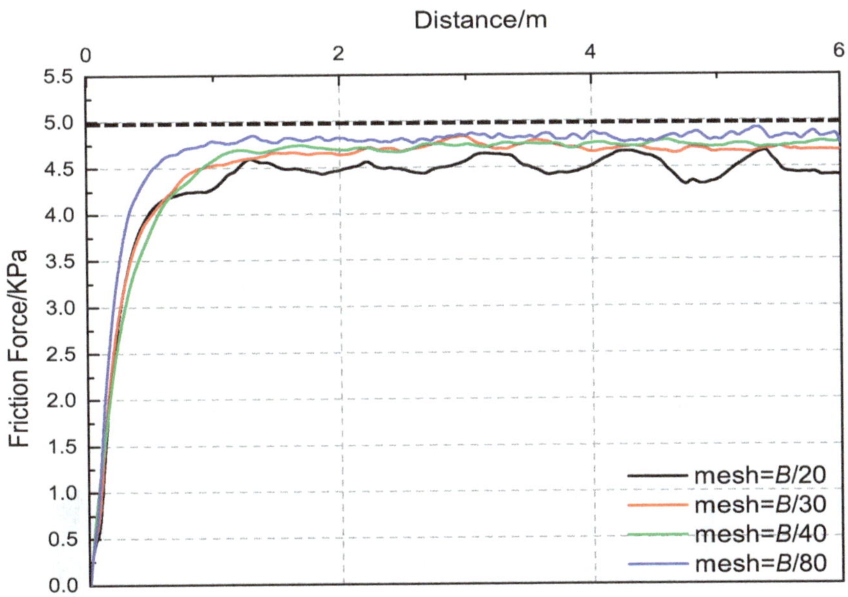

Figure 11. Numerical friction force in test model.

After investigating the behavior of the friction element in CEL, the effects of soil–structure friction on the V H and M of a spudcan penetrating near an existing footprint are carried out. The friction coefficient is set to $\mu = 0$ and 10,000 to represent smooth and rough conditions respectively. The shearing limit value is set as the undrained shear strength of the surrounding clay.

Comparing the smooth and rough cases, it can be seen that some certain friction has a significant effect on H profile, but no obvious effects on V and M, as shown in Figure 12. The friction condition does not affect the location where H_{max} and M_{max} occur. The soil flow mechanism in Figure 13 explains how the friction affects H profile. For the smooth case, H is only from the lateral pushing force of the soil on the right side of the footing. While for the rough case, the friction on the footing bottom also contributes to H.

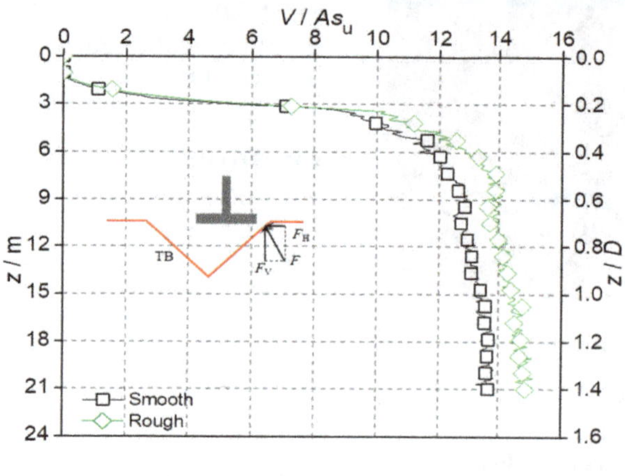

(**a**) Normalized vertical force

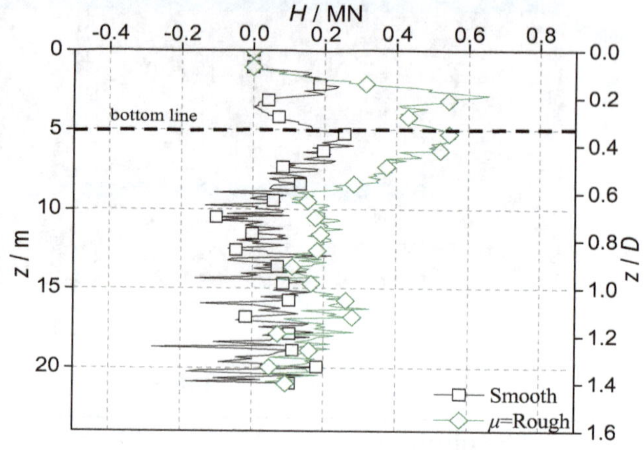

(**b**) Horizontal force

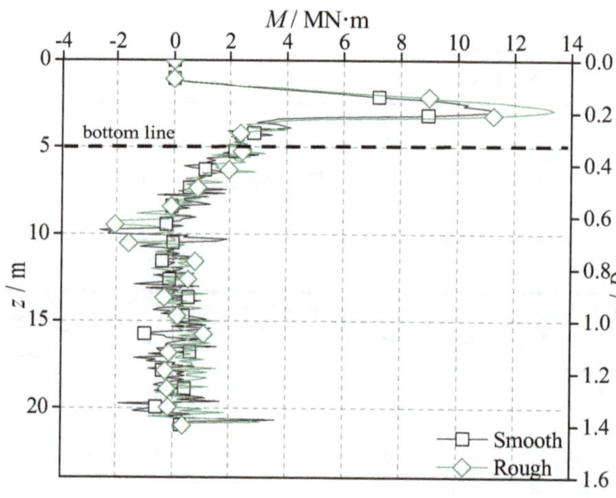

(**c**) Moment

Figure 12. The V, H, and M profile during smooth and rough conditions. (TB−2D−0.25D).

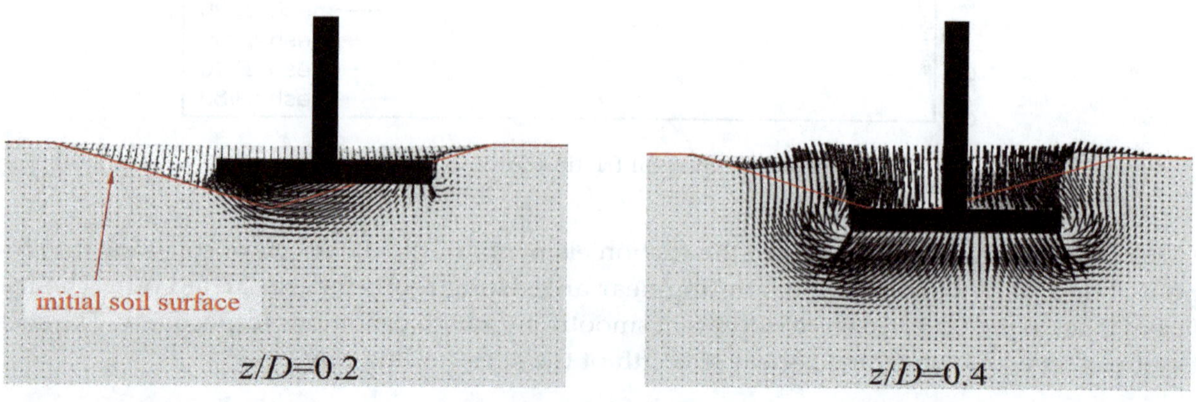

(**a**) smooth

Figure 13. *Cont.*

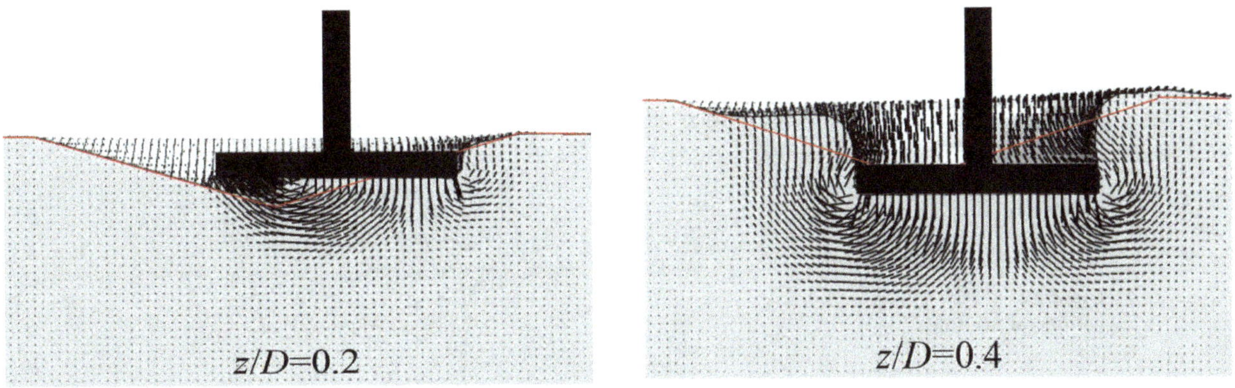

(**b**) rough

Figure 13. The soil flow mechanism of (**a**) smooth and (**b**) rough conditions. (TB–2*D*–0.25*D*).

The maximum normalized values of *H* and *M* of flat base footing are summarized in Figure 14. It can be seen clearly that the maximum *H* value of rough cases is around three times of that of smooth case, while the friction condition has a much smaller effect on M_{max} values (increasing 1.2 to 1.4 times).

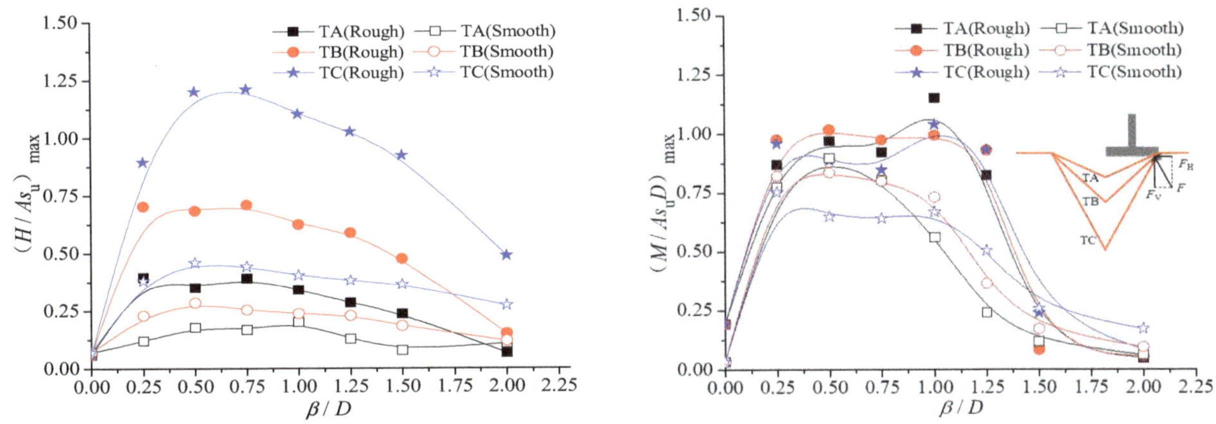

Figure 14. Maximum normalized *H* and *M* values against eccentricity ratio (flat base footing).

3.3. *Effect of the Location of the Reference Point (Working Leg Length)*

For the convenience of discussion, the *V*, *H*, and *M* discussed above are obtained using a reference point (see RP in Figure 4). If the reference point is located at the leg–hull connection section (see RP_0 in Figure 15), an additional moment (M_a) will be mobilized by *H* and its eccentricity (i.e., the working leg length), while the horizontal and vertical forces are not affected by the location of RP. In practical engineering cases, the leg–hull connector section could be the most dangerous section.

Assuming that the top head of the leg is fully fixed and the footing is considered as a rigid body, the additional bending moment (M_a) at RP_0 due to the horizontal force acting on the footing can be calculated as $M_a = H * L_{w\text{-}leg}$. The bending moment at RP_0 (M_{hull}) varies with the working leg length. The total moment at RP_0 (M_{hull}) can, therefore, be calculated as $M_{hull} = M + M_a$. The maximum value of both horizontal force (H_{max}) and bending moment (M_{max}) are taken as the most unfavorable combination of loads to calculate the bending moment on the leg–hull connection at different working leg lengths. As an example, the profile of M_{hull} of the flat base footing reinstalling near the TA footprint is plotted in Figure 16. It can be seen that M_{hull} is within a positive value at a small leg length, which means an anticlockwise moment. With the increasing of the working leg length, M_a increases linearly and, as a result, the total moment M_{hull} decreases. When $L_{w\text{-}leg}$ is less than ~30 m, the total moment is within a negative range (clockwise). With further increasing of $L_{w\text{-}leg}$, the absolute value of the clockwise M_{hull} would be larger than the anticlockwise M_{hull} at $L_{w\text{-}leg} = 0$.

Considering working leg length, the bending moment, M_{hull}, is a combination of the H and M at RP. To simplify the discussions, only the moments at the lower end of the leg (Section 1.1), $M_{1\text{-}1}$, are presented.

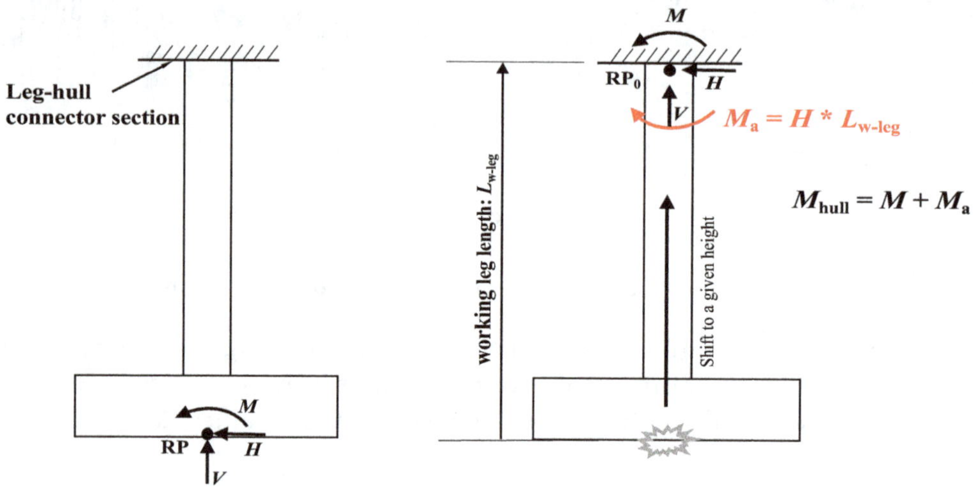

Figure 15. Maximum normalized H and M values against eccentricity ratio (flat base footing).

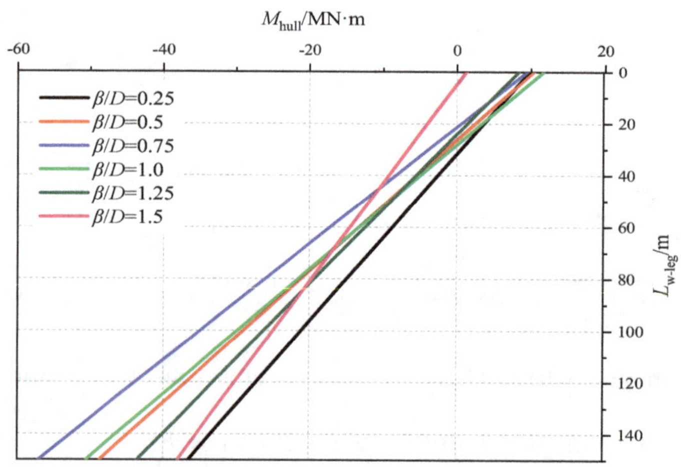

Figure 16. Maximum normalized H and M values against eccentricity ratio (flat base footing).

3.4. *Effect of Footprint Geometry*

The resistance profiles of spudcan penetrating through the edge of footprints TA, TB, and TC are presented in Figure 17, in which the offset distance is 0.75*D*.

As expected, the deeper the footprint is, the more effect it has on the reinstallation resistance profiles. All three H Profiles have the same trend, but the case with steeper slope causes higher H values. The H_{max} value for TC is about 3–5 times higher than that for TA. The deeper the footprint is, the longer it takes for H to reduce to zero.

The bending moment at Section 1.1 ($M_{1\text{-}1}$) can be derived according to the V, H, and M values acting on the Reference Point. The vertical force acting on the RP has no contribution on the bending moment at the section, M has a positive contribution, and H times distance has a negative contribution. The maximum $M_{1\text{-}1}$ values occur at a very shallow depth. With further penetration, the horizontal force becomes larger and plays a leading role in $M_{1\text{-}1}$ value. This results in that the positive $M_{1\text{-}1}$ reduces gradually to negative in Figure 17, with an increasing penetration depth.

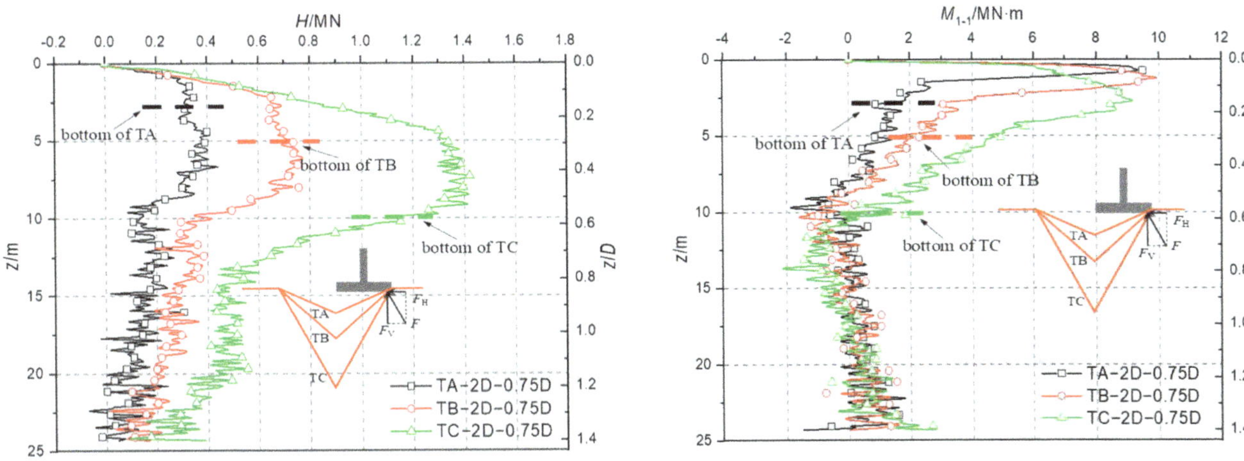

(**a**) Effect of footprint geometry on flat base footing ($\beta = 0.75D$)

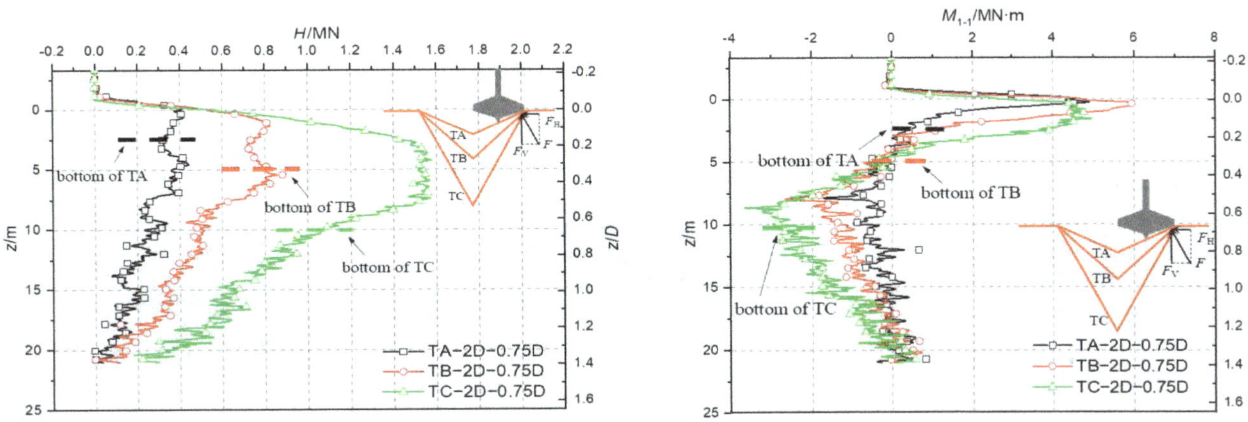

(**b**) Effect of footprint geometry on fusiform spudcan footing ($\beta = 0.75D$)

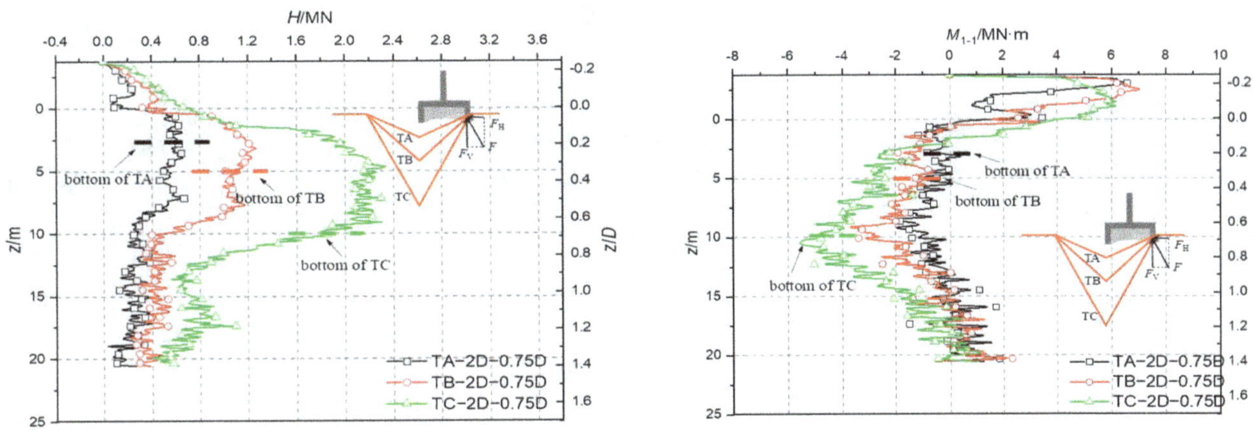

(**c**) Effect of footprint geometry on skirted footing ($\beta = 0.75D$)

Figure 17. Effect of footprint geometry on three types of footings.

3.5. Effect of Footings' Geometry Shape and Offset Distance

The H and M profiles of the three footings reinstalling at selected typical offset distances are shown in Figure 18. The H can be separated into two parts. The first is the horizontal component of normal contact force between the footprint slope and the right side of footing, which is the primary cause of the first peak shown in Figure 19. The second is the lateral pushing force from the right-side soil caused by the asymmetry soil flowing, which is the primary cause of the second peak. For a flat base footing or a skirted footing, the horizontal component of normal contact force is relatively small, since the footing base is horizontal. However, for a fusiform spudcan footing, due to the inverted conical shape, the first part of H force plays a leading role in H profile. After deep penetration, the geometry shape has a minor effect on the resistance, since the soil flow mechanisms are both fully back flow left and right.

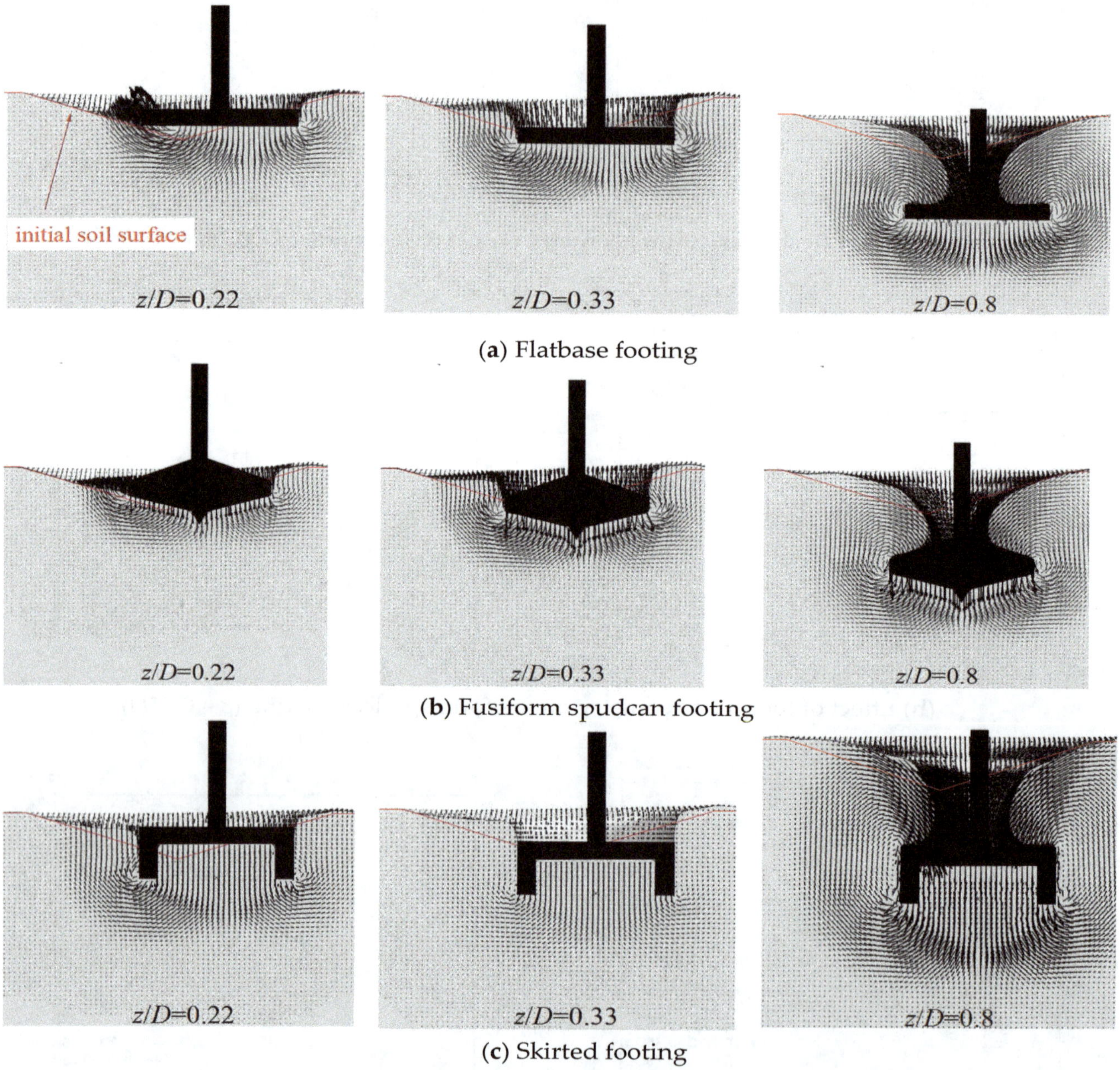

Figure 18. The soil flow mechanism for different footings (TB-2D-0.25D).

The H_{max} and $M_{1\text{-}1max}$ values of all the cases in this study are listed in Table A1 in Appendix A and plotted in Figure 20. For flat base footings and skirted footings, both H_{max} and $M_{1\text{-}1max}$ are significant when $\beta/D = 0.25$ to 1.25. When $\beta/D \geq 1.5$, the value of $M_{1\text{-}1max}$ reduces to zero, while H_{max} still remains at significant values. For fusiform spudcan footings, both H_{max} and $M_{1\text{-}1max}$ are significant when $\beta/D = 0.25$ to 0.5. From the perspective of the footing shape, the flat base footing gives the lowest H_{max} but the largest $M_{1\text{-}1max}$, and the performances of the fusiform spudcan footing and the skirted footing are similar.

It is worthwhile to note that the thickness of the skirt for the skirted footing of the numerical model is higher than in situ skirted footing in order to mitigate numerical divergence. That might cause an overprediction on resistance loads. The effect of the skirt thickness can be ignored when the base level (with the maximum cross-section area) of the skirted footing fully touches the soil.

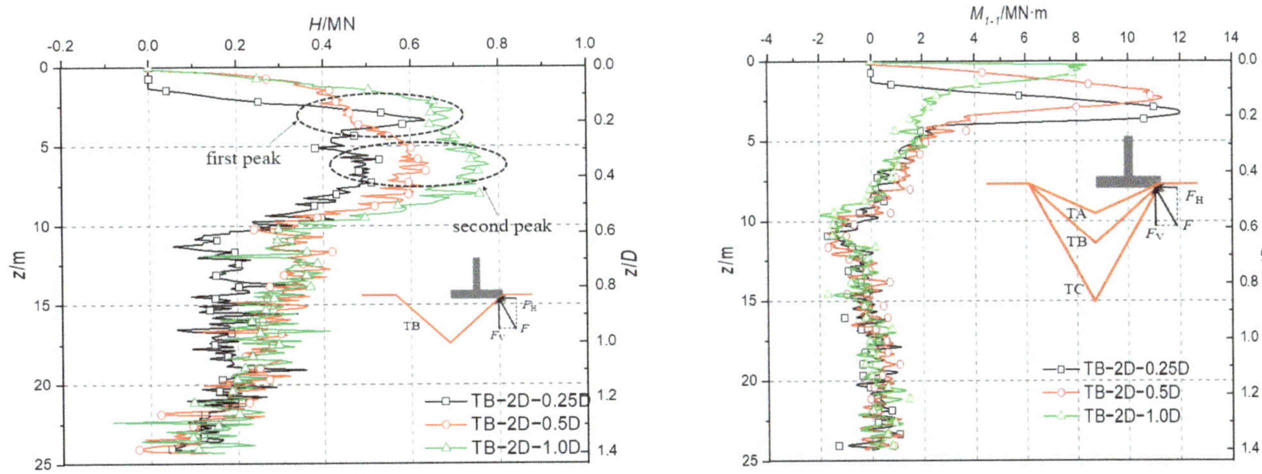

(**a**) The H and $M_{1\text{-}1}$ profile of flat base footing at offsets 0.25D, 0.5D, and 1.0D

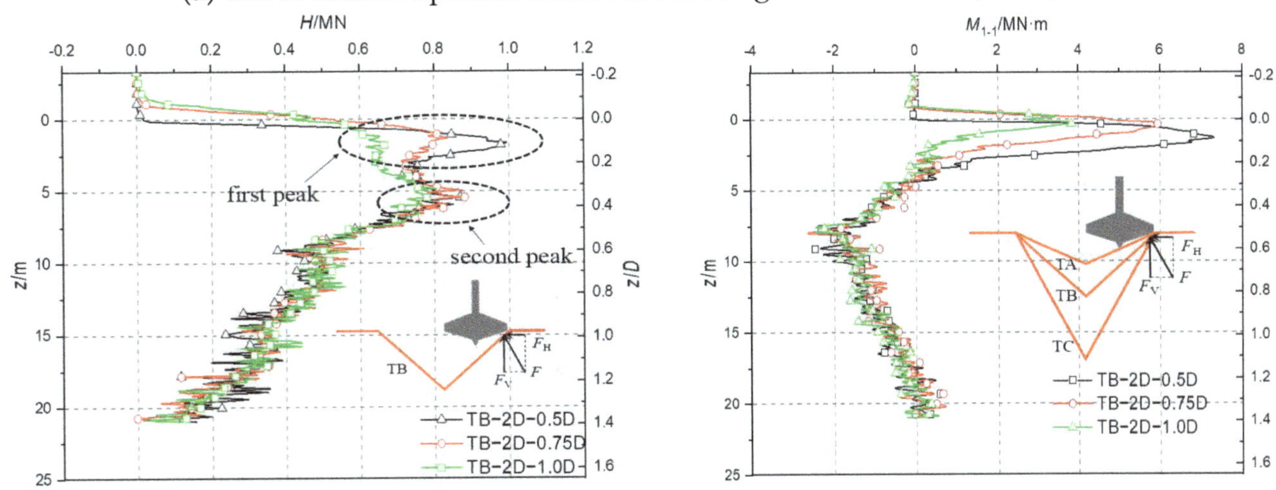

(**b**) The H and $M_{1\text{-}1}$ profile of fusiform spudcan footing at offsets 0.5D, 0.75D, and 1.0D

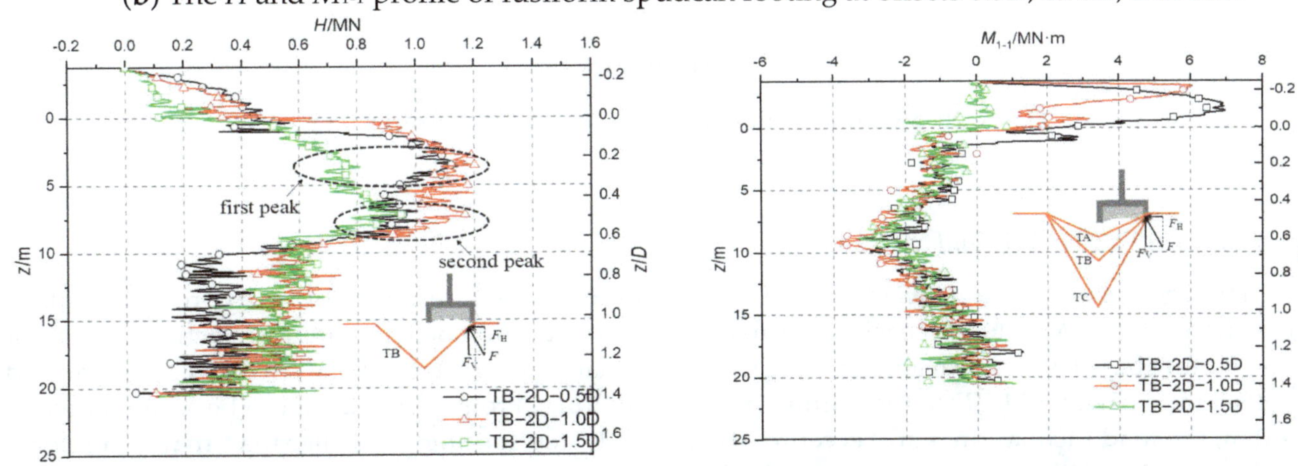

(**c**) The H and M profile of skirted footing at offsets 0.5D, 1.0D, and 1.5D

Figure 19. The H and $M_{1\text{-}1}$ profile of three kind of footings at different offsets.

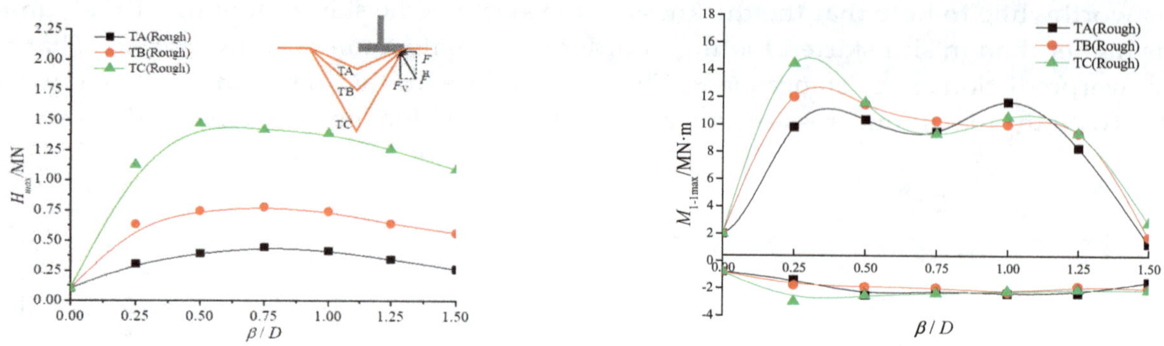

(**a**) Flat base footing

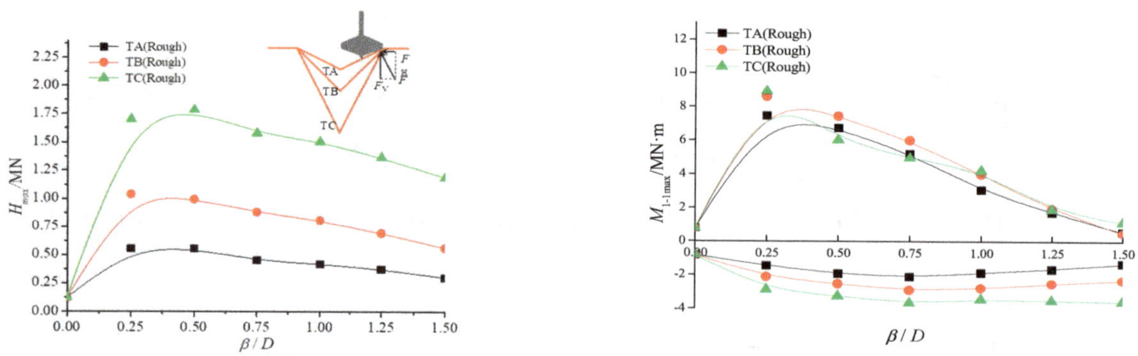

(**b**) Fusifrom Spudcan

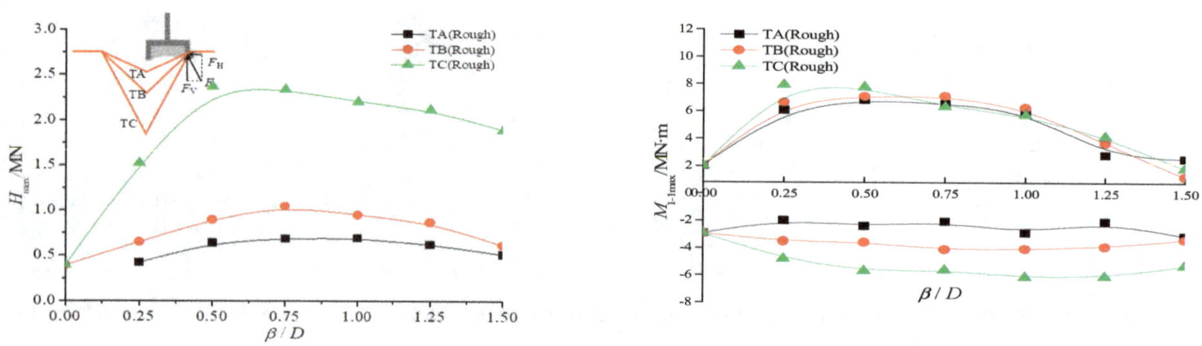

(**c**) Skirted footing

Figure 20. Maximum values of *H* and *M* against eccentricity ratio.

3.6. Resultant Force of V H and M

The above analyses are based on *V H* and *M* values at the reference point. To provide another view, the *V*, *H*, and *M* values of each case can be transformed into one resultant force acting on a point at the footing base level. The resultant force has an inclination of $\alpha = \tan^{-1}(H/V)$ to the vertical line and an offset ratio of $e/D = M/VD$ to the central line of the footing. From Figure 21, it can be seen that when $z > 0$ m, the load inclination α and eccentricity e/D of skirted footing is smaller than that of fusiform spudcan footing. When $z < 0$ m, although both α and e/D of the skirted footing are larger, the vertical force is relatively small and the force acting on the footing may not be sufficient to cause structure failure. That is to say, the skirted footings may have a certain potential in resisting the damage during reinstallation near existing footprints, by comparing with commonly used fusiform spudcan footings.

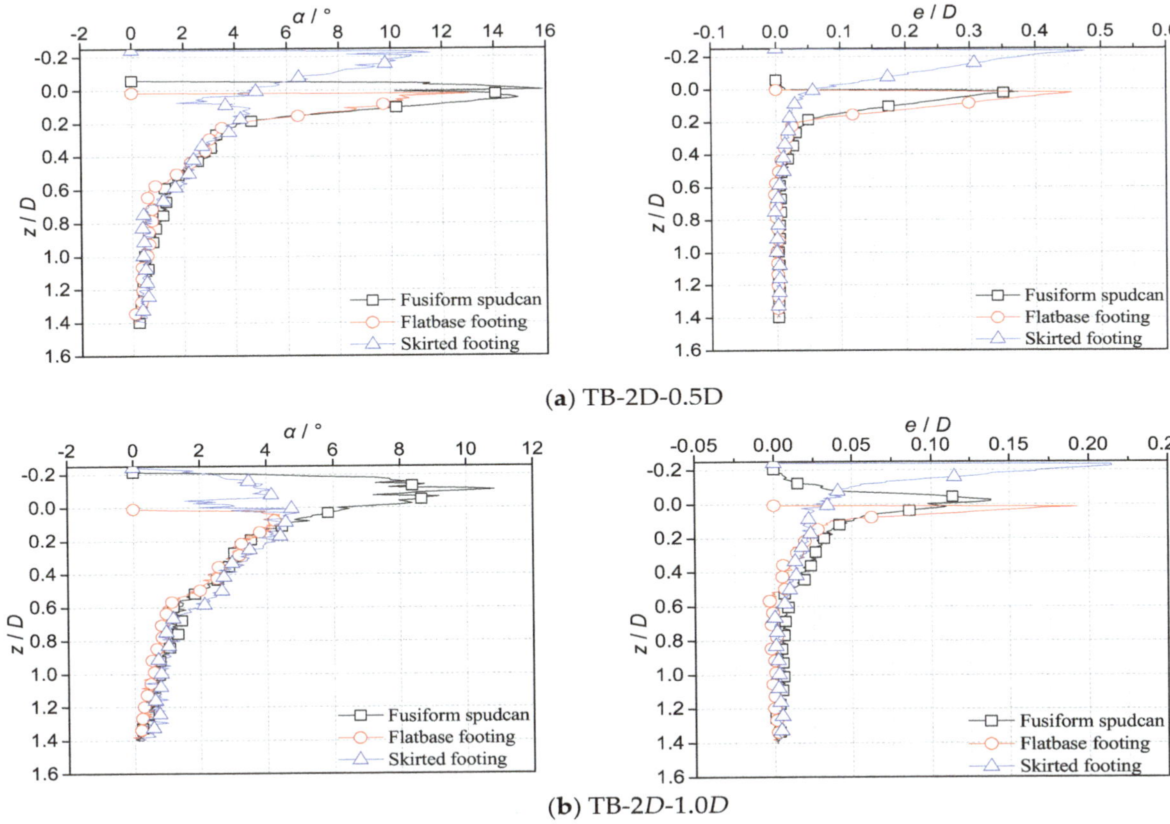

Figure 21. Variations in the (**a**) load inclination and (**b**) load eccentricity during the reinstallation process of the β = 0.5*D*, 1.0*D* cases.

4. Conclusions

This paper carried out large deformation finite element analyses to investigate the effect of an existing footprint on the stability of jack-ups' reinstallation. The following conclusions can be drawn according to the present numerical analyses:

The friction condition of the soil–footing interface has a significant effect on *H* profile but much less effect on *M* profile. The deeper is the footprint, the more effect it has on both *H* and *M* profiles.

The eccentricity ratio is a key factor to evaluate H_{max} and $M_{1\text{-}1max}$. For flat base footings and skirted footings, both H_{max} and $M_{1\text{-}1max}$ are significant when $\beta/D = 0.25$ to 1.25. The value of $M_{1\text{-}1max}$ reduces to zero when $\beta/D \geq 1.5$, while H_{max} still remains at a significant value. For fusiform spudcan footings, both H_{max} and $M_{1\text{-}1max}$ are significant when $\beta/D = 0.25$ to 0.5.

The geometry shape of the footing also has a certain effect on the *V*, *H*, and *M* profiles. The flat base footing gives the lowest H_{max} but the largest $M_{1\text{-}1max}$, and the performances of the fusiform spudcan footing and the skirted footing are similar. From the view of the resultant forces, both α and e/D of the skirted footing are only large before the base level (with the maximum cross-section area) fully touches the soil, which shows a certain potential in resisting the damage during reinstallation near existing footprints by comparing with commonly used fusiform spudcan footings.

The bending moment on the leg–hull connection (M_{hull}) at different working leg lengths ($L_{w\text{-}leg}$) is discussed. When $L_{w\text{-}leg}$ is less than ~30 m, the total moment is within a negative range (clockwise). With further increasing of $L_{w\text{-}leg}$, the absolute value of the clockwise M_{hull} would be larger than the anticlockwise M_{hull} at $L_{w\text{-}leg} = 0$.

In this study, the artificial footprints were adopted to simplify the problem neglecting the disturbance of the soil during initial spudcan penetration. In the further study, the soil profiles, soil properties, geometry of footprints and spudcans, leg details, use of spigots (or not) etc. should be noted as a factor to consider in site-specific analyses.

Author Contributions: Data curation, X.W.; formal analysis, H.Z.; software, J.L.; supervision, L.Y.

Appendix A

Table A1. The list of peak load.

Footprint	β/D	H_{max}/MN			M_{max}/MN·m		
		Fusiform Spudcan Footing	Flat Base Footing	Skirted Footing	Fusiform Spudcan Footing	Flat base Footing	Skirted Footing
TA	0.25	0.55349	0.30883	0.42189	7.44231	9.80965	6.09376
	0.5	0.5524	0.39296	0.6407	6.72754	10.3384	6.84643
	0.75	0.45052	0.44328	0.68789	5.20043	9.48458	6.57694
	1.0	0.4187	0.41402	0.6905	3.06038	11.6203	6.06864
	1.25	0.37315	0.34509	0.62492	1.73015	8.2809	2.8528
	1.5	0.29904	0.26207	0.51236	0.56979	1.28662	2.54351
TB	0.25	1.0387	0.63409	0.77274	8.59083	12.0215	6.65431
	0.5	0.99436	0.74464	1.14443	7.42953	11.4894	7.08853
	0.75	0.8814	0.77806	1.29267	6.00388	10.2634	7.13538
	1.0	0.80606	0.73853	1.2015	3.99285	10.007	6.30153
	1.25	0.69096	0.64062	1.10797	1.96356	9.37414	3.72762
	1.5	0.5602	0.56196	0.9585	0.5096	1.77092	1.25643
TC	0.25	1.701	1.12667	1.52282	8.93624	14.4618	7.94377
	0.5	1.78348	1.47111	2.36442	6.0261	11.5915	7.78829
	0.75	1.57796	1.42199	2.33573	4.96665	9.31258	6.36396
	1.0	1.50321	1.3904	2.203	4.20818	10.5380	5.74945
	1.25	1.36363	1.25953	2.11594	1.90344	9.36561	4.17154
	1.5	1.18654	1.09092	1.89335	1.13969	2.79521	1.85664

References

1. Dier, A.; Carroll, B.; Abolfathi, S. *Guidelines for Jack–up Rigs with Particular Reference to Foundation Integrity*; HSE Books: Hong Kong, 2004.
2. Jack, R.L.; Hoyle, M.J.R.; Smith, N.P.; Hunt, R.J. Jack–up accident statistics—A further update. In Proceedings of the 14th International Conference on the Jack–up Platform Design, Construction and Operation, London, UK, 17–18 September 2013.
3. Cassidy, M.J.; Quah, C.K.; Foo, K.S. Experimental Investigation of the Reinstallation of Spudcan Footings Close to Existing Footprints. *J. Geotech. Geoenviron. Eng.* **2009**, *135*, 474–486. [CrossRef]
4. Gan, C.T. Centrifuge model study on spudcan–footprint interaction. Ph.D. Thesis, National University of Singapore, Singapore, 2 November 2009.
5. Dean, E. A spudcan foundation model with excess pore pressures. Part 1. A principle of effective loads. *Mar. Struct.* **2004**, *17*, 219–243. [CrossRef]
6. Gan, C.T.; Leung, C.F.; Cassidy, M.J.; Gaudin, C.; Chow, Y.K. Effect of time on spudcan–footprint interaction in clay. *Geotechnique* **2012**, *62*, 401–413. [CrossRef]
7. Byrne, B.W.; Villalobos, F.; Houlsby, G.T.; Martin, C.M. Laboratory testing of shallow skirted foundations in sand. In Proceedings of the BGA International Conference on Foundations, Innovations, Observations, Design and Practice, Dundee, UK, 2–5 September 2003; pp. 161–173.
8. Leung, C.; Gan, C.; Chow, Y. Shear strength changes within jack–up spudcan footprint. In Proceedings of the Seventeenth International Offshore and Polar Engineering Conference: International Society of Offshore and Polar Engineers, Lisbon, Portugal, 1–6 July 2007; pp. 1504–1509.
9. Hartono, H.; Tho, K.; Leung, C.; Chow, Y. Centrifuge and Numerical Modelling of Reaming as a Mitigation Measure for Spudcan–Footprint Interaction. In Proceedings of the Offshore Technology Conference–Asia: Offshore Technology Conference, Kuala Lumpur, Malaysia, 25–28 March 2014; pp. 1478–1492.
10. Gan, C.T.; Teh, K.L.; Leung, C.F.; Chow, Y.K.; Swee, S. Behaviour of skirted footings on sand overlying clay. In Proceedings of the 2nd International Symposium on Frontiers in Offshore Geotechnics, Perth, Australia, 8–10 November 2011; pp. 415–420.

11. Hossain, M.S.; Hu, Y.; Ekaputra, D. Skirted foundation to mitigate spudcan punch–through on sand–over–clay. *Geotechnique* **2014**, *64*, 333–340. [CrossRef]
12. Hai–bo, Y. The Optimization of HYSY 944 Jack–up Spudcan Piping Arrangement. *Chin. Offshore Platform.* **2016**, *31*, 23–26.
13. Kong, V.; Cassidy, M.J.; Gaudin, C. Experimental study of effect of geometry on reinstallation of jack–up next to footprint. *Can. Geotech. J.* **2013**, *50*, 557–573. [CrossRef]
14. Kong, V.; Cassidy, M.J.; Gaudin, C. Failure mechanisms of a spudcan penetrating next to an existing footprint. *Theor. Appl. Mech. Lett.* **2015**, *5*, 64–68. [CrossRef]
15. Zhang, W.; Cassidy, M.J.; Tian, Y. 3D Large Deformation Finite Element Analyses of Jack–up Reinstallations Near Idealised Footprints. In Proceedings of the 15th International Conference on the Jack–Up Platform Design, Construction and Operation, London, UK, 15–16 September 2015.
16. Jun, M.; Kim, Y.; Hossain, M.; Cassidy, M.; Hu, Y.; Park, S. Global jack–up rig behaviour next to a footprint. *Mar. Struct.* **2019**, *64*, 421–441. [CrossRef]
17. Jun, M.; Kim, Y.; Hossain, M.; Cassidy, M.; Hu, Y.; Park, S. Optimising spudcan shape for mitigating horizontal and moment loads induced on a spudcan penetrating near a conical footprint. *Appl. Ocean Res.* **2018**, *79*, 62–73. [CrossRef]
18. Jun, M.; Kim, Y.; Hossain, M.; Cassidy, M.; Hu, Y.; Park, S. Physical modelling of reinstallation of a novel spudcan nearby existing footprint. *Phy. Model. Geotech.* **2018**, *1*, 615–621.
19. Jun, M.; Kim, Y.; Hossain, M.; Cassidy, M.; Hu, Y.; Sim, J. Numerical investigation of novel spudcan shapes for easing spudcan–footprint interactions. *J. Geotech. Geoenviron. Eng.* **2018**, *144*, 04018055. [CrossRef]
20. Stewart, D.P.; Finnie, M.S. Spudcan–Footprint Interaction During Jack–Up Workovers. In Proceedings of the 11th International Offshore and Polar Engineering Conference, Stavanger, Norway, 17–22 June 2001; pp. 61–65.
21. Carrington, T.; Hodges, B.; Aldridge, T.; Osborne, J.; Mirrey, J. Jack–Up Advanced Foundation Analysis by Automatic Re–meshing (Large Strain) FEA Methods. In Proceedings of the 9th International Conference on Jack–Up Platform Design, Construction and Operation, London, UK, 23–24 September 2003.
22. Foo, K.; Quah, C.; Wildberger, P.; Vasquez, J. Spudcan footprint interaction and rack phase difference (RPD). In Proceedings of the 9th International Conference on Jack–Up Platform Design, Construction and Operation, London, UK, 23–24 September 2003.
23. Dean, E.R.; Serra, H. Concepts for mitigation of spudcan–footprint interaction in normally consolidated clay. In Proceedings of the Fourteenth International Offshore and Polar Engineering Conference: International Society of Offshore and Polar Engineers, Toulon, France, 23–28 May 2004; pp. 721–728.
24. Jardine, R.; Kovacevic, N.; Hoyle, M.; Sidhu, H.K.; Letty, A. Assessing The Effects On Jack Up Structures Of Eccentric Installation Over Infilled Craters. In Proceedings of the International Conference—Offshore Site Investigation and Geotechnics "Diversity and Sustainability", London, UK, 26–28 November 2002; pp. 307–324.
25. Kong, V.W. Jack–up reinstallation near existing footprints. Ph.D. Thesis, University of Western Australia, Perth, WA, Australia.
26. Liu, J.; Hu, Y.X.; Kong, X.J. Deep penetration of spudcan foundation into double layered soils. *Chin. Ocean Eng.* **2005**, *19*, 309–324.
27. Yu, L.; Hu, Y.; Liu, J.; Randolph, M.F.; Kong, X. Numerical study of spudcan penetration in loose sand overlying clay. *Comp. Geotech.* **2012**, *46*, 1–12. [CrossRef]

11

Pressure Sensing Technique for Observing Seabed Deformation Caused by Submarine Sand Wave Migration

Xiaolei Liu [1,2,*], Xiaoquan Zheng [1], Zhuangcai Tian [1], Hong Zhang [1] and Tian Chen [1]

1 Shandong Provincial Key Laboratory of Marine Environment and Geological Engineering, Ocean University of China, Qingdao 266100, China; zxq9697@stu.ouc.edu.cn (X.Z.); zhuangcaitian@163.com (Z.T.); 18051379505@163.com (H.Z.); chentian@stu.ouc.edu.cn (T.C.)

2 Laboratory for Marine Geology, Qingdao National Laboratory for Marine Science and Technology, Qingdao 266061, China

* Correspondence: xiaolei@ouc.edu.cn

Abstract: Long-term, continuous in-situ observation of seabed deformation plays an important role in studying the mechanisms of sand wave migration and engineering early warning methods. Research on pressure sensing techniques has examined the possibility of using the temporal characteristics of the vertical deformation of the seafloor to identify important factors (e.g., wave height and migration rate) of submarine sand wave migration. Two pressure sensing tools were developed in this study to observe the seabed deformation caused by submarine sand wave migration (a fixed-depth total pressure recorder (TPR_{FD}) and a surface synchronous bottom pressure recorder (BPR_{SS})), based on the principle that as a sand wave migrates under hydrodynamic forcing, the near-bottom water pressure, bottom pressure and total fixed pressure synchronously change with time. Laboratory flume experiments were performed, using natural sandy sediments taken from the beach of Qingdao, China, to better present and discuss the feasibility and limitations of using these two pressure sensing methods to acquire continuous observations of seabed deformation. The results illustrate that the proposed pressure sensor techniques can be effectively applied in reflecting elevation caused by submarine sand wave migration (the accuracy of the two methods in observing the experimental bed morphology was more than 90%). However, an unexpected step-like process of the change in sand wave height observed by BPR_{SS} is presented to show that the sensor states can be easily disturbed by submarine environments, and thus throw the validity of BPR_{SS} into question. Therefore, the TPR_{FD} technique is more worthy of further study for observing submarine sand wave migration continuously and in real-time.

Keywords: sand wave; pressure sensing technique; physical model test; field application

1. Introduction

Submarine sand waves are approximately regular undulating landforms [1] formed by the movement of sandy sediments under various marine hydrodynamic forces, such as ocean currents [2], tidal currents [3] and internal waves [4]. Submarine sand waves are widely distributed across continental slopes [5,6], continental shelves [7,8], straits [9,10], gulfs [11], and other geomorphic units around the world, in shallow to deep seas. Under the action of ocean dynamics, sand waves undergo periodic migration movements, and rates can reach nearly 70 meters per year [12,13]. From this process, vertical deformation may spur the suspension or burial of submarine cables [3,14] and submarine pipelines [8,15], which can seriously damage them. Therefore, the observation and study of the geomorphic morphology of seabed sand waves has attracted the attention of numerous researchers.

At present, the positioning and repeated measurement of water depths is used to observe the migration of seabed sand waves. By analyzing high-precision digital terrain model (DTM) data measured at several time points in a study area, the average rate and direction of sand wave migration for a given period can be obtained [16,17]. DTM data obtained from multi-beam water depth measurements can also be used to calculate the rate of seabed sand wave migration by profile analysis [13,18]. Franzetti [7] used the spatial cross-correlation three-dimensional analysis approach to analyze and calculate the horizontal migration rate and the vertical variation of sand waves. Zhou [19] also used three sets of repeated multi-beam sounding data for 2011–2013 to study the first migration and changes of giant sand wave fields in the Taiwan Shoal in the northern South China Sea. However, because these methods depend on ship operations, they cannot be used to observe continuous changes in height, due to weather, sea conditions and cost constraints.

In-situ integrated observation techniques have been developed to better understand processes and mechanisms of sand wave migration at finer scales since the late 20th Century. Fixed flow meters and small bottom observation platforms, equipped with scan sonar and acoustic backscattering, have been used to observe sand wave surfaces, the dynamic sand wave processes on bottom surfaces [20], variations in sediment velocity on sand wave surfaces [21] and sand ripple variations with rising tides [22]. However, because these acoustic or optical instruments are not only susceptible to the concentration of suspended sediments in near-surface water, but also generate a lot of power consumption during the actual operation, they might not always meet the need for continuous on-site observations in sand wave areas.

Due to their stability and environmental adaptability, pressure sensors have increasingly been used for sea floor height measurements [23–26] and vertical seabed deformation observations [27–29] since the 1990s. Japan's MH21 plan for the exploitation of natural gas hydrates in Japan's seabed involves formation subsidence monitoring at a precision level of 10 mm [30]. In the North Sea, high-precision water pressure measurement technology is used to monitor seabed subsidence [31]. These successful cases highlight possibilities to apply pressure sensing technology in the study of seabed sand wave migration.

In view of this summary of past research, this paper focuses on the application of pressure sensing techniques in the study of sand wave migration. Based on the principle of vertical pressure change caused by sand wave migration, two observation methods were designed and verified by indoor water flume tests. The work presented in this paper can guide the monitoring and early detection of submarine sand wave migration, and be used to better understand mechanisms of submarine sand wave migration.

2. Materials and Methods

2.1. Theory

When using pressure sensing technology to continuously observe the height of the sea floor, the position of the observation point must first be determined. Assuming that the observation point is located at the trough of the initial position of a sand wave (solid line), under hydrodynamic action, the sand wave migrates to the left toward the dotted line (Figure 1). According to the pressure change law acting on the sand wave profile during this process, the bottom pressure (P_B) at the observation point mainly includes the hydrostatic pressure determined by the sea floor surface height (H) and dynamic water pressure created by the waves and current (ΔP_N). The total pressure change (ΔP_T) is mainly shaped by changes in the soil–water pressure ratio's contribution to total pressure created by ΔH, and by changes in dynamic water pressure caused by waves and currents. Therefore, in combining the device arrangement depth of the in-situ observation (b) with the near-bottom pressure sensor mounting height (a), it can be deduced that there is a relationship (shown by Equations (1) and (2)) between ΔH, and ΔP_B and ΔP_T.

$$\Delta H = (\Delta P_B - \Delta P_N)/\rho_w g, \tag{1}$$

$$\Delta H = a \cdot (\Delta P_T - \Delta P_N)/(P_{T0} - P_{N0} - \rho_w g b), \tag{2}$$

$$\Delta P_T = \Delta H \cdot \gamma + \Delta P_N, \tag{3}$$

where ρ_w is the density of seawater, g is gravity acceleration, γ is the buoyant unit weight of the seabed sediment; P_{T0} is the total fixed-depth pressure level at the time of initial recording and P_{N0} is the near-bottom pressure level at the time of initial recording.

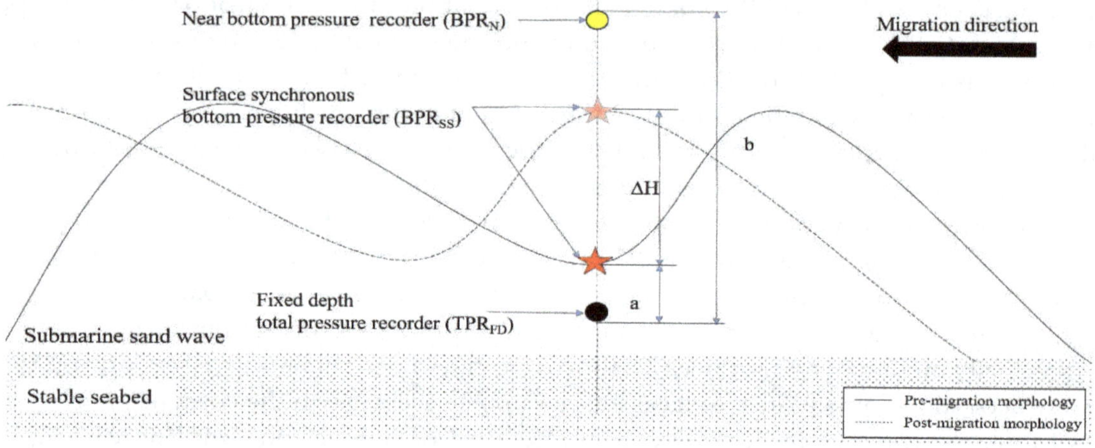

Figure 1. Pressure changes at key points during sand wave migration.

Thus, both P_T and P_N can be used to measure vertical height changes of the sea floor caused by sand wave migration. Based on this, we propose two tools for observing the vertical deformation of the sea floor caused by sand wave migration: (a) a fixed-depth total pressure recorder (TPR_{FD}) kept at a certain depth within a sand wave at the observation point and (b) a surface synchronous bottom pressure recorder (BPR_{SS}) kept at the surface of a sand wave at the point of observation.

2.2. *Experimental Set-Up and Arrangement*

To verify the feasibility of the above two methods, a physical model test was carried out in the wave flume (14.0 m × 0.5 m × 1.3 m) (Figure 2) housed at the Environmental Geotechnical Laboratory of the College of Environmental Science and Engineering, Ocean University of China. The 2.6 m wave-shaped test bed (Figure 2b) used to simulate sand wave terrain was filled with sand samples (Qingdao Beach, China) (Table 1). A wave generator with a wave frequency of 0.2 to 50 Hz was fixed to the right end of the water flume to form a wave with a controllable wave height and frequency (Figure 2a). A permeable slope with a slope of 1:4 was set on the left side of the sink to eliminate the influence of reflected waves (Figure 2a). The two side walls of the sink are made from transparent tempered glass to easily observe test phenomena and markings.

Table 1. Soil sample properties and instrument parameters.

Instrument	Characteristic Parameters	
Sand Sample	Average particle diameter	0.25 mm
	Nonuniformity coefficient	1.47
	Curvature coefficient	1.14
	Buoyant unit weight	16.2 N/cm^3
AA400	Accuracy	1 mm
	Range	0.15–100 m
Ultrasonic terrain scanner	Accuracy	1 mm
	Range	0.1–2 m
Pressure sensor	Accuracy	0.5%
	Range	0–20 kPa
Fiber optic pressure sensor	Accuracy	1‰ F.S
	Range	0–30 kN

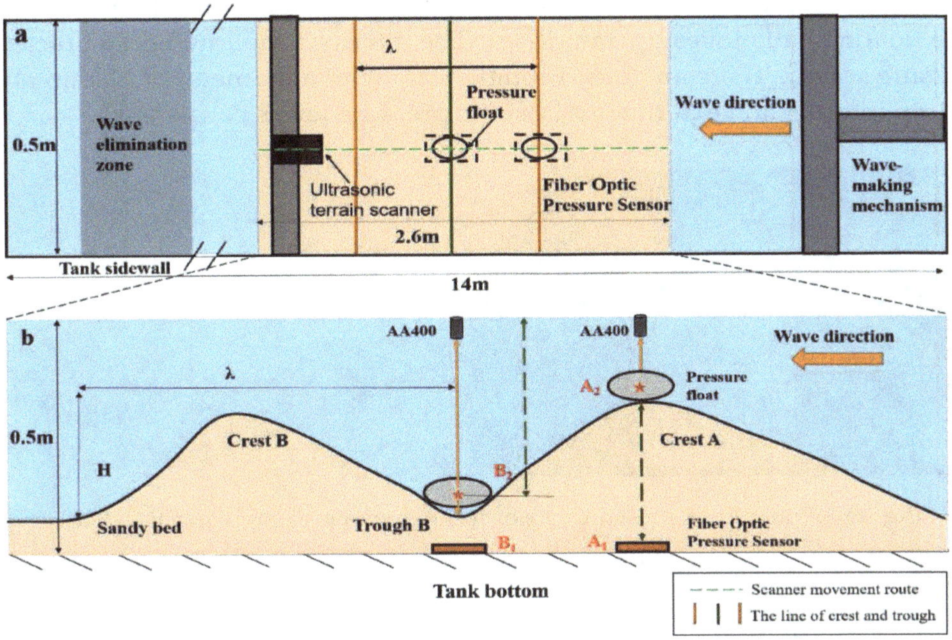

Figure 2. Schematic diagram of the water flume structure and test device. (**a**) Top view of the test device; (**b**) side view of the simulated sand wave bed.

The traditional electrical pressure sensor is easily affected by its own thermal effect and poor linearity, while the optical fiber pressure sensor is small in size, highly sensitive and stable, and responds directly to pressure changes. Therefore, because the measurement performance of a pressure sensor directly determines the accuracy of the ocean observation depth [32], in this experiment, the optical fiber pressure sensor (Suzhou NanZee Sensing Technology Co., Ltd., Suzhou, China) was used to observe the total pressure level. The sensor is 100 mm × 100 mm × 19.2 mm in size, and the interior of it is designed as a "Seesaw" structure (Figure 3). The pressure was measured from the change in the strain wavelength (P_1, P_2) caused by the change in the force of the two fibers installed on the seesaw structure. With no stress, the strain wavelengths of the two fibers were: $P_1 = 1538.22$ nm and $P_2 = 1546.303$ nm. The pressure calculation formula is $W = K_P \times (\Delta P_1 + \Delta P_2)$ where W is the weight of the overlying object, and ΔP_1 and ΔP_2 are the changes in the strain wavelength after being stressed. The NZS-FBG-A01 (M) multi-channel fiber grating sensor demodulation module (Suzhou NanZee Sensing Technology Co., Ltd., Suzhou, China) was used to collect and demodulate the data measured by the fiber sensor. It is a high-resolution Bragg grating sensor demodulation system and a high-precision spectrum analysis system. The resolution of the demodulation wavelength is 0.1 pm and the speed of demodulation is 1 Hz. The central wavelength, peak value and wavelength scanning reflection spectrum of the optical fiber pressure sensor can be output to the computer connected to the demodulation module in real-time through the USB cable. After using standard weights, increasing the number of weights step by step, and combining calculation formulas to calibrate the sensors in the test environment, parameter K_P is 290.3 g/nm, and the linear equation is $W_X = 290.3 \times (\Delta P_{1X} + \Delta P_{2X})$.

Whether the height of the bottom pressure sensor can synchronously change with the height of the sea floor at the observation point is key to whether the BPR_{SS} method can be used for observation. Therefore, a float with a density value between those of the sand bed and water was designed to carry the pressure sensor. Float spheres made from polyamide (PA) have a density (1.14 g/cm^3) slightly greater than that of seawater (1.10 g/cm^3). Because a dish-shaped object has good hydrodynamic features, to reduce measurement errors caused by the movement of the floating ball due to hydrodynamic forces, a dish-shaped floating ball was used in the test. As Figure 4 shows, the float has a diameter of 186 mm on the horizontal axis and a height of 93 mm on the vertical axis. In the middle of the upper structure of the float, a 16 mm diameter penetration hole is reserved. A smooth stainless steel rod is connected to the metal base through the floating ball penetration hole to reduce the left and right sway caused by the wave as the floating ball moves up and down. The interior is hollow, and a bracket is reserved to mount the pressure sensor. There are four through holes with a diameter of 12 mm above and below the sphere to keep internal and external hydrostatic pressure levels consistent.

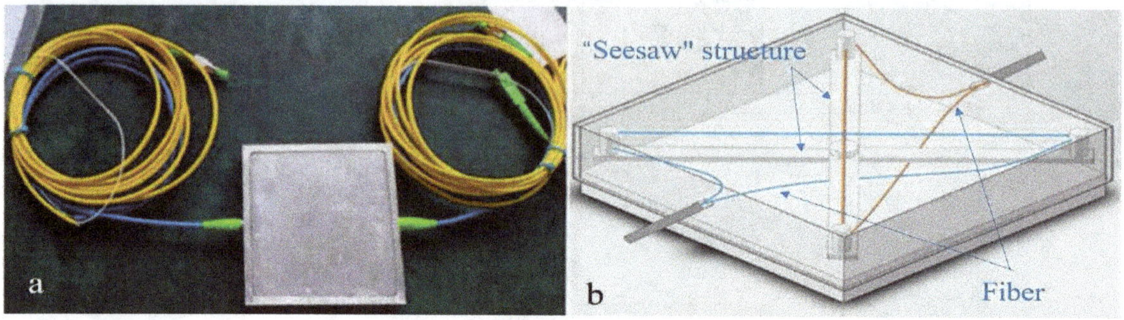

Figure 3. Fiber optic pressure sensor. (**a**) Appearance; (**b**) internal structure.

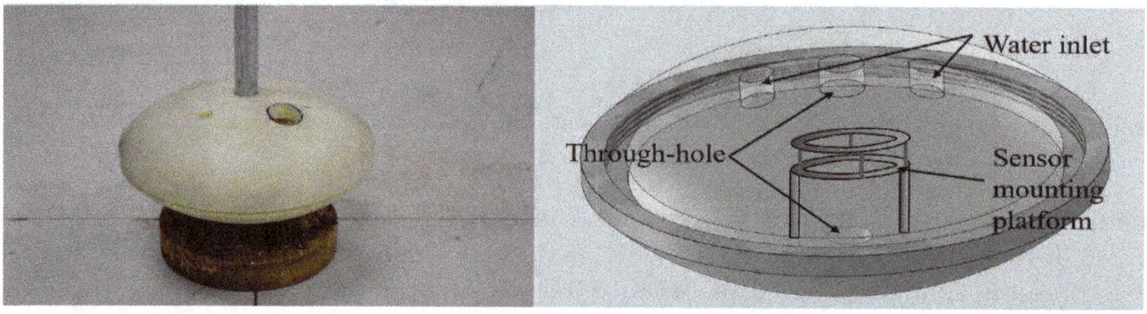

(**a**) (**b**)

Figure 4. Dish-shaped float. (**a**) Appearance; (**b**) internal structure.

The test was applied to TPR_{FD} and BPR_{SS} groups, and observation points were set at the crest (A_1, A_2) and trough (B_1, B_2) (Figure 2) of each group to compare the applicability of the two methods at different positions. With the exception of the pressure sensor, the layout and test conditions of the instruments used in the two groups of tests remained the same. In this test, a hydrodynamic force was applied over two stages. First, waves with a frequency of 34 Hz and a height of 7.9 cm were applied for one hour, and then waves with a frequency of 50 Hz and a height of 12.0 cm were applied for 25 min (in the experiment without any measuring instrument in advance, through direct observation, we found that when the wave action in the second stage reached 25 min, the bed shape had reached a stable state). In addition, to evaluate the accuracy of the two methods, we applied echo ranging, which has been widely used in seabed deformation measurement [33,34], as a control group. A freestyle sonar altimeter (AA400, EofE Ultrasonics Co., Ltd., Goyang-si, Korea) and an ultrasonic terrain scanner were used for the echo ranging group (Table 1). The ultrasonic terrain scanner was used to collect bottom bed morphological data before and after each test, while the AA400 made real-time observations of the bottom bed height at the observation point.

3. Results

3.1. Measurements Made by the Fixed-depth Total Pressure Recorder (TPR_{FD})

Shown by the echo ranging group data and experimental records, the shape of the sandy bed underwent significant migration and deformation after continuous wave loading (Figure 5). The height of the bottom bed surface at observation point A_1 decreased by roughly 7.0 cm, while that at B_1 increased by approximately 7.9 cm. Wave crest A and trough B moved roughly 3.2 cm along the wave propagation direction.

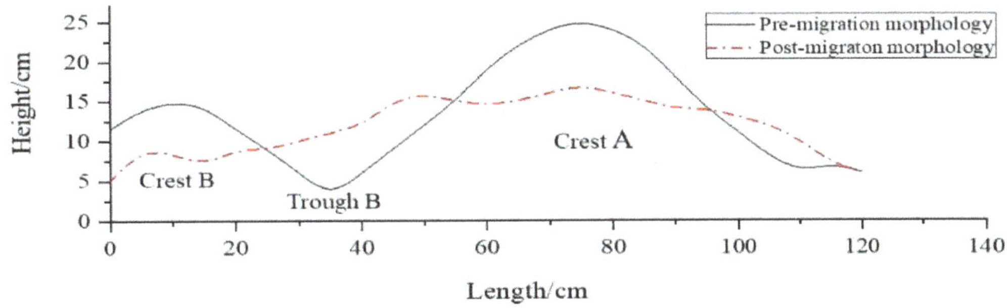

Figure 5. Bed morphology measured by echo ranging for the TPR_{FD} group.

From the measurement results of echo ranging and the TPR_{FD} (Figure 6), overall terrain changes can be divided into two processes: (a) a sharp period of change of 0–20 min, and (b) a slow adjustment period of 20–90 min. The experimental records show that wave crest flattening and wave trough filling processes mainly occurred during the period of sharp change. During this period, the positions of wave crests and troughs moved roughly 2 cm along the wave propagation direction, and the terrain was gentle overall. The height remained basically stable, and the terrain slowly moved in the wave propagation direction and then moved roughly 3.2 cm relative to the original terrain.

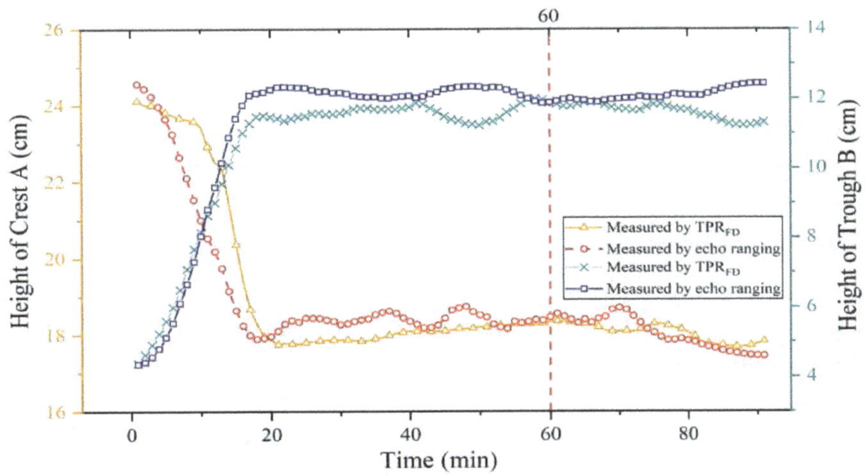

Figure 6. Comparison of results measured by echo ranging and the TPR_{FD}.

3.2. Measurements Made by the Surface Synchronous Bottom Pressure Recorder (BPR_{SS})

The bed height data measured by echo ranging (Figure 7) show that the height of crest A before migration was 20.19 cm, while the trough B reached 11.50 cm. After migration, the height at initial crest A dropped by 8.2 cm, the height of trough B increased by 6.4 cm, and the sand wave migrated roughly 3.4 cm in the wave direction.

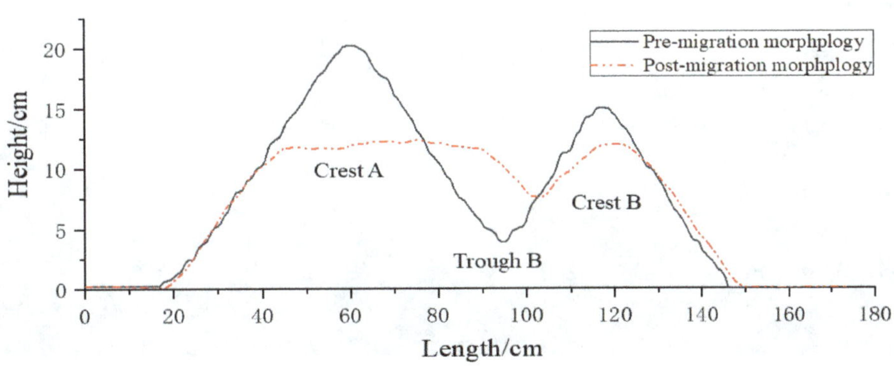

Figure 7. Bed morphology measured by echo ranging for the BPR_{SS} group.

The data measured by the BPR_{SS} and echo ranging (Figure 8) show that the height of the float at the position used in this test continuously rose or descended in a stepwise manner, without an obvious stability period. During the test period of 85 min, heights at observation points A_2 and B_2 underwent four step-like changes. According to the pressure data, the height of the bed at B_2 was increased by roughly 5 cm, falling 1.4 cm below test records. The height of the bed at A_1 dropped by 8.3 cm, which is consistent with test records.

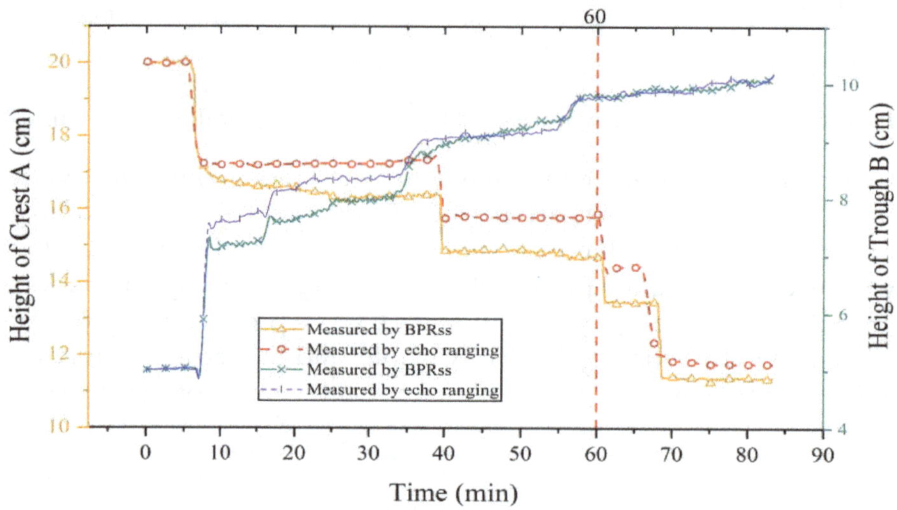

Figure 8. Comparison of results measured by the BPR_{SS} and echo ranging.

4. Discussion

4.1. The comparison of the Results Measured by Pressure Sensor Techniques and Echo Ranging

It is worth noting that there was a significant response delay to the change in elevation within the first 20 min of the TPR_{FD} measurement at crest A. During this period, the results measured by the TPR_{FD} and echo ranging have similar changes in elevation, but there is a delay of about 5 min. From the experimental records, this phenomenon may have occurred because the weight measured by the TPR_{FD} is the total weight of overlying sand in an area of 100 mm × 100 mm, while the height measured by echo ranging is a point height (Figure 9). For this reason, for the point height of sharp terrain changes, rendering the response sensitivity of the fiber optic pressure sensor is insufficient. Therefore, quantitative research into the process of this phenomenon, by increasing the number of acoustic ranging points on the plane where the optical fiber pressure sensor is located or reducing the surface area of the sensor, is needed in the following research.

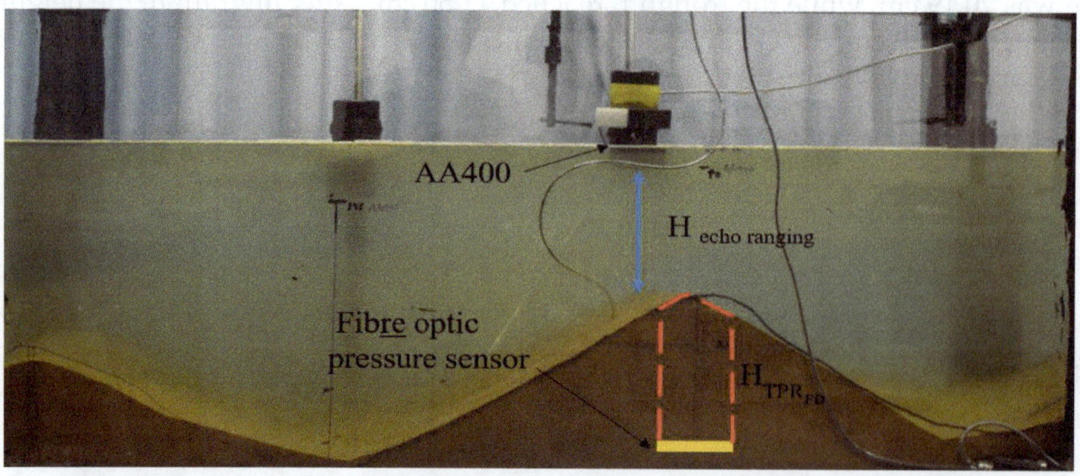

Figure 9. Measurement state of echo ranging and optical fiber pressure sensor.

A comparison of measurement results from the BPR_{SS} and the TPR_{FD} shows that there are two reasons for this stepwise manner: (a) due to the presence of the floating ball, the water flow speed on both sides of the floating ball accelerated, increasing the erosion of the sediment on both sides of the floating ball (Figure 10); (b) because the density of the float is slightly greater than that of seawater, relative to the downward movement at the wave crest, the upward movement of the float at the trough is subjected to more resistance. It should be noted that only the impact of the existing floating ball on the movement of the bottom sediment, and whether and to what extent the movement of the sediment in other areas is affected, still needs to be explored by designing more controlled experiments.

Figure 10. Bottom depression formed by the bottom of the floating ball at the trough.

4.2. The Accuracy of Pressure Sensor Techniques in Reflecting Elevation

A comparison of the two methods with the observation results of the echo ranging group shows that, while the two methods reflect height changes that are consistent with actual height changes of the observation points as a whole, there are significant differences in the accuracy of the observations. Therefore, to quantify the accuracy of the height observations of the BPR_{SS} and TPR_{FD}, we define the observation error and accuracy α as:

$$\text{error} = H_{PR} - H_{\text{echo ranging}}, \tag{4}$$

$$\alpha = \frac{\overline{\text{H}}(\text{error})}{\overline{\text{H}}_{\text{echo ranging}}} \times 100\%, \tag{5}$$

where "H_{PR}" is the height calculated using the pressure observation method, and "$H_{\text{echo ranging}}$" is the height measured by echo ranging.

The two methods produced crest and trough observations with accuracy levels of more than 90%, with the most accurate (98.1%) observation obtained using the pressure floating ball method at the trough (Figure 11). The largest error (2.7 cm) was made in the 17th min of observing the crest using the fiber-optic pressure sensor. With the corresponding delay occurring after 15 min, the overall observation accuracy level decreased to 91.4%, representing the least accurate observation. In addition, the two methods exhibit the following two characteristics in terms of accuracy levels: (a) observations of the trough are more accurate than those of the crest, and (b) the BPR_{SS} method is more accurate than the TPR_{FD} method. These findings show that while seabed elevation can be measured from the weight of overlying sand, this approach is less accurate than using water pressure to reflect elevation.

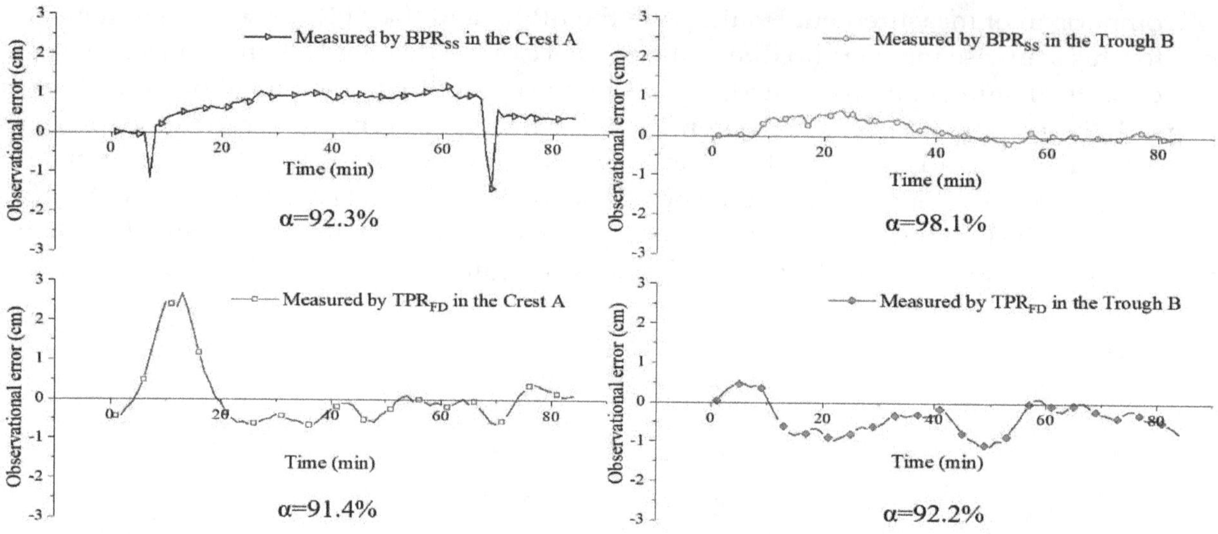

Figure 11. Comparison of the accuracy of the BPR_{SS} and TPR_{FD} methods.

4.3. Applicability of In Situ Observations

A method's reliability is essential to practical observations. Our experiments show that while the BPR_{SS} method, using the floating ball as a carrier, can more accurately reflect bottom bed elevation, it affects the migration of sand waves. When the underflow velocity of active sand wave areas reaches 58 cm/s [16], a floating ball with a density slightly higher than the density of the seawater can generate up and down movement under the action of the current. Increasing the density to reduce this effect may exacerbate the disturbance of the sand wave migration process. Therefore, its optimal density is difficult to determine, making the authenticity of the migration process reflected in observation results difficult to confirm.

In this respect, as a fixed-depth total pressure recorder with an optical fiber pressure sensor is buried in the sea floor, sediment migration on the sea floor surface is less likely to be interrupted. Furthermore, because the sand wave has a common height of approximately 0.4–5 m and a wavelength of approximately 5–100 m [35], it is much larger than that of the bed model used in this experiment. Therefore, the effect of the response delay phenomenon exhibited by the TPR_{FD} method used in this experiment should be greatly reduced. Therefore, the TPR_{FD} method is more applicable for field applications. However, if this technique is to be applied to the field observation, there are still many unresolved problems. How large is this noise caused by bottom current and potential turbulence relative to the measured noise? Further, how to ensure that the TPR_{FD} is always inside the sand wave, and not exposed to current during the sand wave migration process. A designed field observation device, equipped with a TPF_{RD} observation probe and some instruments for measuring hydrodynamics, may help to explore these issues. A multistage penetrating method can also be used to ensure that the TPR_{FD} is always located inside the sand waves [36]. Nevertheless, more relevant subjects in pressure sensing techniques for observing seabed deformation caused by submarine sand wave migration are still expected to be studied and explored.

5. Conclusions

The application of pressure sensing technology in observing changes in the sea floor caused by the migration of submarine sand waves has been demonstrated in this paper. In studying the change law of the near-bottom water pressure, bottom pressure and total fixed pressure during sand wave migration, we propose two methods: BPR_{SS} and TPR_{FD}. Through a simulation experiment with an indoor water flume, the observation effects of the two methods were evaluated. Overall, the main conclusions of this study can be summarized as follows:

(1) Pressure sensor techniques can be used to observe seabed elevation changes (accuracy of above 90%), but for microtopography with severe terrain fluctuations (such as the sand wave crests considered in this experiment), their observation accuracy will decrease.

(2) Reflecting seabed elevation from the weight of overlying sand is feasible, but less accurate than using water pressure to reflect elevation.

(3) The use of a floating ball-mounted pressure sensor as a surface synchronous bottom pressure recorder (BPR_{SS}) to observe sea floor elevation will affect sediment migration on the surface of a sand wave, causing the results to show a step-like change process not observed under real conditions.

(4) Considering the underflow velocity of a sand wave development area, and the actual size of sand waves during field observations, it is more effective to use a fixed-depth total pressure recorder (TPR_{FD}) to observe a trough.

Author Contributions: Conceptualization, X.L.; Methodology, X.Z.; Project administration, X.L.; Resources, X.Z.; Supervision, X.L.; Writing–original draft, X.Z.; Writing–review & editing, X.L., Z.T., H.Z. and T.C. All authors have read and agreed to the published version of the manuscript.

References

1. Allen, J.R.L. Sand waves: A model of origin and internal structure. *Elsevier* **1980**, *26*. [CrossRef]
2. Flemming, B.W. Underwater sand dunes along the southeast African continental margin â Observations and implications. *Elsevier* **1978**, *26*, 177–198. [CrossRef]
3. Morelissen, R.; Hulscher, S.J.M.H.; Knaapen, M.A.F.; Nemeth, A.A.; Bijker, R. Mathematical modelling of sand wave migration and the interaction with pipelines. *Coast. Eng.* **2003**, *48*, 197–209. [CrossRef]
4. Karl, H.A.; Cacchione, D.A.; Carlson, P.R. Internal-Wave Currents as a Mechanism to Account for Large Sand Waves in Navarinsky Canyon Head, Bering Sea. *SEPM J. Sediment. Res.* **1986**, *56*, 706–714. [CrossRef]
5. Reeder, D.B.; Ma, B.B.; Yang, Y.J. Very large subaqueous sand dunes on the upper continental slope in the South China Sea generated by episodic, shoaling deep-water internal solitary waves. *Mar. Geol.* **2011**, *279*, 12–18. [CrossRef]
6. King, E.L.; Boe, R.; Bellec, V.K.; Rise, L.; Skardhamar, J.; Ferre, B.; Dolan, M.F.J. Contour current driven continental slope-situated sandwaves with effects from secondary current processes on the Barents Sea margin offshore Norway. *Mar. Geol.* **2014**, *353*, 108–127. [CrossRef]
7. Franzetti, M.; Le Roy, P.; Delacourt, C.; Garlan, T.; Cancouet, R.; Sukhovich, A.; Deschamps, A. Giant dune morphologies and dynamics in a deep continental shelf environment: Example of the banc du four (Western Brittany, France). *Mar. Geol.* **2013**, *346*, 17–30. [CrossRef]
8. Games, K.P.; Gordon, D.I. Study of sand wave migration over five years as observed in two windfarm development areas, and the implications for building on moving substrates in the North Sea. *Earth Environ. Sci. Trans. R. Soc. Edinb.* **2015**, *105*, 241–249. [CrossRef]
9. Idier, D.; Astruc, D.; Garlan, T. Spatio-temporal variability of currents over a mobile dune field in the Dover Strait. *Cont. Shelf Res.* **2011**, *31*, 1955–1966. [CrossRef]
10. Droghei, R.; Falcini, F.; Casalbore, D.; Martorelli, E.; Mosetti, R.; Sannino, G.; Santoleri, R.; Chiocci, F.L. The role of Internal Solitary Waves on deep-water sedimentary processes: The case of up-slope migrating sediment waves off the Messina Strait. *Sci. Rep.* **2016**, *6*, 36376. [CrossRef]
11. Li, Y.; Lin, M.A.; Jiang, W.B.; Fan, F.X. Process Control of the Sand Wave Migration in Beibu Gulf of the South China Sea. *J. Hydrodyn.* **2011**, *23*, 439–446. [CrossRef]
12. Buijsman, M.C.; Ridderinkhof, H. Long-term evolution of sand waves in the Marsdiep inlet. I: High-resolution observations. *Cont. Shelf Res.* **2008**, *28*, 1190–1201. [CrossRef]
13. van Dijk, T.A.G.P.; Kleinhans, M.G. Processes controlling the dynamics of compound sand waves in the North Sea, Netherlands. *J. Geophys. Res. Earth Surf.* **2005**, *110*. [CrossRef]
14. Besio, G.; Blondeaux, P.; Brocchini, M.; Hulscher, S.J.M.H.; Idier, D.; Knaapen, M.A.F.; Németh, A.A.; Roos, P.C.; Vittori, G. The morphodynamics of tidal sand waves: A model overview. *Coast. Eng.* **2008**, *55*, 657–670. [CrossRef]
15. Németh, A.A.; Hulscher, S.J.M.H.; Van Damme, R.M.J. Simulating offshore sand waves. *Coast. Eng.* **2006**, *53*, 265–275. [CrossRef]

16. Fenster Michael, S.; Fitzgerald Duncan, M.; Bohlen, W.F.; Lewis Ralph, S.; Baldwin Christopher, T. Stability of giant sand waves in eastern Long Island Sound, U.S.A. *Elsevier* **1990**, *91*. [CrossRef]
17. Ingersoll, R.W.; Ryan, B.A. Repetitive surveys to assess sand ridge movement offshore Sable Island. In Proceedings of the Oceans 97 MTS/IEEE Conference, Halifax, NS, Canada, 6–9 October 1997. [CrossRef]
18. Salvatierra, M.M.; Aliotta, S.; Ginsberg, S.S. Morphology and dynamics of large subtidal dunes in Bahia Blanca estuary, Argentina. *Geomorphology* **2015**, *246*, 168–177. [CrossRef]
19. Zhou, J.Q.; Wu, Z.Y.; Jin, X.L.; Zhao, D.N.; Cao, Z.Y.; Guan, W.B. Observations and analysis of giant sand wave fields on the Taiwan Banks, northern South China Sea. *Mar. Geol.* **2018**, *406*, 132–141. [CrossRef]
20. Traykovski, P.; Hay, A.E.; Irish, J.D.; Lynch, J.F. Geometry, migration, and evolution of wave orbital ripples at LEO-15. *J. Geophys. Res.-Oceans* **1999**, *104*, 1505–1524. [CrossRef]
21. Malikides, M.; Harris, P.T.; Tate, P.M. Sediment transport and flow over sandwaves in a non-rectilinear tidal environment: Bass Strait, Australia. *Cont. Shelf Res.* **1989**, *9*, 203–221. [CrossRef]
22. Jo, H.R.; Lee, H.J. Sediment transport processes over a sand bank in macrotidal Garolim Bay, west coast of Korea. *Geosci. J.* **2008**, *12*, 243–253. [CrossRef]
23. Fox, C.G. In situ ground deformation measurements from the summit of Axial Volcano during the 1998 volcanic episode. *Geophys. Res. Lett.* **1999**, *26*, 3437–3440. [CrossRef]
24. Fox, C.G.; Chadwick, W.W.; Embley, R.W. Direct observation of a submarine volcanic eruption from a sea-floor instrument caught in a lava flow. *Nature* **2001**, *412*, 727–729. [CrossRef] [PubMed]
25. Hino, R.; Inazu, D.; Ohta, Y.; Ito, Y.; Suzuki, S.; Iinuma, T.; Osada, Y.; Kido, M.; Fujimoto, H.; Kaneda, Y. Was the 2011 Tohoku-Oki earthquake preceded by aseismic preslip? Examination of seafloor vertical deformation data near the epicenter. *Mar. Geophys. Res.* **2014**, *35*, 181–190. [CrossRef]
26. Wallace, L.M.; Webb, S.C.; Ito, Y.; Mochizuki, K.; Hino, R.; Henrys, S.; Schwartz, S.Y.; Sheehan, A.F. Slow slip near the trench at the Hikurangi subduction zone, New Zealand. *Science* **2016**, *352*, 701–704. [CrossRef]
27. Phillips, K.A.; Chadwell, C.D.; Hildebrand, J.A. Vertical deformation measurements on the submerged south flank of Kīlauea volcano, Hawai'i reveal seafloor motion associated with volcanic collapse. *J. Geophys. Res. Solid Earth* **2008**, *113*. [CrossRef]
28. Ballu, V.; Ammann, J.; Pot, O.; Viron, O.; Sasagawa, G.S.; Reverdin, G.; Bouin, M.-N.; Cannat, M.; Deplus, C.; Deroussi, S.; et al. A seafloor experiment to monitor vertical deformation at the Lucky Strike volcano, Mid-Atlantic Ridge. *J. Geod.* **2009**, *83*, 147–159. [CrossRef]
29. Chadwick, W.W.; Nooner, S.L.; Butterfield, D.A.; Lilley, M.D. Seafloor deformation and forecasts of the April 2011 eruption at Axial Seamount. *Nat. Geosci.* **2012**, *5*, 474–477. [CrossRef]
30. Yokoyama, T.; Shimoyama, M.; Matsuda, S.; Tago, K.; Takeshima, J.; Nakatsuka, Y. Monitoring System of Seafloor Subsidence for Methane Hydrate Production Test. In Proceedings of the SPWLA 18th Formation Evaluation Symposium of Japan, Chiba, Japan, 27 September 2012; p. 6.
31. Stenvold, T.; Eiken, O.; Zumberge, M.A.; Sasagawa, G.S.; Nooner, S.L. High-precision relative depth and subsidence mapping from seafloor water-pressure measurements. *SPE J.* **2006**, *11*, 380–389. [CrossRef]
32. Lobo, F.J.; Hernández-Molina, F.J.; Somoza, L.; Rodero, J.; Maldonado, A.; Barnolas, A. Patterns of bottom current flow deduced from dune asymmetries over the Gulf of Cadiz shelf (southwest Spain). *Mar. Geol.* **2000**, *164*, 91–117. [CrossRef]
33. McKelvey, D.; Honkalehto, T.; Williamson, N.J. *Results of the March 2006 Echo Integration-Trawl Survey of Walleye Pollock (Theragra Chalcogramma) Conducted in the Southeastern Aleutian Basin Near Bogoslof Island, Cruise MF2006-03*; NOAA: Silver Spring, MD, USA, 2006.
34. Bassoullet, P.; Le Hir, P.; Gouleau, D.; Robert, S. Sediment transport over an intertidal mudflat: Field investigations and estimation of fluxes within the "Baie de Marenngres-Oleron" (France). *Cont. Shelf Res.* **2000**, *20*, 1635–1653. [CrossRef]
35. Ashley, G.M. Classification of large-scale subaqueous bedforms; a new look at an old problem. *J. Sediment. Res.* **1990**, *60*, 160–172. [CrossRef]
36. Liu, X.; Zheng, X.; Zhang, B.; Jia, Y.; Tian, Z.; Ji, C. A Multistage Penetrating In-Situ Device and Method to Observe Sand Waves on the Seabed Based on Resistivity Probe. Australia Patent AU2019100321, 2 May 2019.

12

Numerical Investigation of a Two-Element Wingsail for Ship Auxiliary Propulsion

Chen Li [1,2,3,*], **Hongming Wang** [2,3] **and Peiting Sun** [1]

1 Dalian Maritime University, Dalian 116026, China; sunptg@dlmu.edu.cn
2 Jiangsu Maritime Institute, Nanjing 211170, China; lhseptember@163.com
3 Jiangsu Ship Energy-Saving Engineering Technology Center, Nanjing 211170, China
* Correspondence: lichen0987654321@163.com

Abstract: The rigid wingsail is a new type of propulsion equipment which greatly improves the performance of the sailboat under the conditions of upwind and downwind. However, such sail-assisted devices are not common in large ships because the multi-element wingsail is sensitive to changes in upstream flow, making them difficult to operate. This problem shows the need for aerodynamic study of wingsails. A model of two-element wingsail is established and simulated by the steady and unsteady RANS approach with the k-ω SST turbulence model and compared with the known experimental data to ensure the accuracy of the numerical simulation. Then, some key design and structural parameters (camber, the rotating axis position of the flap, angle of attack, flap thickness) are used to characterize the aerodynamic characteristics of the wingsail. The results show that the position of the rotating shaft of the flap has little influence on the lift coefficient at low camber. When stall occurs, the lift coefficient first increases and then decreases as the flap axis moves backward, which also delays the stall angle at a low camber. At the high camber of AOA = 6°, the lift coefficient always increases with the increase of the rotating axis position of the flap; especially between 85% and 95%, the lift coefficient increases suddenly, which is caused by the disappearance of large-scale flow separation on the suction surface of the flap. It reflects the nonlinear coupling effect between camber of wingsail and the rotating axis position of the flap

Keywords: wingsail; aerodynamics; numerical simulation

---

1. Introduction

The rigid wingsail is a common auxiliary propulsion device. Due to its characteristics of environmental protection and energy saving, the rigid wingsail has been applied to various types of ships. In 1980, two rectangular rigid wingsails with a total sail area of 194.4 m^2 on "*Shin Aitoku Maru*" were installed in Japan, which was the world's first modern sail-assisted commercial tanker. After four years of actual sailing, oil tankers can save 8.5% a year compared with conventional ships [1]. Although the international oil price fluctuates greatly, the research on sail-assisted vessels varies from country to country. In 2018, China Shipbuilding Group delivered the world's first Very Large Crude Carrier (VLCC), "Kai Li", with wingsails (Figure 1). The trial results show that the energy-saving efficiency of the VLCC is obvious [2]. At the same time, the wingsails have also been rapidly developed in the field of unpowered navigation. Particularly in 2010, BMW Oracle was powered by a 60-meterhigh, multi-element wingsail, which won the 33rd America's Cup. The flexible size and excellent performance of the wingsail has rekindled the interest of the shipping industry [3]. However, due to the phenomenon of flow separation or stall on the surface of the wingsail, the propulsion performance of the wingsail will be deteriorated, which will seriously affect the stability of the ship. The New Zealand Chiefs [4] suffered a shipwreck in the 35th America's Cup (Figure 2). Therefore, it is very important to find an effective method to control flow separation or delay stall.

Figure 1. "Kaili" VLCC with wingsail.

Figure 2. Capsizing accident of sailboat.

In order to improve the stall characteristics, different flow control methods are used, which can be divided into active control and passive control. The active control method has been applied to the controllable loop sail [5], fluid injection [6,7], turbine sail [8], Magnus sail [9], trailing flap [10,11], leading slat [11,12], etc. The passive method is also applied to different types of technologies, such as Walker sails [13], deformed flaps [14] and leading-edge tubercles [15], etc.

In recent years, the flap setting of the wingsail is considered to be a very feasible active control method to control the flow separation. This idea has been applied in the American Cup Sailing Competition with great success. The rigid wingsail consists of two or three symmetrical wings. There is a gap between them to control the wingsail camber on starboard tack and port tack (Figure 3), so as to improve the propulsion performance and delay stall. Many scholars have also carried out research on the aerodynamic characteristics.

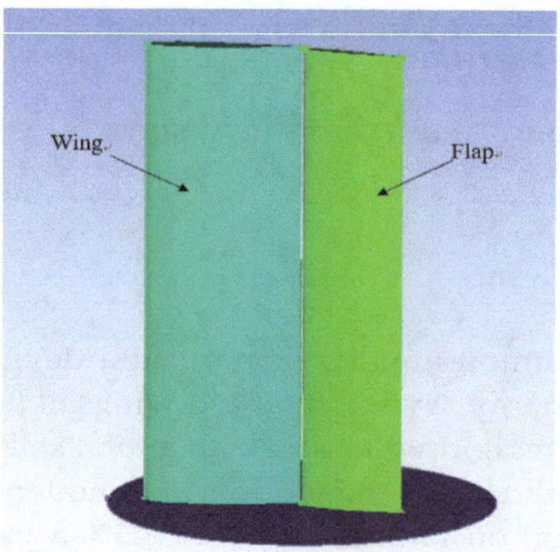

Figure 3. The geometry and simplified configuration of the two-element wingsail.

In 1996, **Daniel** [11] designed a high-performance, three-element wingsail. The experimental results show that the maximum thrust coefficient of the three-element wingsail is increased by 68%, the stall angle is delayed between 4° and 6°, and the thrust in the whole operation area is also improved. At the most efficient wind direction angle of the wingsail, the thrust is increased by 83%. This proves the superiority of multi-element wingsail in propulsion characteristics. In 2003, **Fujiwara** [16,17] switched from a triangular sail to a rectangular one. Its propulsion performance is superior to that of Nojiri's hybrid dynamic sail. In 2015, he cooperated with **Qiao Li** [14] to modify the hybrid sail, replacing the flap with a rigid plate and controlling the rotation of the wing and the plate. Therefore, the sail was also called a variable camber sail(VCS). Through simulation and test, the aerodynamic performance of the VCS was better than that of naca0021 sail and the plate sail. **Vincent** [18] carried out the simulating research of the two-element wingsail with some key design parameters (camber, slot width, angle of attack, flap thickness, etc.) The results show that the transition phenomenon of flap boundary layer occurs due to the existence of laminar separated bubbles and the interaction between the wing and flap boundary layer in the slot region. The stall is related to slot leakage flow with nonlinear coupling and the leakage flow is affected by flap deflection, slot width and flap thickness. Through PIV measurement and numerical simulation, **Alessandro** [19] explained the flow phenomena of rigid wingsails at low and high cambers in detail, especially the flow behavior near the slot between two element wings. He pointed out that the flow field in the gap is the key factor affecting the performance of the wingsail. However, the internal relationship between the change of slot size and the development of wingsail aerodynamic characteristics is still unclear. In order to better understand the aerodynamic characteristics and better control of the rigid wingsail, some numerical studies are proposed in this paper.

2. Methods

2.1. The Geometry of the Wingsail

The geometry is a simplified configuration of a two-element wingsail with 1/20th model-scale (as Figure 3 and Table 1). The nondimensional number for characterizing the flow of fluid Reynolds number is defined by Equation (1):

$$Re = \frac{\rho v c}{\mu} \tag{1}$$

where μ is the viscosity coefficient of air. The results of naca0018 airfoil with $Re = 5 \times 10^5$ (the wind speed is assumed to be 20 m/s) are compared with the existing experiments to ensure the accuracy of the numerical results. The general view of the geometry and its characteristic parameters are as follows:

Table 1. Parameterization of wingsail.

c	0.35 m
h	0.7 m
Re	5×10^5
d	0–25°
α	0–20°
g	2.4%
X_r	75%–95%

The aim of this paper is to understand the influence of geometric parameters on aerodynamic performance. The parametric design of the wingsail is selected and described in Figure 4. In order to simplify the general problem, it has been decided to focus on the flap deflection angle d, the position of flap rotation axis in the direction of the wing chord X_r, and flap thickness e_2/c_2. The angle of attack (α) of the wing represents the angle of attack (AOA) of the wingsail, as seen in Figure 4.

Based on the preliminary simulation and previous research [11,18], the initial wingsail configuration has chord ratio c_1/c_2 = 3:2, wing thickness e_1/c_1 = 18%, flap thickness e_2/c_2 = 15%, one flap rotation axis X_r/c_1 = 85%, slot width g/c_1 = 2.4%. Its name will be r1.5t1815 X85g2.4. Adding the angle of attack (α) and flap deflection angle (d) after this name, and the full configuration name of the wingsail will be r1.5t1815 X85g2.4α6 d15, for $\alpha = 6°$, $d = 15°$.

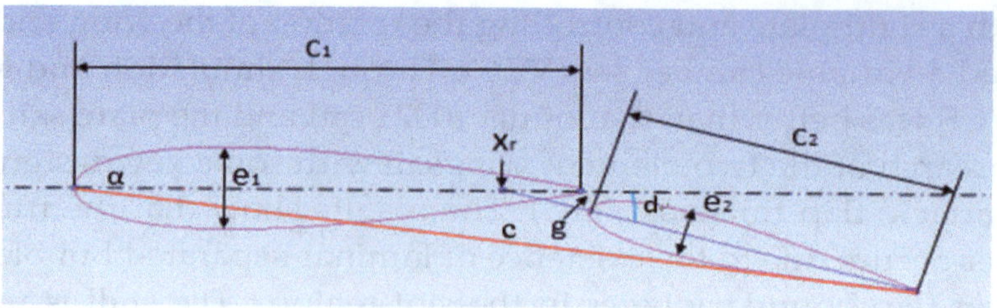

Figure 4. Wingsail geometry and parameterization.

2.2. The Grid Structure

Computational Domain. Before mesh generation, the calculation domain of geometric model should be determined. The computational domain of wingsail is large enough (32c × 30c × 10c) to consider the side effect as negligible (Figure 5). AOA is adjusted by rotating the direction of the airflow. The chord of the wingsail is 0.35 m and the span 0.7 m. In order to simplify the model, the atmospheric boundary layer is not considered. The boundary conditions used are velocity inlets on three sides (inlet, starboard boundary, port boundary) and a static pressure outlet equal to the far-field pressure. The velocity at the inlet is uniform, and the value is the same as the free flow velocity. The wingsail surface and the bottom surface of the computational domain are defined as anti-slip walls.

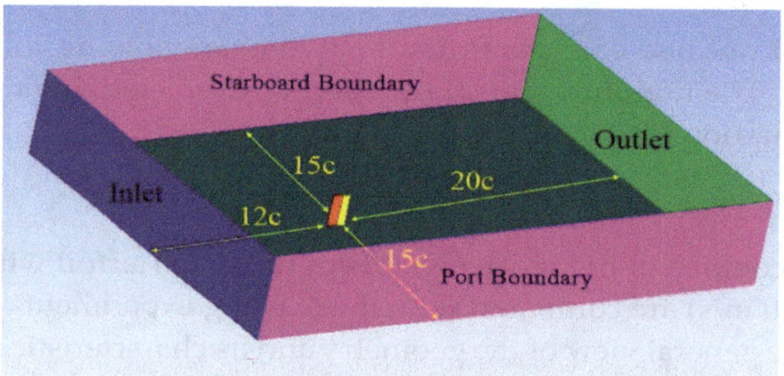

Figure 5. Box domain for the freestream simulation.

Mesh Details. The quality of grid determines the accuracy of numerical results. Particular attention has been paid to refine the mesh in the slot, being a region of interaction between the wake of the wing, flow from the slot and the flap boundary layer. Mesh resolution close to the slot has been encrypted as shown in Figure 6. In addition, the mesh size of the flap leading-edge curve is set as 0.4545% (in the case of mesh number 9.86M). In order to ensure the accurate simulation of the flow field near the wall, the height of the first grid cell closest to the wall should be suitable. The nondimensional wall distance for a wall-bounded flow y^+ is defined by Equation (2):

$$y^+ = \frac{yu_\tau}{\nu} = \frac{\rho y}{\mu}\sqrt{\frac{\tau_{wall}}{\rho}} = y\sqrt{\frac{\rho}{\mu}(\frac{\partial u}{\partial y})}_{y=0} \tag{2}$$

where y is the distance from the first mesh cell to the nearest wall. In this paper, y is specified as 2.871×10^{-5} c on the wingsail surface. In this case, the profile of the dimensionless wall distance y^+ on the surface of the wingsail with AOA = 6° is shown in Figure 7. It can be seen that the value of y^+ on the sail surface is about 1, ensuring the accuracy of numerical calculation results. The growth ratio of the prism layer is 1.2. The total number of nodes in the grid is about 9.86 million.

(a) (b)

Figure 6. The encrypted mesh in (**a**) mid-span and (**b**) bottom wallof the slot.

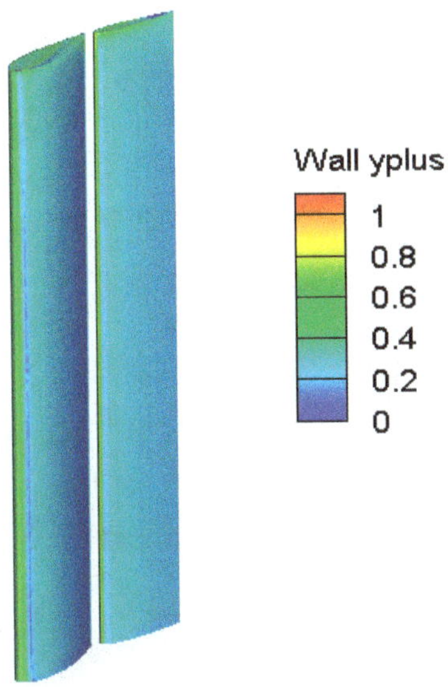

Figure 7. Wall y+ contour on wingsail surface (AOA = 6 deg).

Mesh Independence. It is considered a necessary step to analyze the influence of grid characteristics. When $Re = 5 \times 10^5$, four different grid numbers (including 2.57, 4.22, 9.86 and 14.4M) are used to estimate the grid sensitivity. In the case of AOA = 6°, the grid independence analysis is carried out with steady Reynolds Average N-S equation and AOA = 15° with URANS. Figure 8 reports the lift and drag coefficients of the unmodified sail model. As shown in Figure 8, the number of grids has a great influence on the lift coefficient, and the error of lift coefficient is less than 0.3% between 9.86 and 14.4M. Based on these calculations, the increase in the number of grids (relative to the case of 9.86M) results in a small change in lift. And the drag coefficient, whose change can be considered as acceptable. Therefore, the grid number of 9.86M is used in this study.

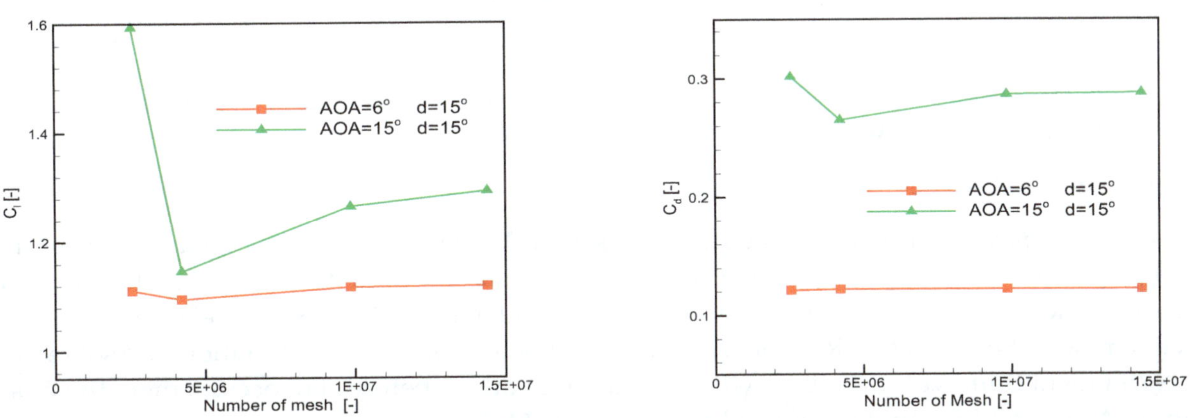

(**a**) (**b**)

Figure 8. Convergence of (**a**) lift and (**b**) drag coefficients as a function of mesh number.

In order to further verify the reliability of the grid with URANS, the influence on the flow field with different grid sizes is also analyzed when $Re = 5 \times 10^5$ and AOA = 15°. For four different grids (including 2.57, 4.22, 9.86 and 14.4M), the velocity vector on the x–y plane is shown in Figure 9. It can be observed that there is a large separation vortex on the suction surface of the flap and a small vortex on the wake of the wing, which is generated by grids of 9.86 and 14.4M, respectively, but there is no obvious flow separation for grids of 2.57 and 4.22M. There was no significant difference in the flow profile between 9.86 and 14.4M. Therefore, the grid number of 9.86M is suitable for the research of wingsail 3D.

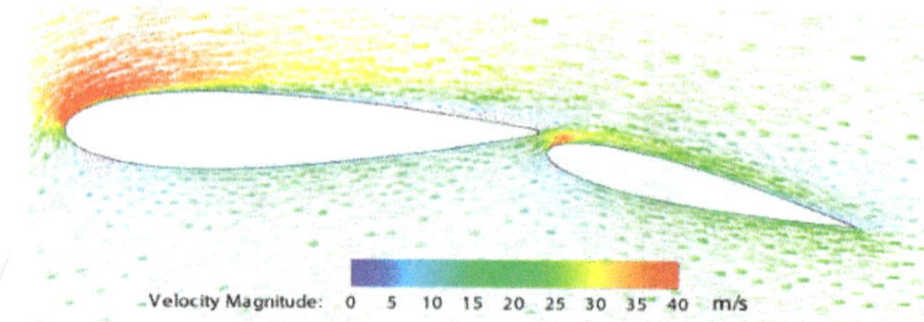

(**a**) 2.57M

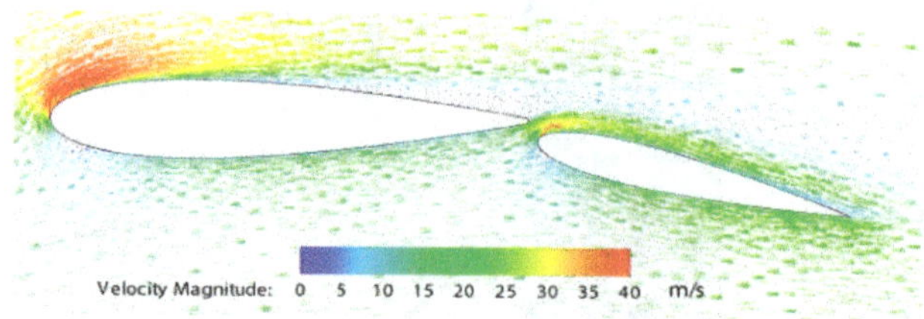

(**b**) 4.22M

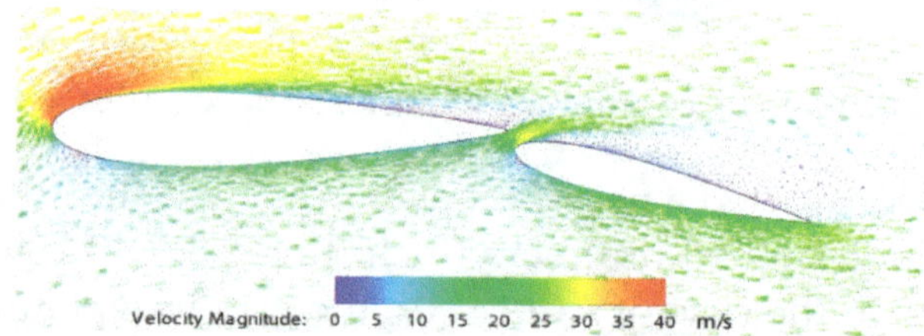

(**c**) 9.86M

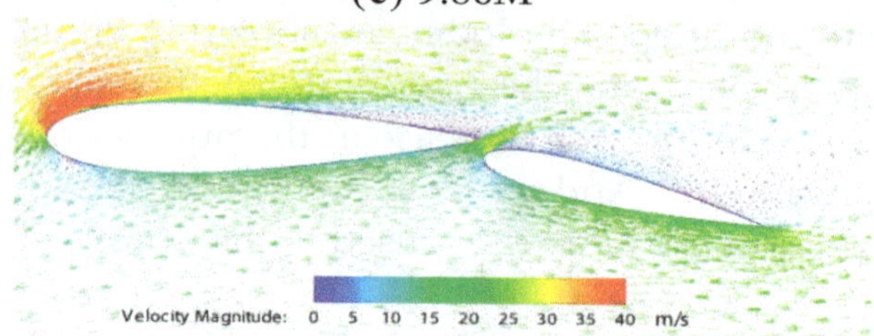

(**d**) 14.4M

Figure 9. Flow profiles on the mid-span of wingsail with different grid numbers of (**a**) 2.57M, (**b**) 4.22M, (**c**) 9.86M and (**d**) 14.4M (AOA = 15 deg).

2.3. Computational Approach

Numericalmethod. **Menter** [20] established the k-ω turbulence model of shear stress transport (SST). **Hassan** [21] compared the numerical predicted results of standard, RNG and realizable K-ε, standard and SST K-ω models with the experimental data, and found that the K-ω SST model is the most accurate prediction model. So the steady Reynolds Averaged Navier–Stokes(RANS) equations are solved thanks to ANSYS Fluent finite-volume solver with low angle of attack before stall occurs; then, the unsteady Reynolds Average N-S equation is used based on the SST K-ω models when physical instabilities exist [22,23]. As the low Reynolds number is based on the chord of the wingsail ($Re = 5 \times 10^5$), the fluid is considered as incompressible (the Mach number of the inlet is about 0.062). These models are assigned solvers which control the number of the time step. The selected time step size is computed in the case of CFL = VΔt/Δx = 1. The impact of time models on the aerodynamic performance of the wingsail is investigated at $Re = 5 \times 10^5$(i.e., V = 20.87 m/s), and the mesh size of the leading-edge curve of the flap is taken as the length interval (i.e., $\Delta x = 5 \times 10^{-4}$ m). Therefore, the magnitude of the time step Δt is set at a value of 2.5×10^{-5} s in this simulation investigation. The turbulence intensity at inlet is 1%. The discrete format is quick. Simple algorithm is adopted for the pressure velocity coupling scheme [24].

Model validation. In order to verify the numerical results, the lift coefficient of the two-dimensional wingsail model in a free flow environment is compared with the experimental results, as shown in Figure 10. Daniel [11] had carried out experimental research on the lift/drag characteristics of naca0018 wing.

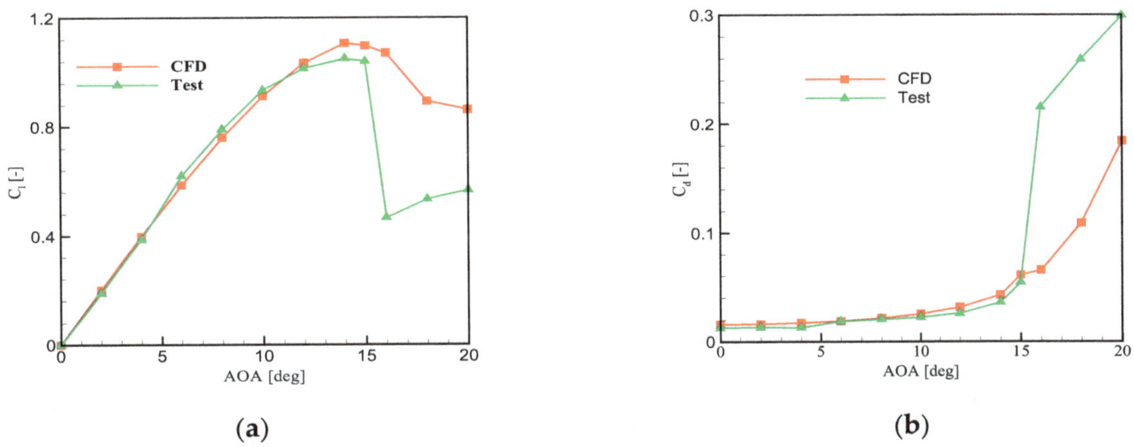

Figure 10. Comparison of lift coefficient (**a**) and drag coefficient (**b**) between test and CFD.

In the case of $Re = 5 \times 10^5$, the RANS method is verified. Before stall (AOA < 15°), the numerical calculation results of the lift coefficient and drag coefficient are close to the experimental values, and the estimation error is less than 5%. At the same time, the prediction of the position of the stall angle is accurate. Although the results of numerical prediction are different from the experimental values after stall, it is acceptable that the 2D numerical calculations used for the qualitative study of the lift/drag characteristics of the wingsail.

3. Numerical Results

3.1. Two-Dimensional Wingsail Configurations Study

The lift force and drag force are converted into dimensionless lift coefficient C_L and drag coefficient C_D, respectively. They are defined as follows:

$$C_L = \frac{L}{0.5\rho v^2 A_R} \tag{3}$$

$$C_D = \frac{D}{0.5\rho v^2 A_R} \tag{4}$$

where L and D are the lift force and drag force of the wingsail, respectively. A_R is the aspect area of wingsail, include the wing and flap. The lift coefficient and drag coefficient characteristics of wingsail with different geometric parameters are compared.

Camber and the rotation axis of the flap effect: as the key parameter, the rotating axis position of the flap in the direction of the wing chord X_r is selected as 75% and 85%; five cambers (d = 5°, 10°, 15°, 20°, 25°) have been selected as Figure 11.

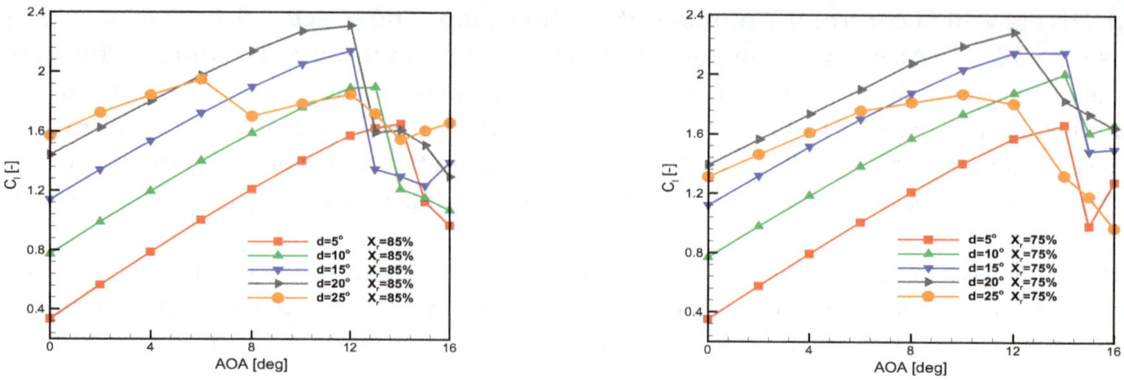

(**a**) X_r=85% (**b**) X_r=75%

Figure 11. Convergence of lift coefficients as a function of AOA and flap deflection angle with (**a**) X_r = 85% and (**b**) X_r = 75%.

When X_r is 85%, the stall does not occur advance with the increase of flap deflection angle at low camber as X_r is 75%. It can be illustrated that the slot width with X_r = 75% exceeds that with X_r = 85% due to the coupling between the camber of the wingsail and the slot width (as Figure 12). The fluid through the slot is enough to supplement the loss of the wing wake, then delay the stall. When the flap deflection angle is more than 15°, the stall angle begins to reduce, which is caused by flow separation of the wing wake, then the flap stalls. The maximum lift coefficient of the wingsail is 2.29 when the flap deflection angle is 20° with AOA = 12°. However, due to the reduction of stall angle, the comprehensive performance of the wingsail with d = 20° is not as good as that with d = 15°. When the flap deflection angle adds to 25°, the lift coefficient decreases in the range of full angles of attack, which may be caused by the stalls. Therefore, it is necessary to consider the rotating axis position of the flap, flap deflection angle and slot width when choosing flap geometry parameters.

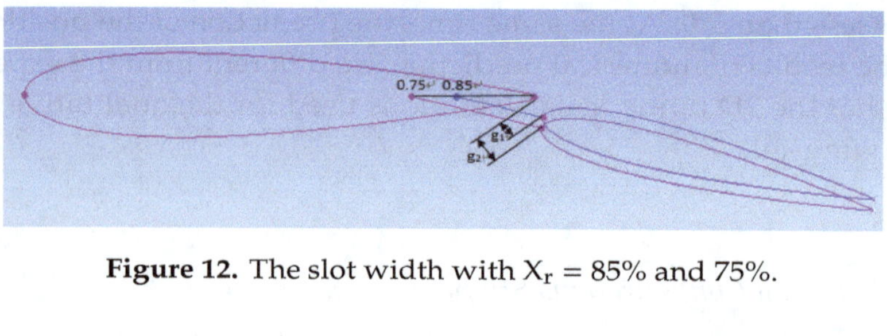

Figure 12. The slot width with X_r = 85% and 75%.

Camber and flap thickness effect: Nine selected simulations are summarized in Table 2 to illustrate the coupling between the camber (flap deflection) and the thickness of the flap. It can be seen that the flap thickness (e_2/c_2) has little effect on the lift and drag coefficients of d = 5° and 15° (low camber), but has significant effect on d = 25° (high camber) which shows the nonlinear coupling between flap thickness and flap deflection angle.

For high camber, the thickening flap leads to an increase of the leading-edge radius which decreases the pressure coefficient suction peak and has postponed the stall of the wingsail. It emphasizes the complexity of the slot flow of a two-element wingsail through the strong coupling between flap deflection, thickness and slot width. The slot effect may be useful to recall some famous papers with **Gentry** [25] and **Smith** [26] which illustrate the complex effect among the wing wake, the leakage flow from the slot and the flap boundary layer as may be seen in Figure 13. Therefore, the slot of the wingsail includes a strong design constraint which include the rotating axis position of the flap, flap deflection, flap thickness and slot width.

Table 2. Results for $\alpha = 6°$ with the SST transitional turbulence model.

Configurations	e2/c2	d	C_L	C_D	L/D
r1.51815x0.85g2.4α6d5	15%	5°	1.005	0.0213	47.16
r1.51815x0.85g2.4α6d15	15%	15°	1.724	0.0338	50.94
r1.51815x0.85g2.4α6d25	15%	25°	1.951	0.06	32.53
r1.51812x0.85g2.4α6d5	12%	5°	1.002	0.0207	48.5
r1.51812x0.85g2.4α6d15	12%	15°	1.73	0.0336	51.47
r1.51812x0.85g2.4α6d25	12%	25°	1.606	0.1628	9.87
r1.51810x0.85g2.4α6d5	10%	5°	1.002	0.0208	48.22
r1.51810x0.85g2.4α6d15	10%	15°	1.74	0.0337	51.59
r1.51810x0.85g2.4α6d25	10%	25°	1.649	0.1649	10

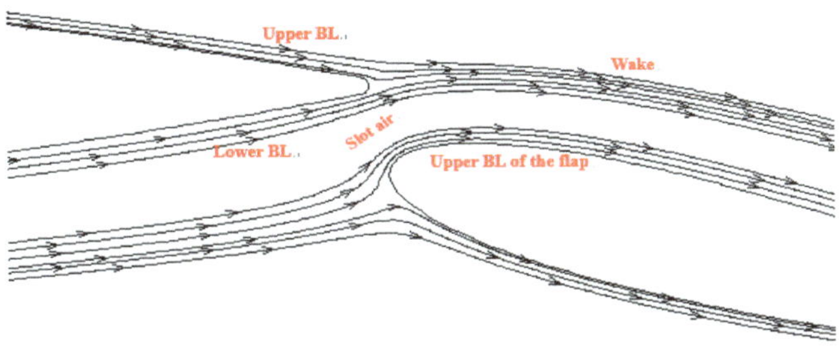

Figure 13. The streamlines in the slot region (URANS).

3.2. Three-Dimensional Study about Effect of Flap Rotation Axis Position

3.2.1. Aerodynamic Performance

The three-dimensional flow around the wingsail at low camber ($d = 15°$) and high camber ($d = 25°$) is studied in detail. In order to better study the effect of the rotating axis position of the flap on the aerodynamic characteristics and delayed stall of the wingsail, we analyzed the lift characteristics of the wingsail when the rotating axis of the flap is located at different positions of the wing chord (X_r = 75%, 80%, 85%, 90% and 95%) at $\alpha = 6°$ and 15°.

Figure 14 shows the lift characteristics for different rotating axis positions of the flap. The position of flap rotating axis has little effect on lift coefficient at $\alpha = 6°$ (stall has not yet occurred) with low camper. The lift coefficient first increases and then reduces with the position of flap rotation axis moving backward; especially from 80% to 85%, it drops suddenly at $\alpha = 15°$ (stall has occurred). It can be illustrated that the change is caused by the flow separation of the wing wake and the flap suction surface from Figure 16. At $\alpha = 6°$ with high camber, the lift coefficient increases with the position of the flap rotating axis moving backward, especially from 85% to 95%. It can be explained from Figure 17b that the flow separation on the suction surface of the flap has disappeared.

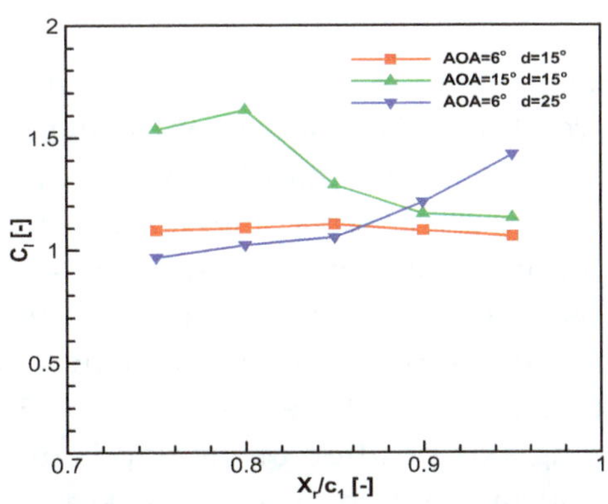

Figure 14. Lift coefficient versus flap rotation axis position of the wingsail in low and high camber configuration.

3.2.2. Streamlines

As for the wingsails with different rotating axis positions of the flap at $\alpha = 15°$ and $d = 15°$, the streamlines on the mid-span of wingsail are depicted in Figure 15. It can be found that a small separation vortex appears in the wake of the wing and a large separation vortex appears in the regions over the suction surface of the flap at $X_r = 90\%$.

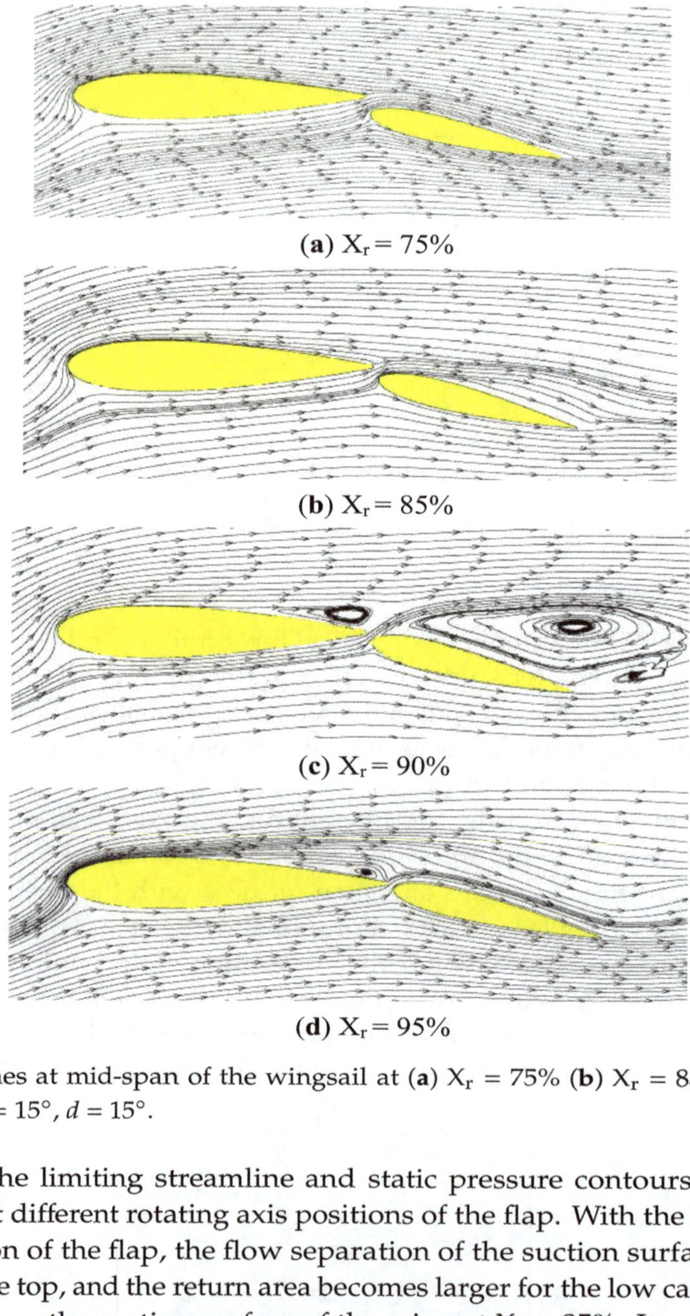

Figure 15. Streamlines at mid-span of the wingsail at (**a**) $X_r = 75\%$ (**b**) $X_r = 85\%$ (**c**) $X_r = 90\%$ and (**d**) $X_r = 95\%$ with $\alpha = 15°$, $d = 15°$.

Figure 16 shows the limiting streamline and static pressure contours on suction surface for two-element wingsail at different rotating axis positions of the flap. With the backward movement of the rotating axis position of the flap, the flow separation of the suction surface of the wing expands from the blade root to the top, and the return area becomes larger for the low camber wingsail. The flow separation line appears on the suction surface of the wing at $X_r = 85\%$. It explains that the forward movement of the rotating axis position of the flap increases the fluid flow through the slot, delays the flow separation of the suction surface of the wing, or delays stall of the wingsail. An obvious separation helix appears on the suction surface of the wing at $X_r = 95\%$, which indicates that the vortex has been formed. When the rotating axis position of the flap is moves backward from 90% to 95%, the flow separation on the suction surface of flap disappears. We guess the fluid flowing through the smaller gap does not supplement the wake of the wing, but flows along the flap boundary layer.

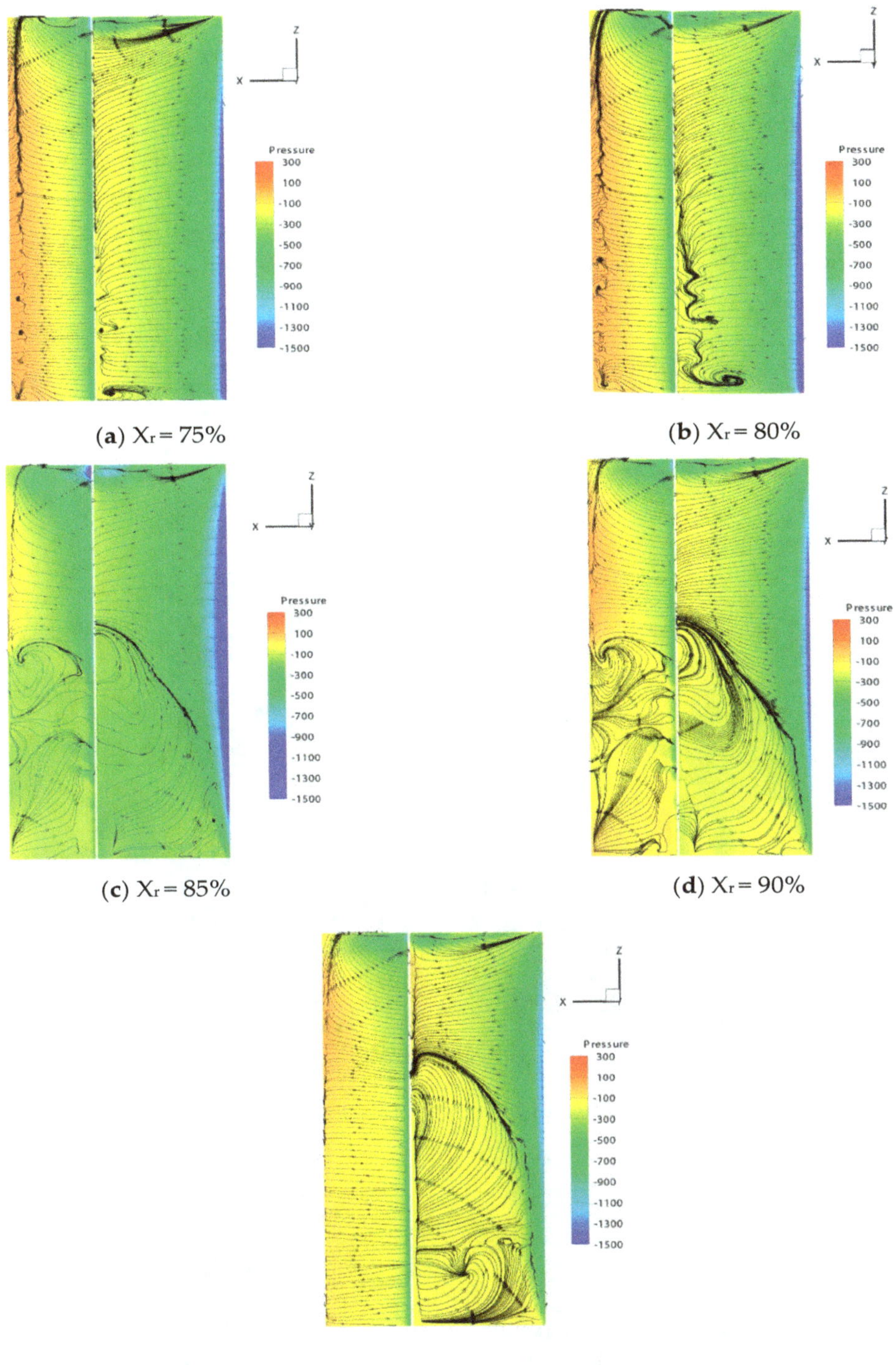

(e) Xr = 95%

Figure 16. Limiting streamline and static pressure contours on suction surface for the two-element wingsail at (**a**) $X_r = 75\%$ (**b**) $X_r = 80\%$ (**c**) $X_r = 85\%$ (**d**) $X_r = 90\%$ and (**e**) $X_r = 95\%$ with $\alpha = 15°$, d = 15°.

3.2.3. Velocity Magnitude Contours

It can be seen from Figure 17 that the suction surface of the wing has a large flow separation at $\alpha = 15°$ with low camber. At $X_r = 85\%$, the fluid flowing through the slot complements the flow separation of the wing wake and the flap suction surface. There is no vortex in the mid-span of the wingsail. The slot jet can be observed clearly at $X_r = 90\%$. As a result of the reduced slot width due to the rotating axis position of the flap backward, the slot jet only divides the vortex of the wing wake and there is a large-scale flow separation on the suction surface of the flap, which causes deep stall of the flap. This phenomenon has also been observed and described in Biber [27] and Chapin [18]. Because of the smaller slot width at $X_r = 95\%$, the slot jet only flows along the boundary layer of the flap, which has little effect on the separation flow of the wing wake. If the slot is further reduced or not set, the flap setting may aggravate the separation of the wing wake, such as the research of the hybrid sail designed by Qiao Li [14]. However, the forward movement of the rotating axis position of the flap is limited by the flap deflection angle. As shown in Figure 18, when the rotating axis position of the flap moves forward from 90% to 85% at $\alpha = 6°$, the large-scale flow separation occurs on the suction surface of the flap, as the phenomenon seen by Fiumara [19] in the two-element sail experiment in 2016.

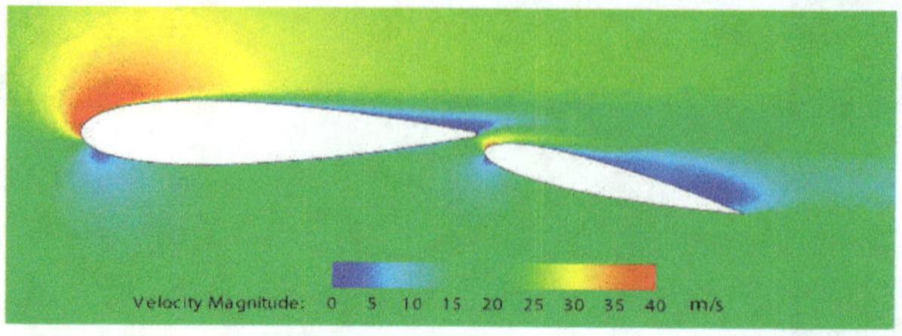

(a) $X_r = 85\%$

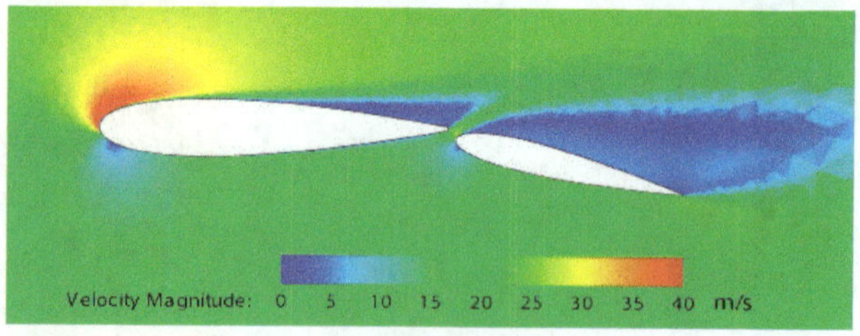

(b) $X_r = 90\%$

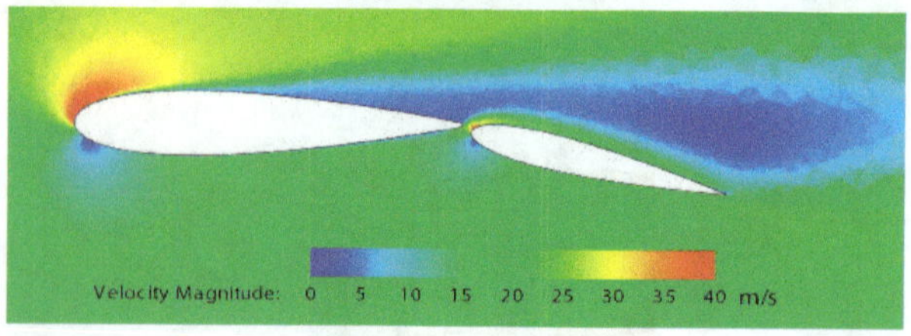

(c) $X_r = 95\%$

Figure 17. Velocity magnitude contours at mid-span of wingsail at **(a)** $X_r = 85\%$ **(b)** $X_r = 90\%$ and **(c)** $X_r = 95\%$ with $\alpha = 15°$, d = 15°

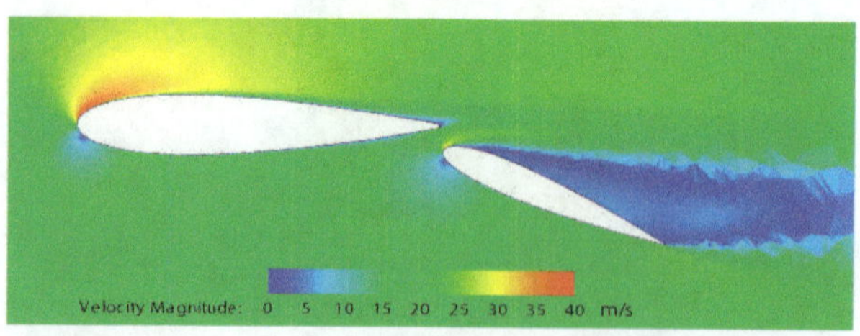

(a) $X_r = 85\%$

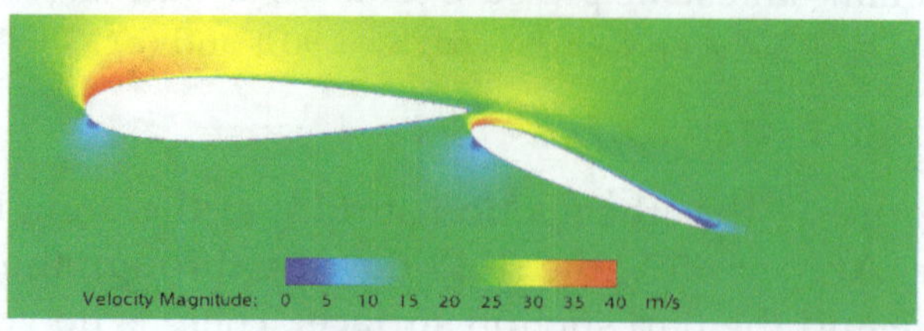

(b) $X_r = 90\%$

Figure 18. *Cont.*

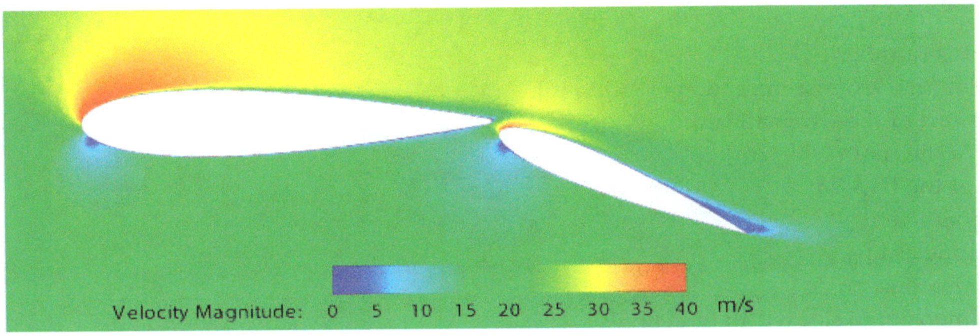

(c) $X_r = 95\%$

Figure 18. Velocity magnitude contours at mid-span of wingsail at (**a**) $X_r = 85\%$ (**b**) $X_r = 90\%$ and (**c**) $X_r = 95\%$ with $\alpha = 6°$, d = 25°.

4. Conclusions

By studying the influence of flap geometric parameters on the aerodynamic characteristics of two-element wingsail under steady and unsteady conditions, two-dimensional and three-dimensional related parameters were simulated using the same Reynolds number ($Re = 5 \times 10^5$).

The 2D simulation results show that, when the rotating axis position of the flap is located at 85% of the wing chord, the thickening flap leads to an increase of the leading-edge radius which decreases the pressure coefficient suction peak and has postponed the stall of the wingsail for high camber. It reflects the nonlinear coupling effect between wingsail camber and flap thickness. When the rotating axis of the flap is located at the 75% of the wing chord, the stall angle is delayed with the increase of the flap deflection angle at low camber. When selecting the geometric parameters of the flap, factors such as the position of the flap rotation axis, the flap deflection angle, and the flap thickness need to be considered comprehensively.

The 3D simulation mainly studies the influence of the flap rotating axis position and the flap deflection angle on the stall characteristics of the wingsail. When stall has not yet occurred with low camber, the rotating axis position of the flap has little effect on the lift coefficient, while stall has occurred, the lift coefficient increases first and then reduces with the rotating axis position of the flap moving backward. The flow separation of the suction surface of the wing expands from the root to the top and the return area becomes larger, especially from the 80% to the 85%, the lift coefficient drops suddenly. This is caused by the flow separation between the wing wake and the flap suction surface. At high camber with AOA = 6°, the lift coefficient always increases with the position of the flap rotation axis, especially from 85% to 95%, the lift coefficient suddenly rises, which is caused by the disappearance of large-scale flow separation of the flap suction surface.

Therefore, the slot width is an important factor affecting the flow separation of the wing wake and the suction surface of the flap, where size is affected by the rotating axis position of the flap and the flap deflection angle. When the flap deflection angle is adjusted to obtain a large lift coefficient, the restriction of the rotating axis position of the flap must be considered to ensure a reasonable stall angle range.

Author Contributions: Conceptualization, C.L. and H.W.; investigation, C.L. and H.W.; methodology, C.L. and P.S.; software, C.L.; writing—original draft preparation, C.L.; writing—review and editing, H.W.; supervision, P.S. All authors have read and agreed to the published version of the manuscript.

Nomenclature

Re	Reynolds number [-]
α	Angle of attack of the main wing (AOA) [°]
c	total chord of the wingsail [m]
c_1	chord of the main wing [m]
c_2	chord of the flap [m]
C_D	Drag coefficient [-]
C_L	Lift coefficient [-]
d	Flap deflection angle [°]
g	non-dimensional slot width (g/c_1) [-]
e_1	thickness of the main wing [m]
e_2	thickness of the flap [m]
A_R	The aspect area of wingsail [m^2]
y^+	Non-dimensional wall distance [-]
ρ	The density of the air [kg/m^3]
z	The height of wingsail in the vertical direction [m]
h	Wingsail height [m]
L	Lift force [N]
D	Drag force [N]
v	The velocity of inflow [m/s]
X_r	The position of flap rotation axis in the direction of the wing chord [-]

References

1. Ishihara, M.; Watanabe, T.; Shimizu, K.; Yoshimi, K.; Namura, H. Prospect of sailequipped motor ship as assessed from experimental ship Daioh. In *Proceedings of the Shipboard Energy Conservation Symposium*; Society of Naval Architects and Marine Engineers: New York, NY, USA, 1980; pp. 181–198.
2. Available online: http://www.sohu.com/a/275284498_100091571 (accessed on 14 November 2018).
3. Blakeley, A.W.; Flay, R.G.J.; Richards, P.J. Design and Optimisation of Multi-Element Wing Sails for Multihull Yachts. In Proceedings of the 18th Australasian Fluid Mechanics Conference, Tasmania, Australia, 3–7 December 2012.
4. Available online: http://k.sina.com.cn/article_6424865154_17ef3a982001001t6p.html (accessed on 20 December 2017).
5. Kind, R.J. An Experimentel Investigation of a Low- Speed Circulation Controlled Airfoil. *Aero. Quart.* **1968**, *19*, 170–182. [CrossRef]
6. Seifert, A.; Bachar, T.; Koss, D.; Shepshelovich, M.; Wygnanski, I. Oscillatory Blowing: A Tool to Delay Boundary-Layer Separation. *AIAA J.* **1993**, *31*, 2052–2060. [CrossRef]
7. Seifert, A.; Darabi, A.; Wyganski, I. Delay of airfoil stall by periodic excitation. *J. Aircr.* **1996**, *33*, 691–698. [CrossRef]
8. Kralj, D.M.; Klarin, B. Wing Sails for Hybrid Propulsion of a Ship. *J. Sustain. Dev. Energy Water Environ. Syst.* **2016**, *4*, 1–13. [CrossRef]
9. Marine, B. E-ShiP1 with Sailing Rotors to Reduce Fuel Costs and to Reduce Emissions. Available online: http://www.marinebuzz.com (accessed on 28 August 2008).
10. Borglund, D.; Kuttenkeuler, J. Active Wing Flutter Suppression Using A Trailing Edge Flap. *J. Fluids Struct.* **2002**, *16*, 271–294. [CrossRef]
11. Daniel, W.A. *The CFD Assisted Design and Experimental Testing of a Wingsail with High Lift Devices*; University of Salford: Salford, UK, 1996.
12. Carr, L.W.; McAlister, K.W. The effect of a leading-edge slat on the dynamic stall of an oscillating airfoil. *AIAA Pap.* **1983**, *83*, 2533.
13. Available online: http://cookeassociates.com/sports/commercial,php (accessed on 6 August 2015).
14. Li, Q.; Nihei, Y.; Nakashima, T.; Iked, Y. A study on the performance of cascade hard sails and sail-equipped vessels. *Ocean Eng.* **2015**, *98*, 23–31. [CrossRef]
15. Fish, F.E.; Battle, J.M. Hydrodynamic design of the humpback whale flipper. *J. Morphol.* **1995**, *225*, 51–60. [CrossRef] [PubMed]

16. Toshifumi, F.; Koichi, H.; Michio, U.; Nimura, T. On aerodynamic characteristics of a hybrid-sail with square soft sail. In Proceedings of the International Offshore and Polar Engineering Conference, International Society of Offshore and Polar Engineers, Honolulu, HI, USA, 25–30 May 2003; pp. 2576–2583.
17. Fujiwara, T.; Hirata, K.; Ueno, M.; Nimura, T. On the development of high-performance sails for an ocean-going sailing ship. In Proceedings of the International Conference on Marine Simulation and Ship Manoeuvrability, MARSIM'03, Kanazawa, Kanazawa, Japan, 25–28 August 2003. RC-23-1–9.
18. Chapin, V.; Gourdain, N.; Verdin, N. Aerodynamic Study of a Two-elements Wingsail for High Performance Multihull Yachts. In Proceedings of the 5th High Performance Yacht Design Conference Auckland, Auckland, New Zealand, 10–12 March 2015.
19. Fiumara, A.; Gourdain, N.; Chapin, V.; Senter, J.; Bury, Y. Numerical and experimental analysis of the flow around a two-element wingsail at Reynolds number 0.53 × 106. *Int. J. Heat Fluid Flow* **2016**, *62*, 538–551. [CrossRef]
20. Menter, F.R. Two-equation eddy-viscosity turbulence models for engineering applications. *AIAA J.* **1994**, *32*, 1598–1605. [CrossRef]
21. Hassan, G.E.; Hassan, A.; Youssef, M.E. Numerical Investigation of Medium Range Re Number Aerodynamics Characteristics for NACA0018 Airfoil. *CFD Lett.* **2014**, *6*, 175–187.
22. Malan, P.; Suluksna, K.; Juntasaro, E. Calibrating the γ-Reθ Transition Model for Commercial CFD. In Proceedings of the 47th AIAA Aerospace Science Meeting, Orlando, FL, USA, 5–8 January 2009.
23. Douvi, C.E.; Tsavalos, I.A.; Margaris, P.D. Evaluation of the Turbulence Models for the Simulation of the Flow Over a National Advisory Committee for Aeronautics (NACA) 0012 Airfoil. *Mech. Eng. Res.* **2012**, *4*, 100–111.
24. Rhee, S.H.; Kim, H. A suggestion of gap flow control devices for the suppression of rudder cavitation. *J. Mar. Sci. Technol.* **2008**, *13*, 356–370. [CrossRef]
25. Gentry, A. The Aerodynamic of Sail Interaction. In Proceedings of the 3th AIAA Symposium on the Aero/Hydronautics of Sailing, Redondo Beach, CA, USA, 20 November 1971.
26. Smith, A.M.O. High-Lift Aerodynamics. *J. Aircr.* **1975**, *12*, 501–530. [CrossRef]
27. Biber, K.; Zumwalt, G. Experimental studies of a two-element airfoil with large separation. In Proceedings of the 30th Aerospace Sciences Meeting and Exhibit, Reno, NV, USA, 6–9 January 1992. [CrossRef]

13

Constitutive Relationship Proposition of Marine Soft Soil in Korea using Finite Strain Consolidation Theory

Sang Hyun Jun [1,*] and Hyuk Jae Kwon [2]

1 Infra Division, POSCO E&C, Incheon 22009, Korea
2 Department of Civil Engineering, Cheongju University, Chungbuk 28503, Korea; hjkwon@cju.ac.kr
* Correspondence: clays@poscoenc.com

Abstract: This paper proposes representative constitutive relationship equations of dredging and reclamation soft soil in Korea. The marine soft soils were sampled at 23 dredged-reclaimed construction sites in the Busan, Gwangyang, and Incheon regions in Korea; then, laboratory tests were carried out. The consolidation property was classified as LL = 60% for Busan and Gwangyang marine soft soil and LL = 30% for Incheon marine soft soil by conducting basic physical property tests and consolidation tests. Busan soft soil showed a slightly higher consolidation settlement property than Gwangyang soft soil. Incheon soft soil showed the lowest consolidation settlement property among the three regions. In particular, 77 consolidation simulations were carried out at a high void ratio using the centrifugal experiment to realize high water content and in-field stress conditions. The constitutive relationship equations of each of the 23 specimens were analyzed with regard to the void ratio–effective stress and void ratio–permeability coefficient through the back analysis of finite consolidation theory from the experimental results. The constitutive relationship equation for Korean soft soil was determined to be a reasonable power function equation. The representative constitutive relationships for soft soils in the three regions were estimated using six equations, which were classified by physical and consolidation properties. The representative constitutive equations were compared to those in previous studies on high void ratio conditions of marine soft soil, and the results showed a similar range.

Keywords: marine soft soil; dredging and reclamation; constitutive relationship; centrifugal experiment; void ratio–effective stress; void ratio–permeability coefficient

---

1. Introduction

The demand for land is addressed efficiently through the reclamation of foreshores, which include vast lands for industrial use, dwellings, airports, and harbors. Superior soil, suitable for making grounds through the reclamation of the foreshore, has been depleted due to resource exhaustion and environmental preservation, and marine soft soil is used to substitute soil to be reclaimed from nearby sites. Each year, over 30 million m^3 of dredged soil is used for the environmental reformation of offshores and estuaries, harbor construction, and maintenance of waterways. Recently, dredged soil has been reclaimed to dumping areas only because the 1996 Protocol to the London Convention has not permitted ocean dumping from January 2009. The dredged marine soft soil is reclaimed to dump areas due to demands for new ground and the treatment of a huge amount of dredged soil.

In Korea, most of the dredged marine soft soil is reclaimed due to its high water content; however, the difference between construction methods as well as the very weak strength and large settlement of the soft soil is a problem. Settlement is the main geotechnical issue related to the stability and estimation of the amount of dredged soil.

The difference between actual settlement and Terzaghi's one-dimensional consolidation theory is well explained by Gibson et al. [1] and Cargill [2]. Finite strain consolidation theory induces a large strain consolidation phenomenon without numerical inconsistency and, specifically, it applies nonlinear compressibility and permeability. Numerical analysis conducted using the equation that is governed by the finite strain consolidation theory is known to estimate a large consolidation settlement [3–7]. Finite strain consolidation theory is not commonly applied to the design and construction in Korea. The general reasons are the lack of understanding of the theory, and difficulty in estimating input data and analyzing the numerical input or result.

The studies on void ratio-effective stress and void ratio-permeability have mainly focused on the exponential or power equation. Somogyi [8,9] proposed a constitution equation as a power law. Carrier et al. [10] proposed a constitution equation of modified power laws by analyzing the relationship between physical character (Atterberg limits and Activity) and the coefficient of the constitution equation. Gibson et al. [11] proposed an exponential constitution equation to simplify the consolidation governing equation.

Non-linear relationships between void ratio and effective stress, and void ratio and permeability, which directly affect settlement behavior, are observed in the high water content range of dredged soil. The non-linear constitution equation in a high water content range is difficult to estimate by conducting direct experiments. To overcome this problem, consolidation tests and settlement tests are carried out, and the results of these tests are back-analyzed through numerical analysis based on the finite strain consolidation theory. Therefore, in this study, representative relationship equations of void ratio-effective stress and void ratio-permeability of marine soft soil in Korea are proposed using finite strain consolidation theory.

2. Characteristics of Dredged and Reclaimed Soil

2.1. Methods of Analysis

A huge amount of dredged soil has been reclaimed in disposal areas that are mostly developed at new ports. There are three main disposal areas of dredged soil at Busan New Port, five at Incheon port, and three at Pyeongtaek-Dangjin Port, in addition to several others in various locations. The demand for the disposal area is continuously increasing around the areas developed at new ports, and the expansion of the existing disposal area is covered due to the expropriation of dredged soil. In this study, dredged soil was sampled at 23 dredged-reclaimed construction sites and laboratory tests were carried out using 23 samples. The sampling locations and construction sites are shown in Table 1 and Figure 1. The samples were separated into three regions: the Busan, Gwangyang, and Incheon regions.

Table 1. Sampling locations.

Region	Mark	Locations in Korea	Region	Mark	Locations in Korea
	bs-a	Angol-dong, Changwon		gy-c	Hwanggil-dong, Gwangyang
	bs-b	Cheonseong-dong, Busan		gy-d	Doi-dong, Gwangyang
	bs-c	Gamcheon-dong, Busan		gy-e	Hwanggeum-dong
	bs-d	Ung-dong, Changwon	Gwang-yang (9 sites)	gy-f	Doi-dong, Gwangyang
Busan (10 sites)	bs-e	Ung-dong, Changwon		gy-g	Hwachi-dong, Yeosu
	bs-f	Seongbuk-dong, Busan		gy-i	Doi-dong, Gwangyang
	bs-g	Haeun-dong, Changwon		gy-j	Jeongryang-dong, Yeosu
	bs-i	Ung-dong, Changwon		ic-a	Songdo-dong, Incheon
	bs-j	Youngdang-dong, Busan	Incheon (4 sites)	ic-b	Oryu-dong, Incheon
	bs-k	Gapo-dong, Changwon		ic-c	Unbuk-dong, Incheon
Gwang-yang	gy-a	Jung-dong, Gwangyang		ic-d	Songdo-dong, Incheon
	gy-b	Hwanggeum-dong	Total	23 dredged-reclaimed sites	

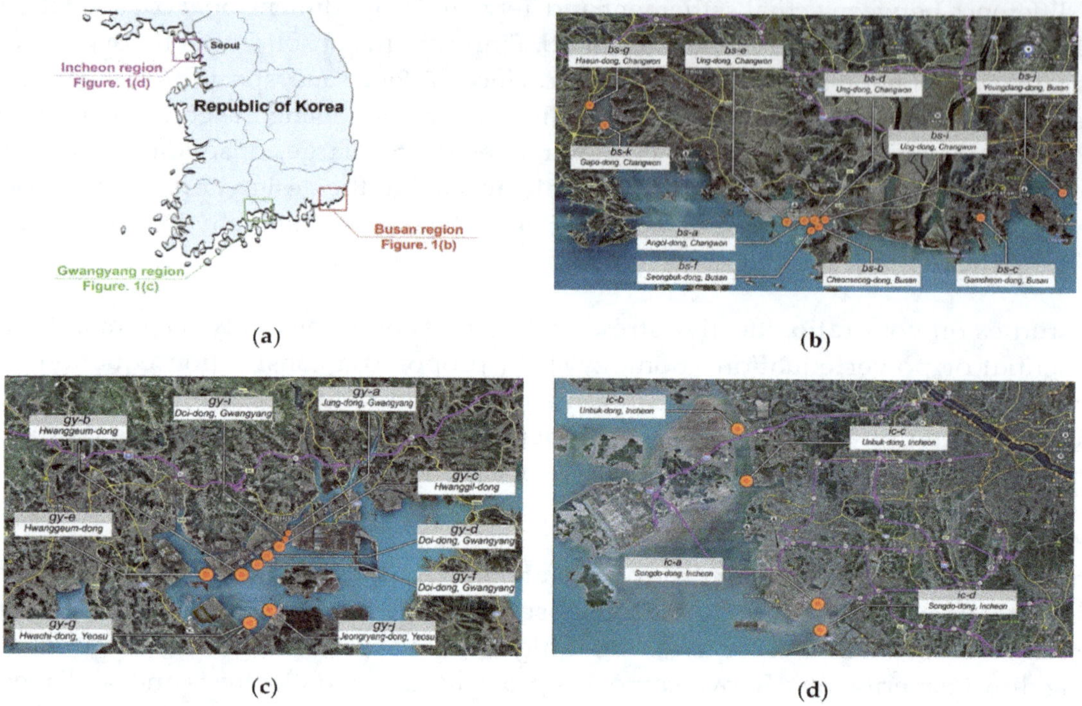

Figure 1. Sampling locations: (**a**) the regions in Korea; (**b**) the Busan region; (**c**) the Gwangyang region; (**d**) the Incheon region.

Dredged soil is disposed of under disturbed conditions using different dredging methods. In particular, the pump-dredged method mostly used in Korea moves dredged soil with water into the disposal area after being totally disturbed. Therefore, in this study, experiments on engineering properties and centrifugal tests used disturbed samples. The sieve analysis, particle size analysis, soil classification system, and Atterberg limit test for 23 samples were performed by ASTM C136, ASTM D422, ASTM D2487, and ASTM D4318, respectively. The oedometer one-dimensional consolidation test and the constant rate of the strain (CRS) consolidation test followed ASTM D2435-11 and ASTM D4186-06. The CRS test is performed as strain rates of 0.04–0.001% per minute depending on the liquid limit of the soil, and the result is analyzed using the consolidation solution provided by Wissa et al. [12].

2.2. *Characteristics of Dredged Soil*

2.2.1. Busan Region

The characteristics of dredged soil in the Busan region were analyzed using 10 samples, and are shown as plasticity and activity charts in Figure 2a,b, respectively. The specific gravity, plastic limit, liquid limit, plastic index, and activity of the samples were 2.706–2.731 (average 2.718), 20.88–39.31 (avg. 28.97%), 40.10–94.26% (avg. 60.74%), 15.56–54.96% (avg. 31.77%), and 0.8–2.29 (avg. 1.63), respectively. The specimens bs-g, bs-i, bs-j, and bs-k followed the A-line on the plasticity chart, and other specimens showed liquid limits (LL) in the range of 40–55% (LL < 60%). The specimens in the Busan region can be separated into bs-a to bs-f (LL < 60%) and bs-g to bs-k (LL ≥ 60%), according to the liquid limit. It can be observed from Figure 2b that specimens with LL less than 60% represent a contented clay mineral with a PI (plasticity index) range of 10–20%, and the specimens with LL greater than 60% represent activated clay with a PI range of 33–55%. According to the sieve and particle size analysis tests, the sand particle content (less than the No. 200 sieve, 0.074 mm), silt particle content, and clay particle content were 0.62–5.07%, 72.90–82.93%, and 16.05–24.53%.

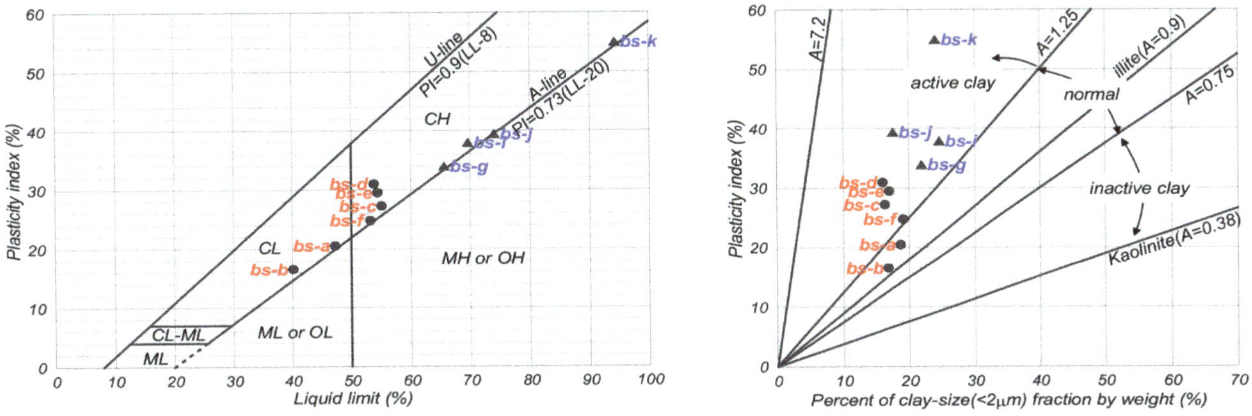

(a) (b)

Figure 2. Physical properties of Busan region soil: (**a**) Plasticity chart; (**b**) Activity chart.

The specimens showed the compression index (C_c) of 0.316–0.883 and permeability change index (C_k) of 0.209–1.328 during the oedometer [incremental loading (IL)] test, and C_c of 0.530–1.011 and C_k of 0.731–1.799 during the CRS consolidation test where the compression index (C_c) and the permeability change index are defined as $\frac{\Delta e}{\log p/p_0}$ and $\frac{\Delta e}{\log k/k_0}$. The consolidation properties of the Busan region soils as per the liquid limit are shown in Figure 3. Specimens (bs-a, b, d) with LL < 60% showed small compression and high permeability than the specimens with LL ≥ 60%. Therefore, the separation of the Busan region soil using the criteria of liquid limit (LL = 60%) can be considered reasonable because the physical and consolidation properties are different. The clay with LL < 60% was named the Busan-L marine clay and the clay with LL ≥ 60% was named the Busan-H marine clay.

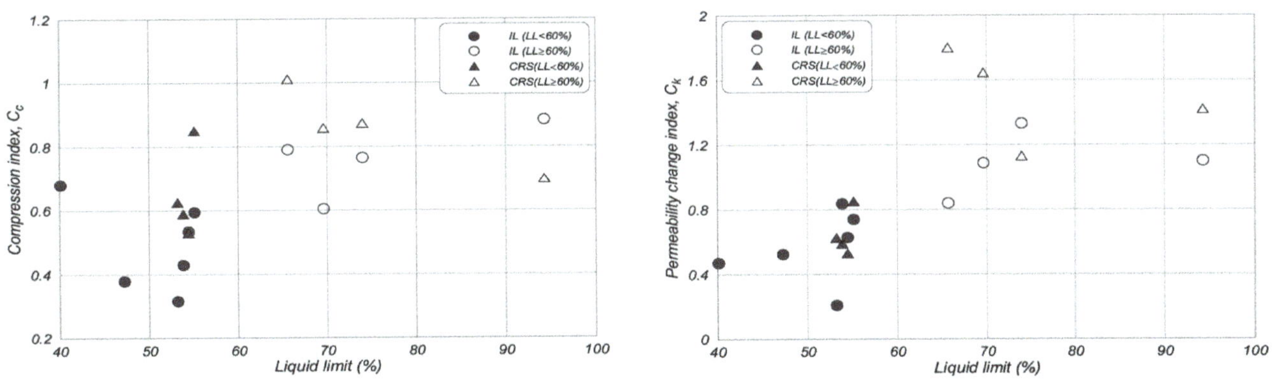

(a) (b)

Figure 3. Consolidation properties of Busan region soil: (**a**) liquid limit (LL)-compression index (C_c); (**b**) liquid limit (LL)-permeability change index (C_k).

2.2.2. Gwangyang Region

The soil in the Gwangyang region, which was analyzed using nine samples, had the specific gravity, plastic limit, liquid limit, and plastic index of 2.702–2.745 (average 2.718), 14.78–36.37% (avg. 24.13%), 33.77–82.71% (avg. 55.77%), and 18.99–50.50% (avg. 31.63%), respectively, as shown in Figure 4. It had an activity of 0.94–3.79 (avg. 2.60) and was classified into CL and CH as per the unified soil classification system (USCS). The plastic and liquid limits of the Gwangyang region soil were slightly less than those of the Busan region soil. According to the sieve and particle size analysis tests, the sand particle content, silt particle content, and clay particle content were 0.47–10.30%, 67.41–87.14%, and 6.9–22.81%, respectively. The Busan region soil contained slightly finer particles than the Gwangyang region soil as per the result of the plasticity, sieve, and particle size analysis tests. However, clay minerals might be different due to the higher activity of the Gwangyang region soil than the Busan region soil.

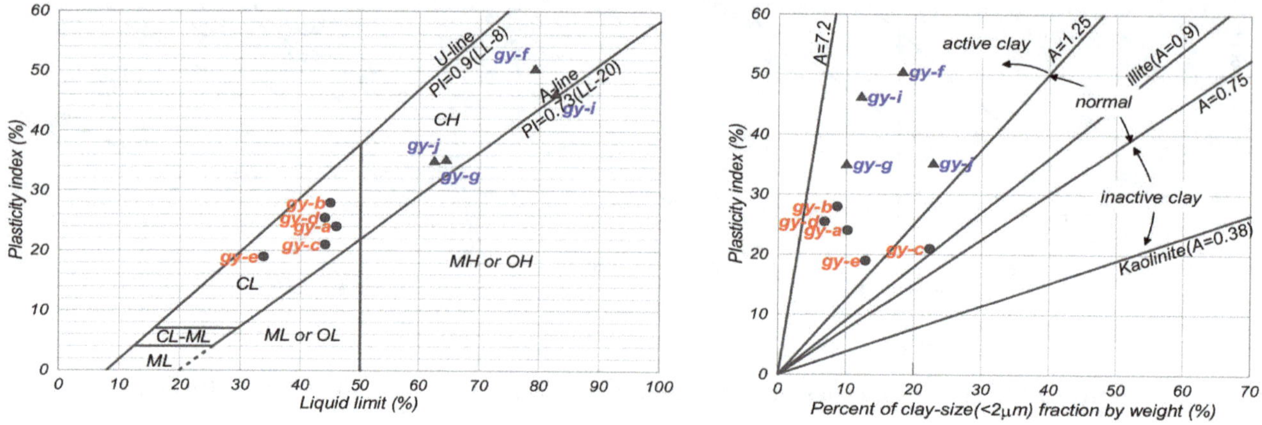

(a) (b)

Figure 4. Physical properties of Gwangyang region soil: (**a**) Plasticity chart; (**b**) Activity chart.

The clay showed a C_c of 0.231–0.939 and C_k of 0.478–0.840 during the IL test, and C_c of 0.242–1.273 and C_k of 0.518–1.371 during the CRS test, as shwon in Figure 5. Hence, the Gwangyang region soil showed small ranges of C_c and C_k. This might be the reason why the Busan clay included more fine clay than the Gwangyang clay.

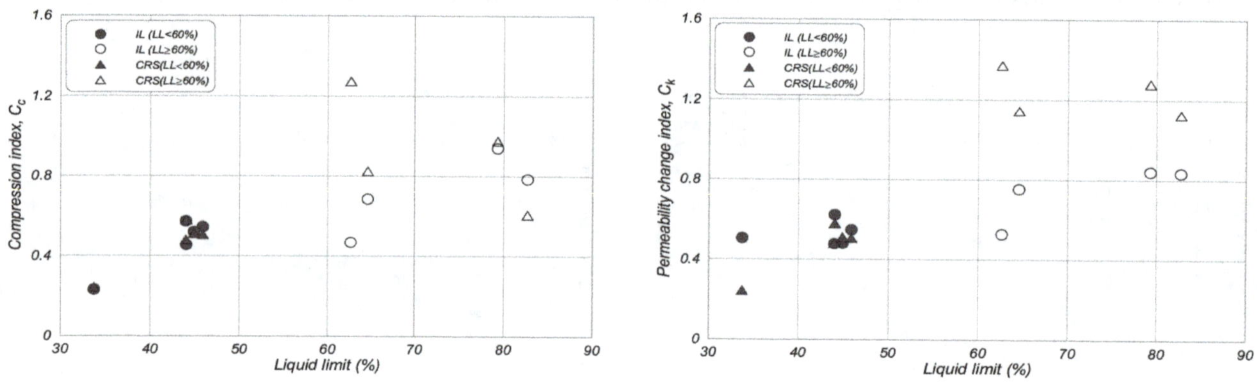

(a) (b)

Figure 5. Consolidation properties of Gwangyang region soil: (**a**) liquid limit (LL)-compression index (C_c); (**b**) liquid limit (LL)-permeability change index (C_k).

In the Gwangyang region soil, specimens such as gy-a to gy-e with LL > 60% showed high clay particle content, high plastic index, high compression, and low permeability than the specimens with LL < 60%. Therefore, the Gwangyang region dredged soil can be separated reasonably using the criteria of liquid limit (LL = 60%) in the same way as the Busan region dredged soil. The clay with LL < 60% was named the Gwangyang-L marine clay and the clay with LL ≥ 60% was named the Gwangyang-H marine clay.

2.2.3. Incheon Region

The Incheon region soil analyzed from four sites shown in Table 1 had the specific gravity, plastic limit, liquid limit, and plastic index of 2.693–2.713 (avg. 2.703), 12.48–22.74% (avg. 16.48%), 24.86–38.34% (avg. 30.62%), and 11.84–16.78% (avg. 14.15%), respectively. The sand particle content, silt particle content, and clay particle content for the Incheon region soil were 4.59–8.78%, 79.38–84.41%, and 10.79–14.42%. The activity with a range of 1.04–1.22 and average of 1.13 lies in the normal area, as shown in Figure 6. The consolidation properties were analyzed with $C_c = 0.167$–0.406 and $C_k = 0.246$–0.628 during the oedometer test, and $C_c = 0.204$–0.493 and $C_k = 0.380$–0.791 during the CRS test, as shown in Figure 7.

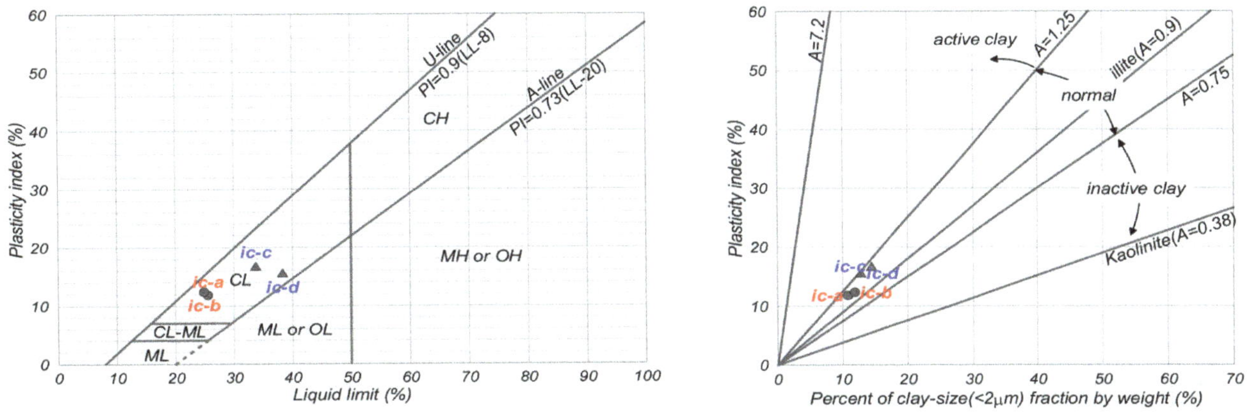

(a) (b)

Figure 6. Physical properties of the Incheon region soil: (**a**) Plasticity chart; (**b**) Activity chart.

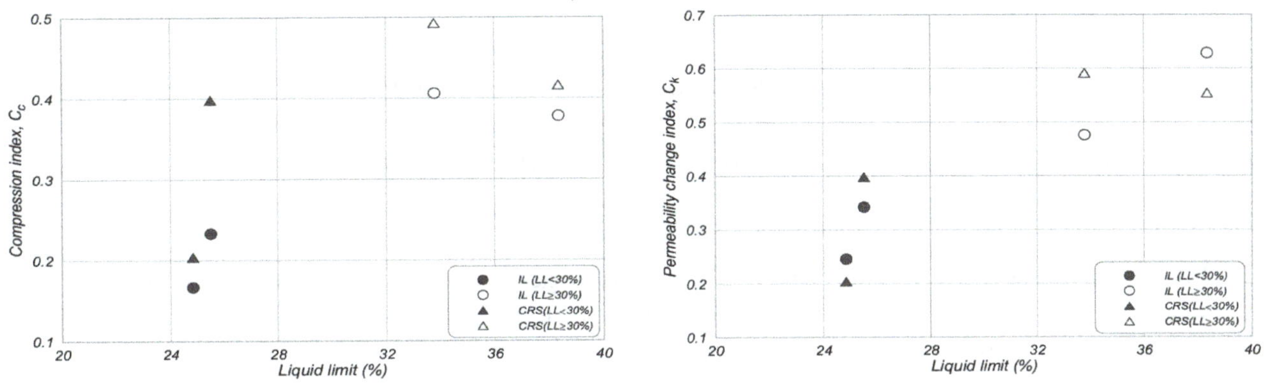

(a) (b)

Figure 7. Consolidation properties of Incheon region soil: (**a**) liquid limit (LL)-compression index (C_c); (**b**) liquid limit (LL)-permeability change index (C_k).

The Incheon region soil had lower liquid and plastic limits than the Busan and Gwangyang region soils. The liquid limit was 30.12% less than that of Busan clay and 25.15% less than that of the Gwangyang clay. The plastic index was less than 17.5 compared to the Busan and Gwangyang clays. The results of the oedometer test show that Incheon clay exhibited a $C_c = 0.31$ and $C_k = 0.246$–0.628, which were less than those of the Busan and Gwangyang clays. That is, the Incheon region soil clearly showed low plasticity, small compressibility, and large permeability than the Busan and Gwangyang clays.

Furthermore, the Incheon region soil had different physical and compressive properties that were separated by the criteria of LL = 30% compared to the Busan and Gwangyang clays (LL = 60%). The ic-c and ic-d specimens with LL ≥ 30% had large fine particles, large compressibility, and small permeability compared to the ic-a and ic-b specimens with LL < 30%. Therefore, the Incheon region dredged soil can be separated reasonably using the criteria of liquid limit (LL = 30%). The clay with LL < 30% was named the Incheon-L marine clay and the clay with LL ≥ 30% was named the Incheon-H marine clay. Engineering properties of the marine soft soil were classified by liquid limits and are summarized in Figure 8.

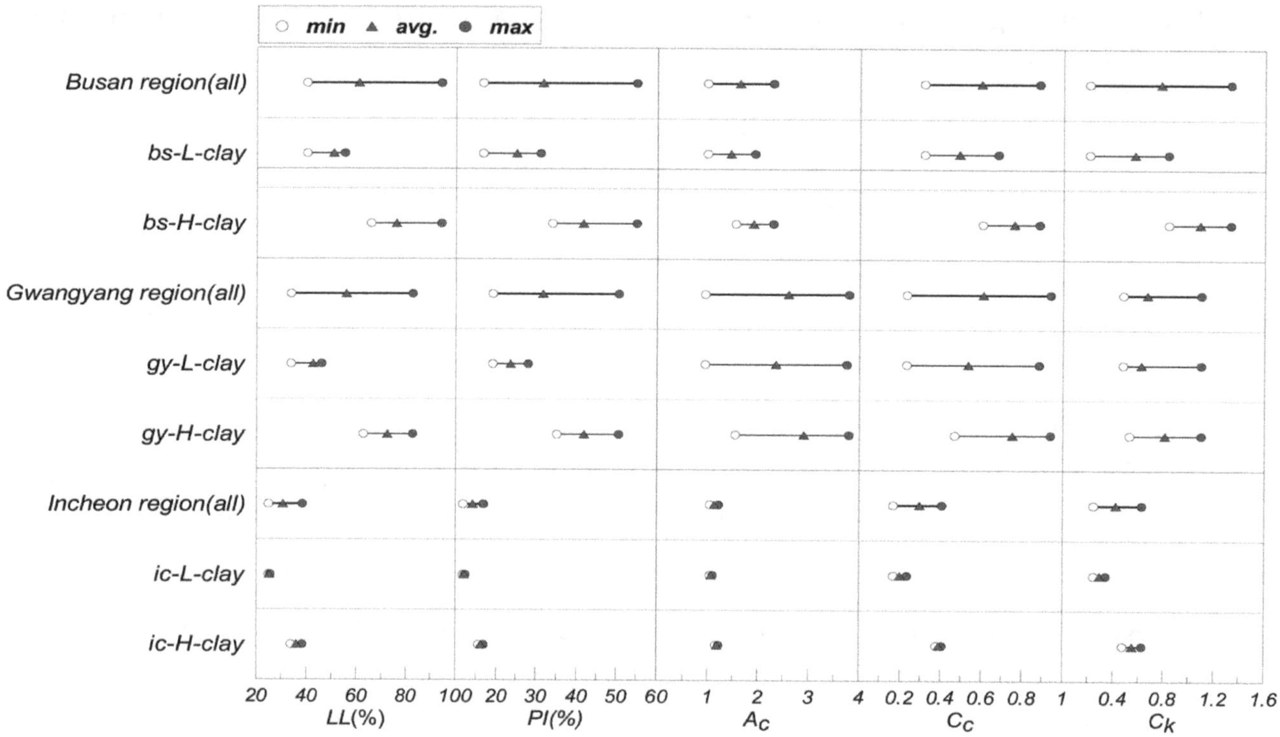

Figure 8. Engineering properties for the classified marine soft soil by liquid limits.

3. Relationships between Void Ratio–Effective Stress and Permeability

3.1. Estimation of Non-Linear Constitution Equation

The dredged-reclaimed clay with a high void ratio should be estimated using finite strain consolidation theory due to huge deformations. The finite strain consolidation is suitable to analyze large deformations, considering the variation in the compressibility and permeability of clay as per effective stress. The variation in compressibility and permeability depend on the relationship void ratio (e)–effective stress (σ′)–permeability coefficient (k) during analysis, and this relationship is called the constitutive relationship of finite strain consolidation theory [13,14]. The constitutive relationship between void ratio–effective stress and void ratio–permeability is examined in this section to choose a suitable constitutive relationship equation in Korea.

The constitutive relation equations were mostly the proposed exponential function and power function. Somogyi [8,9] proposed Equations (1) and (2) of the power function using the empirical data of void ratio–effective stress and void ratio–permeability:

$$e = A\sigma'^{B} \tag{1}$$

$$k = Ce^{D} \tag{2}$$

where A, B, C, and D are the decided coefficients including the material properties observed during the test or empirical study. These equations were used in this study due to the following three considerations. First, there should be an equation that was proposed from research on Korean clay. Second, there should be an equation that represents a reasonable relationship between the void ratio–effective stress and void ratio–permeability coefficient in the high void ratio range. Finally, there should be a simple equation to propose a design chart.

3.2. *Back Analysis of Constitutive Relationship Equation through Centrifugal Experiment*

The centrifugal test is applied to carry out a one day long self-weight consolidation of dredged/reclaimed clay in foreign and domestic scenarios because it can surmount the limitations of consolidation time and initial height through a back-analysis study using the time-settlement results of the test. The back analysis using the centrifugal experiment is explained as a flowchart in Figure 9. The equipment used for the centrifuge test in this study was from the Kangwon National University in Korea. The centrifugal experiment was carried out 2–4 times under different initial void ratio and height conditions for each specimen, and then a constitutive relationship equation was estimated through the back analysis from the experimental results. The experimental conditions such as test number, gravity acceleration, initial void ratio, and constitutive relationship equations analyzed by the back analysis for each region specimen are shown in Table 2. Numerical analysis for the self-weight consolidation test in the flowchart was performed by the microcomputer program 'Primary Consolidation, Secondary Compression, and Desiccation of Dredged Fill' (PSDDF) by Stark et al. [3,4].

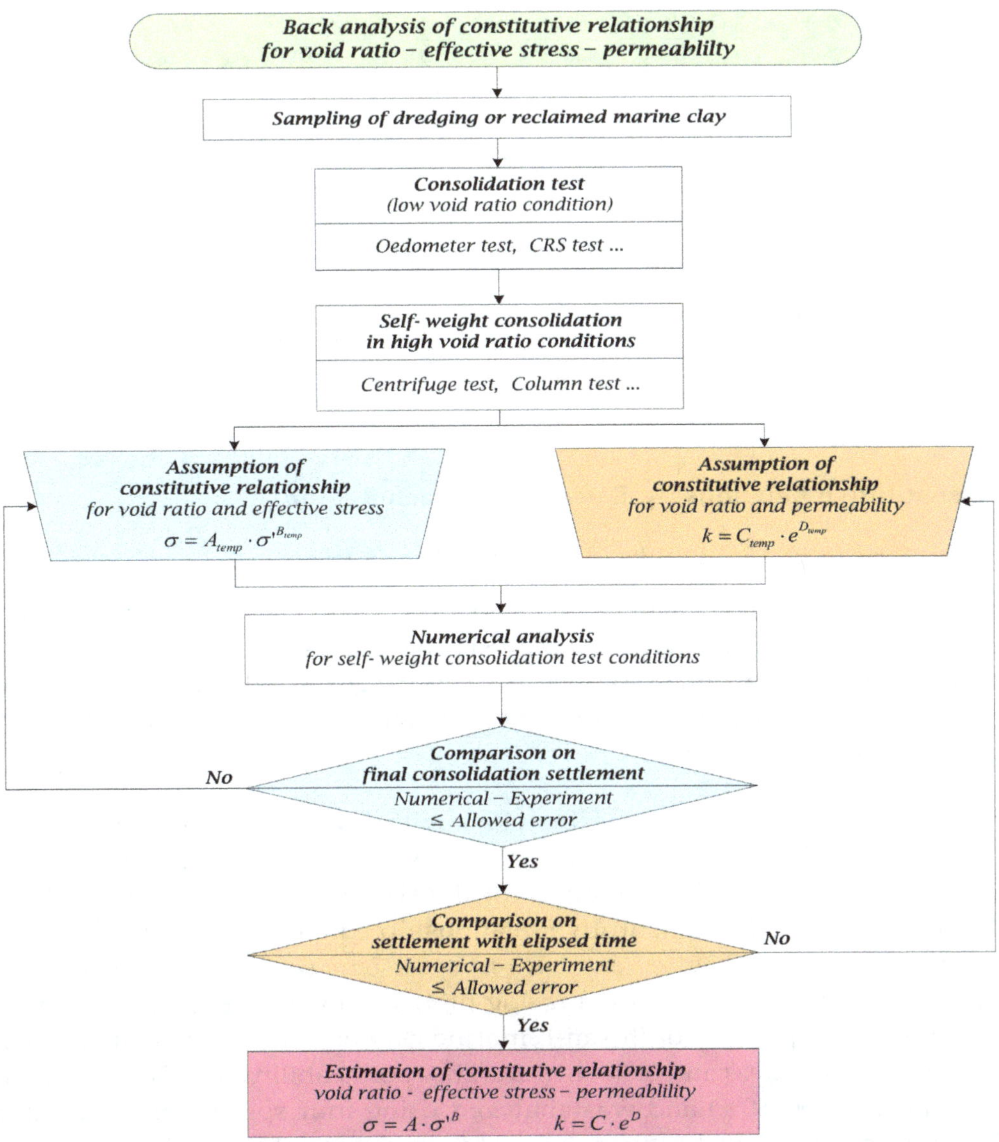

Figure 9. Flowchart of back analysis of constitutive relationship using the centrifugal experiment.

Table 2. Centrifugal experimental conditions and results.

Symbol	Mark	Accel. of Gravity, n (g)	Initial Void Ratio e_0	Initial Height of Column H_0 (mm)	Atterberg Limits		Constitutive Relationship	
					LL (%)	PI	e=σ′ (kPa)	e-k (m/day)
Busan -L-clay (bs-L-clay)	bs-a	20~30	4.08~6.80	75~200	47.2	20.5	$e = 3.44\,\sigma'^{-0.241}$	$k = 1.42 \times 10^{-5}\,e^{5.22}$
	bs-b	30	4.07~6.78	100~200	40.1	16.6	$e = 2.85\,\sigma'^{-0.121}$	$k = 1.17 \times 10^{-6}\,e^{8.22}$
	bs-c	40	5.42~6.78	100	55.1	27.3	$e = 3.49\,\sigma'^{-0.207}$	$k = 1.74 \times 10^{-5}\,e^{4.06}$
	bs-d	50~60	4.07~5.43	100~150	53.8	31.0	$e = 2.32\,\sigma'^{-0.171}$	$k = 1.13 \times 10^{-5}\,e^{5.17}$
	bs-e	50~60	4.34~5.42	150~200	54.4	29.5	$e = 2.84\,\sigma'^{-0.175}$	$k = 6.39 \times 10^{-6}\,e^{6.33}$
	bs-f	30~40	4.06~5.41	100	53.2	24.8	$e = 2.84\,\sigma'^{-0.187}$	$k = 9.91 \times 10^{-6}\,e^{5.34}$
Busan -H-clay (bs-H-clay)	bs-g	30~40	5.42~6.78	100	65.6	33.9	$e = 4.20\,\sigma'^{-0.174}$	$k = 9.30 \times 10^{-6}\,e^{4.01}$
	bs-i	50~60	4.08~5.44	120~220	69.6	37.9	$e = 4.33\,\sigma'^{-0.222}$	$k = 3.58 \times 10^{-6}\,e^{5.38}$
	bs-j	30~40	4.10~5.46	100~200	74.0	39.4	$e = 4.16\,\sigma'^{-0.254}$	$k = 1.03 \times 10^{-6}\,e^{6.03}$
	bs-k	30~50	5.44~10.88	100~200	94.3	55.0	$e = 4.58\,\sigma'^{-0.180}$	$k = 1.68 \times 10^{-5}\,e^{3.56}$
Gwangyang -L-clay (gy-L-clay)	gy-a	30~40	4.08~5.43	100	45.9	24.0	$e = 2.97\,\sigma'^{-0.197}$	$k = 2.67 \times 10^{-6}\,e^{5.52}$
	gy-b	30~40	4.12~5.49	100	44.9	28.0	$e = 2.66\,\sigma'^{-0.176}$	$k = 8.88 \times 10^{-6}\,e^{6.17}$
	gy-c	30~40	4.07~5.42	100	44.1	21.0	$e = 2.81\,\sigma'^{-0.175}$	$k = 5.32 \times 10^{-6}\,e^{5.26}$
	gy-d	30~40	4.11~5.48	100	44.0	25.5	$e = 3.03\,\sigma'^{-0.244}$	$k = 9.72 \times 10^{-6}\,e^{6.10}$
	gy-e	40~50	4.05~8.11	180~225	33.8	19.0	$e = 3.16\,\sigma'^{-0.174}$	$k = 4.25 \times 10^{-6}\,e^{7.50}$
Gwangyang -H-clay (gy-H-clay)	gy-f	30~50	4.08~6.80	100~200	79.3	50.5	$e = 3.96\,\sigma'^{-0.294}$	$k = 1.69 \times 10^{-5}\,e^{4.25}$
	gy-g	30~40	4.17~6.95	100~200	62.6	35.1	$e = 4.62\,\sigma'^{-0.277}$	$k = 1.15 \times 10^{-5}\,e^{3.75}$
	gy-i	30~40	4.07~5.42	100~200	82.7	46.3	$e = 3.77\,\sigma'^{-0.179}$	$k = 8.09 \times 10^{-6}\,e^{5.91}$
	gy-j	40	3.32~5.45	100~125	64.5	35.3	$e = 3.71\,\sigma'^{-0.177}$	$k = 1.26 \times 10^{-5}\,e^{3.78}$
Incheon -L-clay (ic-L-clay)	ic-a	60	2.71~4.06	200	25.5	11.8	$e = 1.76\,\sigma'^{-0.150}$	$k = 1.09 \times 10^{-4}\,e^{5.94}$
	ic-b	50	2.69~4.04	100~200	24.8	12.4	$e = 1.50\,\sigma'^{-0.161}$	$k = 3.12 \times 10^{-4}\,e^{4.46}$
Incheon -H-clay (ic-H-clay)	ic-c	30~40	2.71~4.34	100~200	33.8	16.8	$e = 2.15\,\sigma'^{-0.139}$	$k = 6.69 \times 10^{-5}\,e^{4.90}$
	ic-d	30~40	4.05~5.40	100	38.3	15.6	$e = 2.97\,\sigma'^{-0.197}$	$k = 2.67 \times 10^{-5}\,e^{5.52}$

4. Proposed Constitutive Relationship Equation for Each Region

4.1. Representative Constitutive Relationship Equation for Each Region

The constitutive relationship equations for a total of 23 dredged clay specimens from three regions and six groups were obtained through back analysis and the results of the centrifugal experiments. The equation of the power function is shown on a log–log scale and is the deduced representative equation for each of the six groups. The units of the constitutive equation are kPa of effective stress and m/day of permeability coefficient.

The Busan-L marine clay (Busan marine clays with low liquid limit) with a liquid limit in the 40–60% range in the Busan region dredged clay was analyzed using six specimens (bs-a, b, c, d, e, f). A constitutive relationship equation for the six specimens is shown in Figure 10 of the log–log scale. The representative constitutive relationship equation (bs-L-clay) of Busan-L marine clay can be expressed as $e = 3.1\sigma'^{-0.19}$ and $k = 9 \times 10^{-6}e^{5.5}$ ($A = 3.1$, $B = -0.19$, $C = 9 \times 10^{-6}$, $D = 5.5$) using the arithmetic mean on the log–log plot, as shown in Figure 10. The representative constitutive relationship equations of Busan-L marine clay had the coefficient of determination of 0.87 on void ratio–effective stress and 0.84 on void ratio–permeability. The Busan-H marine clay had a high liquid limit (LL = 60–80%) and was analyzed using four specimens (bs-g~k), as shown in Figure 11. The representative constitutive relationship equations of Busan-H clay (bs-H-clay) can be expressed as $e = 4.3\sigma'^{-0.20}$ and $k = 6 \times 10^{-6}e^{4.5}$, and had the coefficient of determination of 0.97 on void ratio–effective stress and 0.88 on void ratio–permeability.

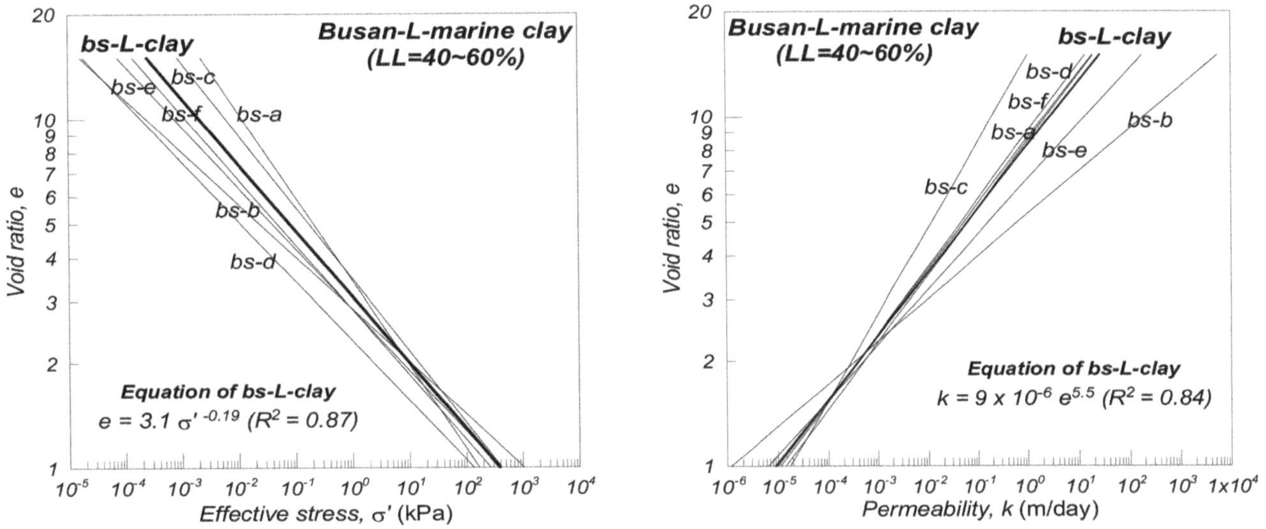

(a) (b)

Figure 10. Constitutive relationship of Busan L marine clay (bs-L-clay): (**a**) void ratio vs. effective stress; (**b**) void ratio vs. permeability.

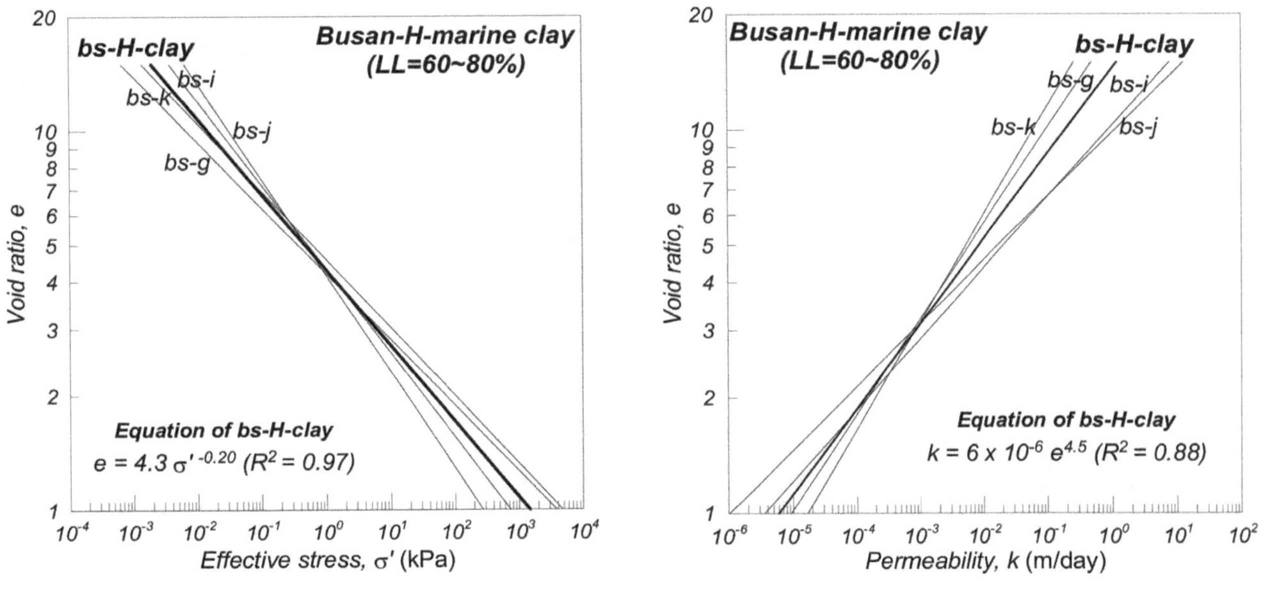

(a) (b)

Figure 11. Constitutive relationship of Busan H marine clay (bs-H-clay): (**a**) void ratio vs. effective stress; (**b**) void ratio vs. permeability.

The Gwangyang-L marine clay with LL = 40–60% and the representative constitutive equation (gy-L-clay) of $e = 2.9\sigma'^{-0.18}$ and $k = 9 \times 10^{-6}e^{6.0}$ was analyzed using five specimens (gy-a~e), as shown in Figure 12. The equations of gy-L-clay had the coefficient of determination of 0.95 on void ratio–effective stress and 0.86 on void ratio–permeability. The Gwangyang-H marine clay with a high liquid limit of 60–80% and the representative constitutive equation (gy-H-clay) of $e = 3.9\sigma'^{-0.20}$ and $k = 8 \times 10^{-6}e^{4.5}$ was analyzed using four specimens (gy-f~j), as shown in Figure 13. The equations of gy-H-clay had the coefficient of determination of 0.88 on the e-σ′ relationship and 0.77 on the e-k relationship.

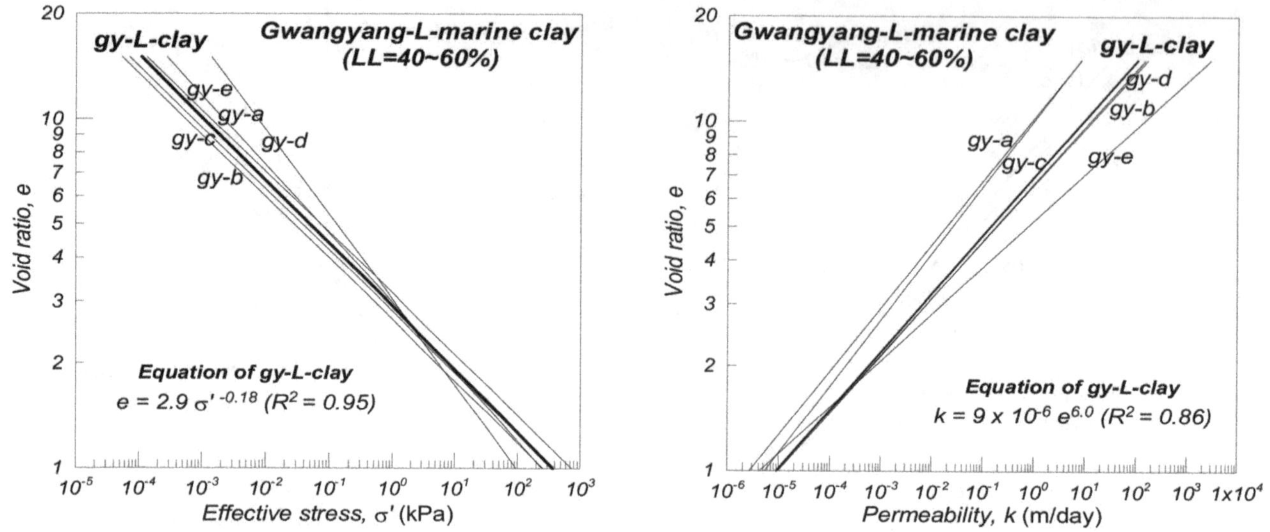

(a) (b)

Figure 12. Constitutive relationship of Gwangyang L marine clay (gy-L-clay): (**a**) void ratio vs. effective stress; (**b**) void ratio vs. permeability.

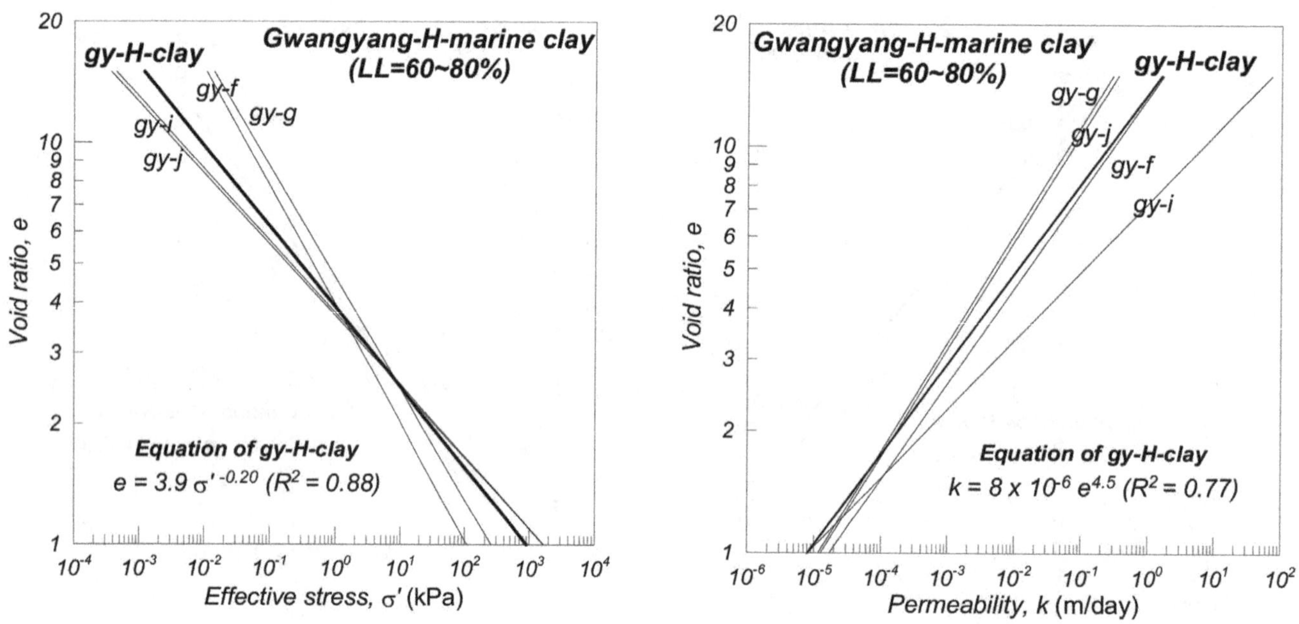

(a) (b)

Figure 13. Constitutive relationship of Gwangyang H marine clay (gy-H-clay): (**a**) void ratio vs. effective stress; (**b**) void ratio vs. permeability.

The Incheon-L marine clay with LL = 20–30% and the representative constitutive equation (ic-L-clay) of $e = 1.7\sigma'^{-0.15}$ and $k = 1\times10^{-4}e^{5.5}$ was analyzed using ic-b and ic-c specimens, as shown in Figure 14. The equations of ic-L-clay had the coefficient of determination of 0.98 on the e-σ′ relationship and 0.92 on the e-k relationship. The representative constitutive equation (ic-H-clay) of the Incheon-H marine clay was $e = 2.2\sigma'^{-0.17}$ and $k = 5\times10^{-5}e^{5.5}$, which was deduced from the ic-c and ic-d specimens, as shown in Figure 15. The equations of ic-H-clay had the coefficient of determination of 0.85 on the e-σ′ relationship and 0.94 on the e-k relationship.

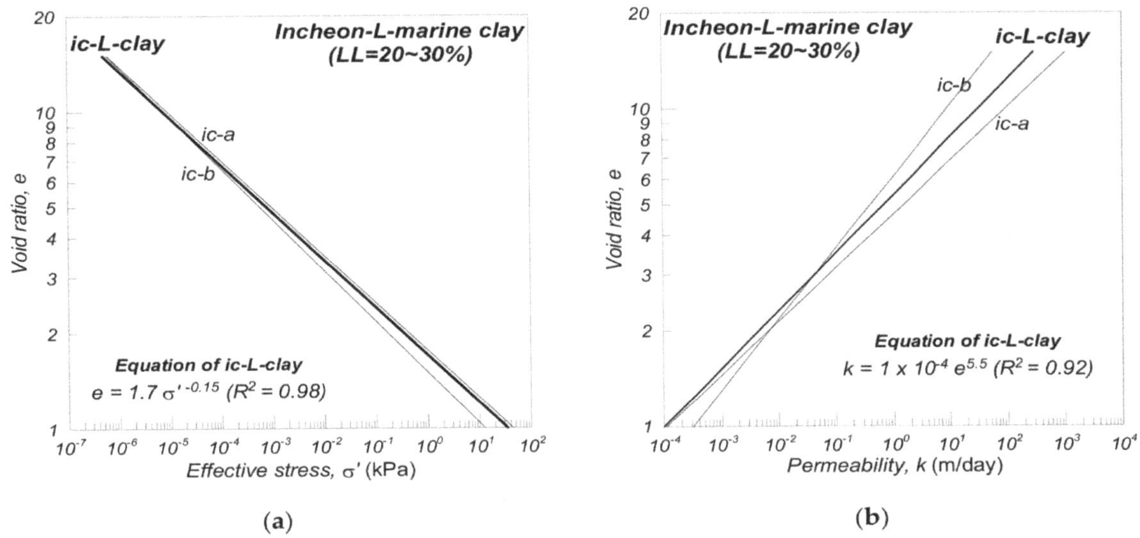

Figure 14. Constitutive relationship of Incheon L marine clay (ic-L-clay): (**a**) void ratio vs. effective stress; (**b**) void ratio vs. permeability.

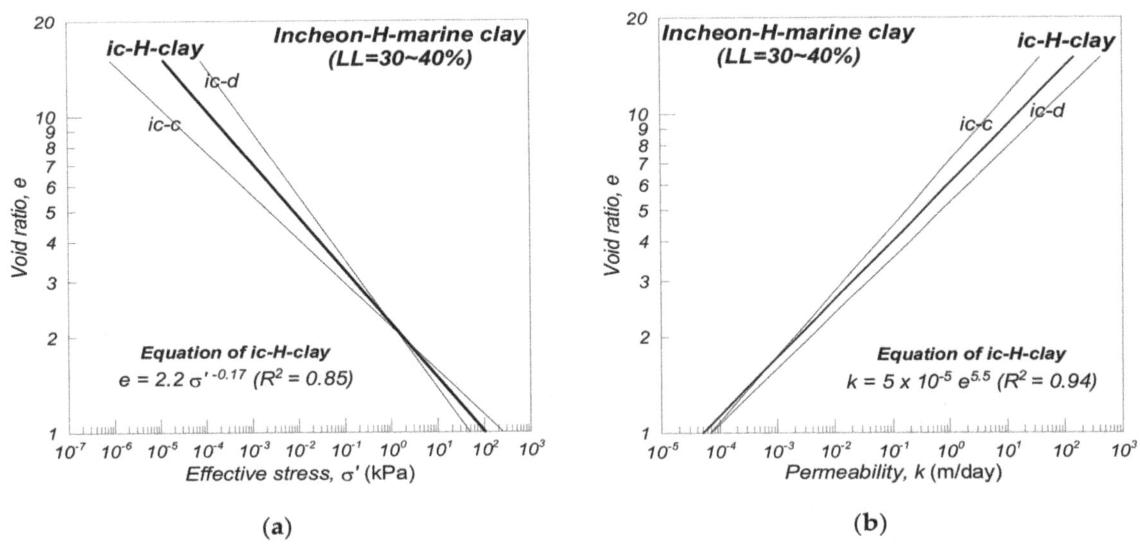

Figure 15. Constitutive relationship of Incheon H marine clay (ic-H-clay): (**a**) void ratio vs. effective stress; (**b**) void ratio vs. permeability.

The representative constitutive relationship equations for each region are summarized in Table 3.

Table 3. Representative constitutive relationship equation.

Region	Symbol	Representative Constitutive Relationship		Applicable Range of Liquid Limits
		Void Ratio-Effective Stress (kPa)	Void Ratio-Permeability (m/day)	
Busan	bs-L-clay	$e = 3.1\ \sigma'^{-0.19}$	$k = 9 \times 10^{-6}\ e^{5.5}$	40~60%
	bs-H-clay	$e = 4.3\ \sigma'^{-0.20}$	$k = 6 \times 10^{-6}\ e^{4.5}$	60~80%
Gwangyang	gy-L-clay	$e = 2.9\ \sigma'^{-0.18}$	$k = 9 \times 10^{-6}\ e^{6.0}$	40~60%
	gy-H-clay	$e = 3.9\ \sigma'^{-0.20}$	$k = 8 \times 10^{-6}\ e^{4.5}$	60~80%
Incheon	ic-L-clay	$e = 1.7\ \sigma'^{-0.15}$	$k = 1 \times 10^{-4}\ e^{5.5}$	20~30%
	ic-H-clay	$e = 2.2\ \sigma'^{-0.17}$	$k = 5 \times 10^{-5}\ e^{5.5}$	30~40%

4.2. *Comparison with Precedent Studies*

The representative constitutive relationship equations were compared to previous studies that had similar basic physical properties such as a classification of soil and Atterberg limits. The constitutive equations proposed in this paper and those presented by Carrier et al. [10] were plotted, as shown in Figure 16. The relationships of void ratio–effective stress were plotted in the low effective stress range at the same void ratio. The equations were similar to Horton, Naumee River, Todedo and Craney Island clays, which had similar physical properties. The permeability of Korean clay had a comparatively high range at the same void ratio. The constitutive relationship equations were similarly analyzed for aluminum red mud (LL = 41–46%, PI = 7–9), FGD (Flue-Gas Desulfurization) sludge (LL = 65%, PI = 17), and Craney Island clay.

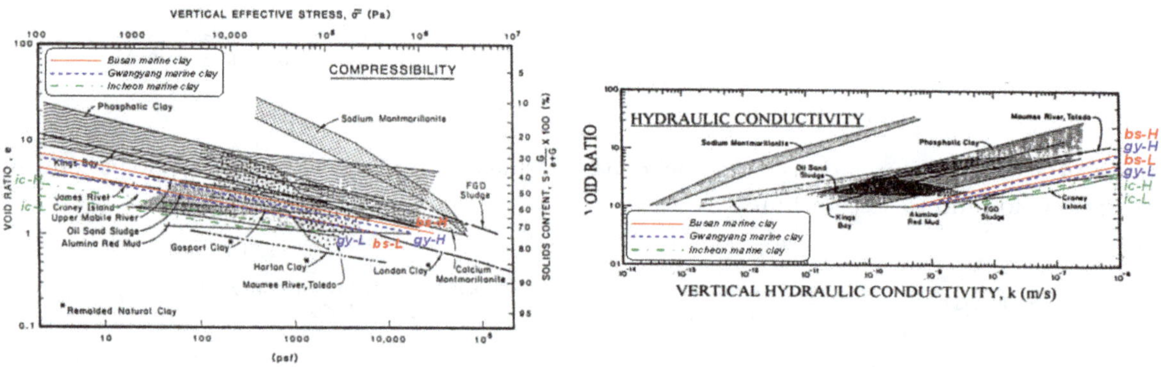

(a) (b)

Figure 16. Comparison of our study and the study by Carrier et al. [10]: (**a**) void ratio vs. effective stress; (**b**) void ratio vs. permeability.

Yamagami et al. [15] proposed the power function relationship equations of void ratio–effective stress and void ratio–permeability coefficient for each stage of settling and consolidation during the estimation of settling and consolidation characteristics from back analysis. The equations proposed here were plotted and compared with the constitutive equations for mud A (PI = 40.1) and mud B (PI = 22.5) presented by Yamagami et al., as shown in Figure 17. The equation for mud A was similar to that of the e-σ′-k of bs-L-clay and gy-L-clay; mud B showed low compressibility and high permeability at high void ratio ranges. The representative constitutive equations in this study were compared to the existing research data that had similar physical properties as a Korean marine soft soil, and the results showed a similar range.

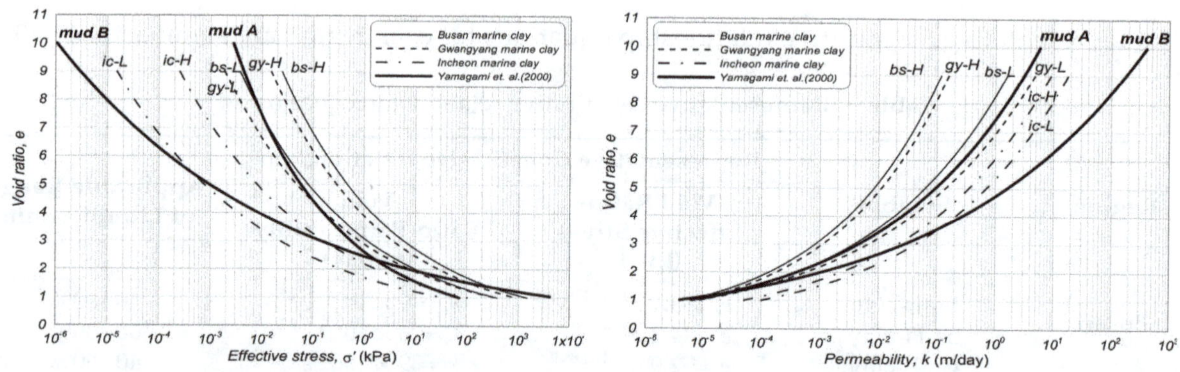

(a) (b)

Figure 17. Comparison of our study and the study by Yamagami et al. [15]: (**a**) void ratio vs. effective stress; (**b**) void ratio vs. permeability.

5. Conclusions

In this study, the marine soft soil that was sampled at dredging and reclamation sites in Korea was analyzed, and the representative constitutive relationship equations were proposed as follows:

The consolidation property was classified according to LL = 60% for Busan and Gwangyang marine clay and according to LL = 30% for Incheon marine clay by conducting basic physical property tests and consolidation tests for 23 marine soft soils in the Busan, Gwangyang, and Incheon regions. Busan clay showed a slightly higher consolidation settlement property than Gwangyang clay. Incheon clay showed the lowest consolidation settlement property among the three regions. The void ratio–effective stress and void ratio–permeability coefficient were estimated using the back analysis and centrifugal experiment with high void ratios, and the constitutive relationships were shown to be similar in terms of region and properties. This is a reasonable analysis procedure to carry out the consolidation experimental high void ratios using the centrifugal experiment, and then to estimate the constitutive relationship equation by back analysis from the result of the centrifugal experiment. Furthermore, the constitutive relationship equation for Korean clay was analyzed to be a reasonable power function equation. The representative constitutive relationship equations of each region were analyzed by estimating each constitutive equation for each specimen using the back analysis and centrifugal experiment, and then the specimens were classified on the basis of LL criteria, reflecting the engineering properties. For future study, a design chart for estimating the consolidation phenomenon of marine soft soil will be proposed using the constitutive relationship equations.

Author Contributions: Conceptualization, S.H.J. and H.J.K.; Methodology, S.H.J.; Investigation, S.H.J.; Writing—original draft preparation, S.H.J.; Writing—review and editing, H.J.K. All authors have read and agreed to the published version of the manuscript.

References

1. Gibson, R.E.; England, G.L.; Hussey, M.J.L. The Theory of One-dimensional Consolidation of Saturated Clays, I. Finite Non-linear Consolidation of Thin Homogeneous Layers. *Geotechnique* **1967**, *17*, 261–273. [CrossRef]
2. Cargill, K.W. *Consolidation of Soft Layers by Finite Strain Analysis*; U.S. Army Engineer Waterways Experimentation Station: Vicksburg, MS, USA, 1982.
3. Stark, T.D.; Choi, H.; Schroeder, P.R. Settlement of Dredged and Contaminated Material Placement Areas. I: Theory and Use of Primary Consolidation, Secondary Compression, and Desiccation of Dredged Fill. *J. Waterw. Port Coast. Ocean Eng.* **2005**, *131*, 43–51. [CrossRef]
4. Stark, T.D.; Choi, H.; Schroeder, P.R. Settlement of Dredged and Contaminated Material Placement Areas. II: Primary Consolidation, Secondary Compression, and Desiccation of Dredged Fill Input Parameters. *J. Waterw. Port Coast. Ocean Eng.* **2005**, *131*, 52–61. [CrossRef]
5. Liu, J.C.; Griffiths, D.V. A general solution for 1D consolidation induced by depth-and time-dependent changes in stress. *Geotechnique* **2015**, *65*, 66–72. [CrossRef]
6. Hu, A.F.; Xia, C.Q.; Cui, J.; LI, C.X.; Xie, K.H. Nonlinear consolidation analysis of natural structured clays under time-dependent loading. *Int. J. Geomech.* **2018**, *18*, 04017140. [CrossRef]
7. Liu, W.; Shi, Z.; Zhang, J.; Zhang, D. One-dimensional nonlinear consolidation behavior of structured soft clay under time-dependent loading. *Geomech. Eng.* **2019**, *18*, 299–313.
8. Somogyi, F. *Analysis and Prediction of Phosphatic Clay Consolidation: Implementation Package*; Bromwell & Carrier Engineering Inc.: Lakeland, FL, USA, 1979.
9. Somogyi, F. Large Strain Consolidation of Fine Grained Slurries. In Proceedings of the Canadian Society for Civil Engineering, Winnipeg, Manitoba, 29–30 May 1980.
10. Carrier, W.D., III; Bromwell, L.G.; Somogyi, F. Design Capacity of Slurried Mineral Waste Ponds. *J. Geotech. Eng.* **1983**, *109*, 699–716. [CrossRef]
11. Gibson, R.E.; Schiffman, R.L.; Cargill, K.W. The Theory of One-dimensional Consolidation of Saturated Clay II: Finite Non-linear Consolidation of Thick Homogeneous Layers. *Can. Geotech. J.* **1981**, *18*, 280–293. [CrossRef]
12. Wissa, A.E.Z.; Christian, J.T.; Davis, E.H.; Heiberg, S. Consolidation at Constant Rate of Strain. *J. Soil Mech. Found. Div.* **1971**, *97*, 1393–1413.

13. Schiffman, R.L.; Pane, V.; Gibson, R.E. The Theory of One-dimensional Consolidation of Saturated Clays IV: An Overview of Nonlinear Finite Strain Sedimentation and Consolidation. In Proceedings of the A Symposium on Sedimentation/Consolidation Models: Predictions and Validation; Townsend: San Francisco, CA, USA, 1984; pp. 1–29.
14. Morris, P.H. Analytical Solutions of Linear Finite-Strain One-dimensional Consolidation. *J. Geotech. Geoenvironmental Eng.* **2002**, *128*, 319–326. [CrossRef]
15. Yamagami, T.; Jiang, J.C.; Ueno, K. Back Analysis for Determination of Sedimentation and Consolidation Properties. In Proceedings of the International Symposium on Coastal Geotechnical Engineering in Practice, Yokohama, Japan, 20–22 September 2000; pp. 217–222.

Permissions

All chapters in this book were first published in MDPI; hereby published with permission under the Creative Commons Attribution License or equivalent. Every chapter published in this book has been scrutinized by our experts. Their significance has been extensively debated. The topics covered herein carry significant findings which will fuel the growth of the discipline. They may even be implemented as practical applications or may be referred to as a beginning point for another development.

The contributors of this book come from diverse backgrounds, making this book a truly international effort. This book will bring forth new frontiers with its revolutionizing research information and detailed analysis of the nascent developments around the world.

We would like to thank all the contributing authors for lending their expertise to make the book truly unique. They have played a crucial role in the development of this book. Without their invaluable contributions this book wouldn't have been possible. They have made vital efforts to compile up to date information on the varied aspects of this subject to make this book a valuable addition to the collection of many professionals and students.

This book was conceptualized with the vision of imparting up-to-date information and advanced data in this field. To ensure the same, a matchless editorial board was set up. Every individual on the board went through rigorous rounds of assessment to prove their worth. After which they invested a large part of their time researching and compiling the most relevant data for our readers.

The editorial board has been involved in producing this book since its inception. They have spent rigorous hours researching and exploring the diverse topics which have resulted in the successful publishing of this book. They have passed on their knowledge of decades through this book. To expedite this challenging task, the publisher supported the team at every step. A small team of assistant editors was also appointed to further simplify the editing procedure and attain best results for the readers.

Apart from the editorial board, the designing team has also invested a significant amount of their time in understanding the subject and creating the most relevant covers. They scrutinized every image to scout for the most suitable representation of the subject and create an appropriate cover for the book.

The publishing team has been an ardent support to the editorial, designing and production team. Their endless efforts to recruit the best for this project, has resulted in the accomplishment of this book. They are a veteran in the field of academics and their pool of knowledge is as vast as their experience in printing. Their expertise and guidance has proved useful at every step. Their uncompromising quality standards have made this book an exceptional effort. Their encouragement from time to time has been an inspiration for everyone.

The publisher and the editorial board hope that this book will prove to be a valuable piece of knowledge for researchers, students, practitioners and scholars across the globe.

List of Contributors

Myeong-Ho Song and Taek-Kun Nam
Division of Marine Engineering, Mokpo National Maritime University, Mokpo 58628, Korea

Jae-ung Lee
Division of Marine Information Technology, Korea Maritime and Ocean University, Busan 49112, Korea

Evangelos S. Filippas, George P. Papadakis and Kostas A. Belibassakis
School of Naval Architecture & Marine Engineering, National Technical University of Athens, 15780 Zografos, Greece

Shuang Han and Dong-Sheng Jeng
School of Engineering and Built Environment, Griffith University Gold Coast Campus, Queensland 4222, Australia

Chia-Cheng Tsai
Department of Marine Environmental Engineering, National Kaohsiung University of Science and Technology, Kaohsiung 80778, Taiwan
Department of Marine Environment and Engineering, National Sun Yat-Sen University, Kaohsiung 80424, Taiwan
Center of Excellence for Ocean Engineering, National Taiwan Ocean University, Keelung 20224, Taiwan

Xiuhai Wang and Hongjun Liu
College of Environmental Science and Engineering, Ocean University of China, Qingdao 266000, China
Shandong Provincial Key Laboratory of Marine Environment and Geological Engineering, Qingdao 266000, China
Laboratory for Marine Geology, Qingdao National Laboratory for Marine Science and Technology, Qingdao 266000, China

Chaoqi Zhu
College of Environmental Science and Engineering, Ocean University of China, Qingdao 266000, China
Shandong Provincial Key Laboratory of Marine Environment and Geological Engineering, Qingdao 266000, China
Laboratory for Marine Geology, Qingdao National Laboratory for Marine Science and Technology, Qingdao 266000, China
Shandong Provincial Key Laboratory of Marine Ecology and Environment & Disaster Prevention and Mitigation, Qingdao 266000, China

Yanling Zhang and Qi Wu
School of Civil Engineering, Chongqing University, Chongqing 400045, China

Xuanming Ding and Zhixiong Chen
School of Civil Engineering, Chongqing University, Chongqing 400045, China
National Joint Engineering Research Center of Geohazards Prevention in the Reservoir Areas, Chongqing 400045, China

Deqiang Chen
College of Civil Engineering and Architecture, Guangxi University, Nanning 530004, China

Junhui Luo, Xianlin Liu, Decai Mi and Longwang Xu
Guangxi Communication Design Group Co., LTD, Nanning 530029, China

Changqi Zhu, Mingjian Hu, Xinzhi Wang and Haifeng Liu
State Key Laboratory of Geomechanics and Geotechnical Engineering, Institute of Rock and Soil Mechanics, Chinese Academy of Sciences, Wuhan 430071, China

Xiang Cui
State Key Laboratory of Geomechanics and Geotechnical Engineering, Institute of Rock and Soil Mechanics, Chinese Academy of Sciences, Wuhan 430071, China
University of Chinese Academy of Sciences, Beijing 100049, China

Zhen Guo
College of Civil Engineering, Tongji University, Shanghai 200092, China

Adnan Aydin
Department of Geology and Geological Engineering, The University of Mississippi, University, MS 38677, USA

Mei Xue
State Key Laboratory of Marine Geology, Tongji University, Shanghai 200092, China

Xu Han
Department of Geotechnical Engineering, College of Civil Engineering, Tongji University, Shanghai 200092, China

Yu Huang
Department of Geotechnical Engineering, College of Civil Engineering, Tongji University, Shanghai 200092, China
Key Laboratory of Geotechnical and Underground Engineering of the Ministry of Education, Tongji University, Shanghai 200092, China

Long Yu, Heyue Zhang, Jing Li and Xian Wang
State Key Laboratory of Coastal and Offshore Engineering, Dalian University of Technology, Dalian 116024, China

Xiaoquan Zheng, Zhuangcai Tian, Hong Zhang and Tian Chen
Shandong Provincial Key Laboratory of Marine Environment and Geological Engineering, Ocean University of China, Qingdao 266100, China

Xiaolei Liu
Shandong Provincial Key Laboratory of Marine Environment and Geological Engineering, Ocean University of China, Qingdao 266100, China
Laboratory for Marine Geology, Qingdao National Laboratory for Marine Science and Technology, Qingdao 266061, China

Peiting Sun
Dalian Maritime University, Dalian 116026, China

Hongming Wang
Jiangsu Maritime Institute, Nanjing 211170, China
Jiangsu Ship Energy-Saving Engineering Technology Center, Nanjing 211170, China

Chen Li
Dalian Maritime University, Dalian 116026, China
Jiangsu Maritime Institute, Nanjing 211170, China
Jiangsu Ship Energy-Saving Engineering Technology Center, Nanjing 211170, China

Sang Hyun Jun
Infra Division, POSCO E&C, Incheon 22009, Korea

Hyuk Jae Kwon
Department of Civil Engineering, Cheongju University, Chungbuk 28503, Korea

Index

www.ingramcontent.com/pod-product-compliance
Lightning Source LLC
LaVergne TN
LVHW060533210926
848485LV00005B/1189